PHYSICAL METHODS IN AGRICULTURE

APPROACH TO PRECISION AND QUALITY

PHYSICAL METHODS IN AGRICULTURE
APPROACH TO PRECISION AND QUALITY

Edited by

Jiří Blahovec
Czech University of Agriculture
Prague, Czech Republic

and

Miroslav Kutílek
Czech Technical University
Prague, Czech Republic

Kluwer Academic / Plenum Publishers
New York, Boston, Dordrecht, London, Moscow

Library of Congress Cataloging-in-Publication Data

Physical methods in agriculture : approach to precision and quality / edited by Jirí Blahovec and Miroslav Kutílek.
p. cm.
Includes bibliographical references (p.).
ISBN 0-306-47430-1
1. Precision farming. I. Blahovec, Jirí, 1942- II. Kutílek, Miroslav. III. International Conference on Physical Methods in Agriculture--Approach to Precision and Quality (2001 : Prague, Czech Republic)

S494.5.P73 P58 2002
631--dc21

2002034141

Based on the Proceedings of the International Conference on Physical Methods in Agriculture—Approach to Precision and Quality, held August 27-30, 2001, in Prague, Czech Republic.

ISBN 0-306-47430-1

233 Spring Street, New York, New York 10013

http//www.wkap.nl/

10 9 8 7 6 5 4 3 2 1

A C.I.P. record for this book is available from the Library of Congress.

Printed in the United States of America

Contributors

Nawaf Abu-Khalaf, AgroTechnology, Department of Agricultural Sciences, The Royal Veterinary and Agricultural University (KVL), Højbakkegård Alle-2, DK-2630 Taastrup, Denmark, *e-mail: nak@kvl.dk*

Vladimir Badenko, Agrophysical Research Institute, Russian Academy of Agricultural Sciences, Grazhdansky Prospect 14, 195220 St. Petersburg, Russia, *e-mail: badenko@venture.spb.ru*

Bent S. Bennedsen, AgroTechnology, Department of Agricultural Sciences, The Royal Veterinary and Agricultural University (KVL), Højbakkegård Alle-2, DK-2630 Taastrup, Denmark, *e-mail: bsb@kvl.dk*

Horst Beuche, Department of Technology-Assessment and Substance Flow, Institute of Agricultural Engineering Bornim (ATB), Max-Eyth-Allee 100, 14469 Potsdam, Germany, *e-mail: hbeuche@atb-potsdam.de*

Jiri Blahovec, Department of Physics, Czech University of Agriculture, 16521 Prague 6 – Suchdol, Czech Republic, e-mail: *blahovec@tf.czu.cz*

Wioletta Błaszczak, Institute of Animal Reproduction and Food Research, Polish Academy of Sciences, Tuwima 10, 10-747 Olsztyn, Poland, *e-mail: viola@pan.olsztyn.pl*

Jaroslav Buchar, Department of Physics, Faculty of Agronomy, Mendel University of Agriculture and Forestry, Zemedelska 1, 613 00 Brno, Czech Republic, *e-mail: buchar@mendelu.cz*

Victor Y. Chertkov, Faculty of Agricultural Engineering, Technion, Haifa 32000, Israel, *e-mail: agvictor@tx.technion.ac.il*

Paul Darius, Katholieke Universiteit Leuven, Department of Agro Engineering and –Economics, Laboratory for Agro- Machinery and Processing, Kasteelpark Arenberg 30, B-3001 Leuven, Belgium, *e-mail: Paul.Darius@agr.kuleuven.ac.be*

Josse De Baerdemaeker, Katholieke Universiteit Leuven, Department of Agro Engineering and –Economics, Laboratory for Agro- Machinery and Processing, Kasteelpark Arenberg 30, B-3001 Leuven, Belgium, *e-mail: Josse.Debaerdemaeker@agr.kuleuven.ac.be*

Michele Desmet, Katholieke Universiteit Leuven, Laboratory/Flanders Center for Postharvest Technology, W. de Croyelaan 42, B-3001 Leuven, Belgium, *e-mail: Michele.Desmet@agr.kuleuven.ac.be*

Bohdan Dobrzański, jr., Polish Academy of Sciences, B. Dobrzański Institute of Agrophysics, ul. Doświadczalna 4, 20-290 Lublin 27, *e-mail: bdob@demeter.ipan.lublin.pl*

Józef Fornal, Institute of Animal Reproduction and Food Research, Polish Academy of Sciences, Tuwima 10, 10-747 Olsztyn, Poland, *e-mail: fornal@pan.olsztyn.pl*

Richard J Godwin, Cranfield University at Silsoe, Silsoe, Bedford MK17 9DY, UK, *e-mail: r.godwin@cranfield.ac.uk*

Stanisław Grundas, Institute of Agrophysics, Polish Academy of Sciences, Doświadczalna 4, 20-290 Lublin, Poland, *e-mail: grundas@demeter.ipan.lublin.pl*

Hans J. Hellebrand, Department of Technology-Assessment and Substance Flow, Institute of Agricultural Engineering Bornim (ATB), Max-Eyth-Allee 100, 14469 Potsdam, Germany, *e-mail: jhellebrand@atb-potsdam.de*

Józef Horabik, Institute of Agrophysics, Polish Academy of Sciences, Doświadczalna 4, 20-290 Lublin, Poland, *e-mail: jhorabik@demeter.ipan.lublin.pl*

Pál T. Jancsók, Katholieke Universiteit Leuven, Department of Agro Engineering and –Economics, Laboratory for Agro- Machinery and Processing, Kasteelpark Arenberg 30, B-3001 Leuven, Belgium, *e-mail: pal.jancsok@agr.kuleuven.ac.be*

Tomasz Jeliński, Institute of Animal Reproduction and Food Research, Polish Academy of Sciences, Tuwima 10, 10-747 Olsztyn, Poland, *e-mail: tomi@pan.olsztyn.pl*

Miroslav Krejča, School of Technological Education, Komenského 86, 39711 Písek, Czech Republic

Jana Kupcová-Vlašimská, Konstruktiva, Kryštofova 57/27, 14900 Prague 4, Czech Republic

Dmitry Kurtener, Agrophysical Research Institute, Russian Academy of Agricultural Sciences, Grazhdansky Prospect 14, 195220 St. Petersburg, Russia, *e-mail: Courtene@DK3516.spb.edu*

Miroslav Kutílek, Nad Paťankou 34, Prague 6, Czech Republic, *e-mail: kutilek@ecn.cz*

Martin Libra, Department of Physics, Czech University of Agriculture, 16521 Prague 6 – Suchdol, Czech Republic, *e-mail: libra@tf.czu.cz*

Veerle Van linden, Katholieke Universiteit Leuven, Department of Agro Engineering and –Economics, Laboratory for Agro- Machinery and Processing, Kasteelpark Arenberg 30, B-3001 Leuven, Belgium, *e-mail: Veerle.Vanlinden@agr.kuleuven.ac.be*

Manfred Linke, Department of Horticultural Engineering, Institute of Agricultural Engineering Bornim (ATB), Max-Eyth-Allee 100, 14469 Potsdam, Germany, *e-mail: mlinke@atb-potsdam.de*

Svatopluk Matula, Department of Soil Science and Geology, Czech University of Agriculture, Prague, Kamycka 129, 165 21 Prague 6 – Suchdol, Czech Republic, *e-mail: Matula@af.czu.cz*

Paolo Menesatti, Agricultural Mechanization Research Institute (ISMA), Via della Pascolare, 16 - 00016 Monterotondo (Rome), Italy, *e-mail: pmenesatti@tin.it, p.menesatti@ingegneriaagraria.it*

Marek Molenda, Institute of Agrophysics, Polish Academy of Sciences, Doświadczalna 4, 20-290 Lublin, Poland, *e-mail: mmolenda@demeter.ipan.lublin.pl*

Miloš Ondřej, Institute of Plant Molecular Biology, Academy of Sciences of Czech Republic, Branišovská 31, 370 05 České Budějovice, Czech Republic, *e-mail: ondrej@umbr.cas.cz*

Graziella Paglia, Agricultural Mechanization Research Institute (ISMA), Via della Pascolare, 16 - 00016 Monterotondo (Rome), Italy, *e-mail: g.paglia@tin.it*

Marcello Pagliai, Istituto Sperimentale per lo Studio e la Difesa del Suolo, Piazza M. D'Azeglio 30, 50121 Firenze, Italy, *e-mail: pagliai@issds.it*

Vladislav Poulek, Poulek Solar Co. Ltd., Kaštanová 1481, 25001 Brandýs nad Labem, Czech Republic, *e-mail: poulek@telecom.cz*

Herman Ramon, Katholieke Universiteit Leuven, Department of Agro Engineering and –Economics, Laboratory for Agro- Machinery and Processing, Kasteelpark Arenberg 30, B-3001 Leuven, Belgium, *e-mail: Herman.Ramon@agr.kuleuven.ac.be*

Rafał Rybczyński, Polish Academy of Sciences, B. Dobrzański Institute of Agrophysics, ul. Doświadczalna 4, 20-290 Lublin 27, *e-mail: rryb@demeter.ipan.lublin.pl*

Jadwiga Sadowska, Institute of Animal Reproduction and Food Research, Polish Academy of Sciences, Tuwima 10, 10-747 Olsztyn, Poland, *e-mail: jaga@pan.olsztyn.pl*

Naoki Sakurai, Faculty of Integrated Arts and Sciences, Hiroshima University, 739-8521 Japan, *e-mail: nsakura@hiroshima-u.ac.jp*

Libor Severa, Department of Physics, Faculty of Agronomy, Mendel University of Agriculture and Forestry, Zemedelska 1, 613 00 Brno, Czech Republic

Jana Simeonovová, Department of Physics, Faculty of Agronomy, Mendel University of Agriculture and Forestry, Zemedelska 1, 613 00 Brno, Czech Republic, *e-mail: simeon@mendelu.cz*

Silvia Solaini, Agricultural Mechanization Research Institute (ISMA), Via della Pascolare, 16 - 00016 Monterotondo (Rome), Italy, *e-mail: ssolain@tin.it*

Sebnem Tavman, Ege University Food Engineering Department, 35100 Bornova-Izmir, Turkey, *e-mail*: *tavman@textil.ege.edu.tr*

Bert E. Verlinden, Flanders Centre of Postharvest Technology, W. de Croylaan 42 B-3001 Leuven, Belgium, *e-mail: Bert.Verlinden@agr.kuleuven.ac.be*

Nadia Vignozzi, Istituto Sperimentale per lo Studio e la Difesa del Suolo, Piazza M. D'Azeglio 30, 50121 Firenze, Italy, *e-mail*: *vignozzi@issds.it*

Victor Yakushev, Agrophysical Research Institute, Russian Academy of Agricultural Sciences, Grazhdansky Prospect 14, 195220 St. Petersburg, Russia

Perihan Yolci, TUBITAK, The Scientific and Technical Research Council of Turkey, Marmara Research Center, Food Science and Technology Research Institute, P.K. 21, 41470 Gebze- Kocaeli, Turkey, *e-mail : Perihan.Yolcu@posta.mam.gov.tr*

Angelo Zanella, Research Centre for Agriculture and Forestry Laimburg, 39040 AUER-ORA (Bolzano) Laimburg 6, Pfatten-Vadena, Italy, *e-mail: angelo.zanella@provinz.bz.it*

Acknowledgements

The conference “Physical Methods in Agriculture – Approach to Precision and Quality” was co-sponsored by CIGR (represented by Bill Stout), IUSS (International Union of Soil Science, represented by W.E.H. Blum), and ISASE (International Society of Agromaterials Science and Engineering, represented by J. Blahovec). The conference was sponsored by Czech University of Agriculture in Prague.

The conference boards and committees. *Honorary Board* : B. Szot, I. Huszár, R. Řezníček, H.J. Hellebrand, K.H. Kromer, and J. Glinski. *Scientific Board*: J. Blahovec (chairman), W.E.H. Blum, J. De Baerdemaeker, R. Horn, H.D. Kutzbach, J.P. Melcion, J. Pecen, Y. Sarig, B. Stout, C. Studman, S.A. Thompson, R. Walczak, V.P. Yakushev, and Wen-Jei Yang. *Local Organising Committee*: J. Blahovec, M. Houška, M. Libra, S. Matula, J. Mareček, Z. Pastorek, J. Pecen (chairman), J. Petr, V. Rataj, R. Žitný.

The editing process of the book depended on help of different anonymous referees who helped to improve the text of majority of papers.

Preface

The first attempts to use physical methods in agriculture can be found in nineteenth century as a necessary component of farm and food machinery. There were mechanics, electricity and physical chemistry that were the first physical disciplines used in agriculture and food industry. In the same time period the studies on physical properties of soils started to be one of main topics of soil science. The twentieth century was a century of research on physical properties of agromaterials. The physical properties of agromaterials have been studied e.g. in the USA, where a big role has been played by ASAE (1907), and in the Soviet Union where the special Institute of Agrophysics was founded (1932) by Academician Ioffe. The ASAE's activity was enlarged in 1960s and 1970s, especially with the role playing by the Mohsenin's group and its followers. At that time the Institute of Agrophysics of Polish Academy of Sciences was founded in Lublin and conferences on physical methods in agriculture began to be organised.

The participants of the last conference - "Physical Methods in Agriculture - Approach to Precision and Quality", held on August 27-30, 2001, have prepared the basis for this book. Part of the conference participants decided to enlarge their conference papers to be more general and more instructive in relation to further development of the science. New papers prepared under this decision were reviewed, discussed and revised, repeatedly, to be presented in this book. We assume that the papers of the book represent the scientific discipline in its wholeness in spite of the fact that not all important problems, methods, and theories are included into this book.

The philosophy hidden behind the book tittle is formulated in the front essay "On precision and quality" by one of us (M.K.). The papers are divided into three groups. The first group entitled "Soil and Allocation Agriculture" contains papers dealing with soil properties on one side and allocation agriculture on the other side. Pagliai (direct methods of studies on soil pore systems) represents the first part of papers as it reviews the basis for more accurate models of soil characteristics. Following authors are dealing with individual soil parameters: Chertkov (theory of soil cracking and hydraulic consequences), Kutilek (statistical distribution of soil physical parameters), Matula (soil tillage and hydraulic properties of soils). The agricultural allocation is studied in relation to solar collectors (M. Libra and V. Poulek), precision farming (R.J. Godwin), and geo-information system (V. Yakushev, D. Kurtener, V. Badenko).

The second group: "Properties and Quality" can be subdivided into two parts. In the first part, papers with experimental and theoretical improvements in product testing were included. Theoretical basis for experimental studies is in papers of J. Blahovec (on stress relaxation in soft agro-products), J. Buchar (eggshell strength), B. Dobrzanski (cracking of legume grains), J. Horabik and M. Molenda (grain behaviour en masse), T. Jelinski, J. Sadowska, and J. Fornal (digital image analysis applied to quality of cereals), and S. Tavman, and P. Yolci (thermal conductivity in granular material – macaroni products). The second part of the second group is a reaction to new perspectives in agriculture, especially in relation to the present development of genetic disciplines. Introduction into these new activities is given in the paper of M. Ondřej. More detailed information on transgenic potato based on cereals are in paper of J. Fornal, W. Blaszczak and S. Grundas.

The third group of papers ("Fruits") contains the papers related to properties and testing the fruits. The paper of N. Sakurai represents complex information on fruit structure and chemical composition in relation to texture. J. Baerdemaeker, P.T. Jancsok, and B.E. Verlinden give an overview on fruit texture. Precise studies of specified physical parameters are in the texts on temperature (H.J. Hellebrand,, H. Beuche, and M. Linke), and on interactions with electromagnetic rays (B. Dobrzanski and R. Rybczinski for light), (N. Abu-Khalaf and B. Bennedsen for near infrared rays). P. Menesatti, G. Paglia, S. Solaini, and A. Zanella deal with logistic model used for damage description in falling test. V. Van Linden, M. Desmet, P. Darius, J. De Baerdemaeker, and H. Ramon present study on tomato impacts in the handling chain. They used – artificial sphere for registration of the real impacts.

Prague, July, 2002

Jiří Blahovec,
Miroslav Kutílek

Contents

On Precision and Quality

MIROSLAV KUTÍLEK
Professor Emeritus, Czech Technical University, Prague, Czech Republic

1. ON APPLIED AND EMPIRICAL SCIENCES

My discussion is related to the theory of scientific discovery and it is based mainly upon my personal experience in soil science and in its branches, in soil physics and soil hydrology.

The title of our conference is Physical Methods in Agriculture. It means that we are dealing with applications of physical methods and theories to agriculture. The question arises whether we are allowed to use the term applied physics in this connotation. Is the question correct? It was already Pasteur 130 years ago, who has stated that "there does not exist a category of science to which one can give the name applied science. There are science and the application of science, bound together as the fruit to the tree which bears it" (Nichols, 1999). And Popper (1994) says it explicitly: It is the method, not subject matter that distinguishes science. Insofar as we are dealing with applications to a certain domain, to agriculture, we are developing science related to this domain provided that we are using the tools of science. Application of a theory is always the process of testing of that theory in a domain not yet considered by the original authors of that theory.

The deductive procedure of testing a theory consists of four steps (Popper, 1994):

1. Study of the internal consistency and incontestability of the theory. It is performed inter alia in comparing the consequences of the theory.
2. Study of the logical forms.

3. Comparison to other (linked) theories.
4. Testing of the theory by applications.

Application is therefore an inseparable part of a scientific theory. Proposals on classification of sciences into categories of basic and applied sciences lack in logic if we are accepting methods of the Popper´s school. We come to the same conclusion even if we follow epistemological deductions of the early neopositivism.

Likewise applied sciences we find empirical sciences also assorted into a special category separated from the category of theoretical sciences in some classification schemes of sciences. Is there a demarcation line separating "empirical science" from "theoretical science"? Does "empirical science" exist as a special category of sciences? The absence of logic in such categorization of sciences is similar to the discussed problem of detached category of "applied science". There exists a widespread confusion on empirical nature of some branches of natural sciences, as if the "empirical science" was subordinate to "theoretical science". However, without empirical testing we do not verify the theory, the empirical procedure is incorporated into each valid (= verified) theory. Popper (1994) concludes in his discussion on demarcation that theoretical physics is a complete realization of the empirical science.

Let us turn our attention to the principal thesis of neopositivism formulated before Popper. According to it, the science is based upon experiments and description of studied objects and processes. The results of observed phenomena offer a "given" (das Gegebene, Schlick, 1938). The main aim of neopositivism is the study of fundamental concepts and theorems and their reducibility to the empirical basis and vice versa, to what extend the theorems are deducible from the empirical basis.

Even the recent philosophy does not exclude empirical methods from "basic sciences". E.g. the bootstrap method of justification shows that there is no priority of data in relation to theoretical assumptions but that a sort of symbiosis between data and theories is supposed (Fajkus, 1997).

Philosophy on methodology of sciences objectifies the view that empiricism can not have a connotation of inferiority and that the empirical part of sciences is inseparable from objectification performed by construction of a theory.

Soil science belongs to the set of branches of sciences frequently denoted as empirical sciences. Mankind had knowledge on practical use of soil just from the first settlement and from the start of simple agriculture. This was a real empiricism and I would prefer the term technology for this early period of practical studies on soil. The term technology will be discussed later on. Only with the objectification leading to the abstraction and to the formulation of first laws or theories on soil genesis and on soil processes we are allowed to speak on soil science. It does not matter that the first impulse to formulate those laws originates in agricultural practices. The birth of soil science as a real science is clearly evident.

Separation of special categories for applied sciences and empirical sciences in classification schemes of sciences is therefore irreconcilable with the logic of scientific methods.

Corollary: The same critics as used against false employment of terms "empirical" or "applied" science are appropriate when we meet the qualification scheme on the degree of excellence of scientists, where the scientists are qualified according to the "class of science" they are working in. The procedure is used frequently by the bureaucrats at the institutions governing scientific institutions. Once, they prefer scientists working in "empirical" or "applied" sciences and this is the case of ministries of industry, agriculture etc. Others at the presidium of Academy or at a ministry for development of sciences tend to classify "theoretical scientists" as the top scientists. The bureaucratic quasi-classification in all instances of ministries and at regional or national governments is then relevant not only regarding the salaries, but the financial support for the related disciplines is not negligible. The decision makers have then in their hands the future of whole scientific and technological disciplines. Absurdity of doing so is obvious.

2. ON PRECISION IN SCIENCES

Precision means accuracy, and precise is equal to accurately expressed, scrupulous in observance of rules (Fowler and Fowler, 1964). The same source says that quality is degree of excellence, e.g. we speak on good, high, poor quality. Thus, it looks that the two terms denote two different characteristics and that they can not be mutually interchanged. Is there a common attribute to them? The requirement on the increase of precision may lead to improved measuring devices, and thus to the introduction of more precise entry data and to newly modified computer programs. The pragmatic concept of precision could be then defined as a relative simple technical requirement leading to an automatic progress of science. Is the problem so simple?

Precision stands for conformity with a standard, or with truth. The next question is posed: What is the scientific truth, what are the rules for a definition of a standard in science?

We are generally accepting reality that the antique ideal on absolute flawless science and knowledge (episteme) characterized by a total provability is a sort of an unattainable idol. The requirement on scientific objectivity leads inevitably to the statement that each theory has features of temporality. This is what our experience says, too. The scientist is not the owner of an irrefutable truth, but he is searching and formulating hypothesis, testing them with the aim of their refutation. Popper speaks on a well corroborated theory. It can be overcome only by theory of a higher level of universality. It means that the new theory can be better tested and the condition is that it includes the older well corroborated theory, or at least its good approximation. Our research activity should not defend old well corroborated theories. Quite opposite, it should try to elaborate new hypothesis by overcoming the old ones and the new hypothesis is tested with the aim of a good corroboration. The procedure is called deduction.

Corollary: Let us confront this Popperian statement with the reality of our journals. There is a great majority of papers where the researchers practically confirm well corroborated theories. If there is a grain of new finding, then it is restricted to modification of empirical parameters or to description of physical characteristics of materials slightly different from the known and earlier well studied materials. If we accept the arguments of the discussion on science and scientific discovery, we are hesitating to denote such papers as scientific, and the term technological will fit better, as I am demonstrating later on. Even if new measuring tools are used I would doubt to call the procedures as bringing an increased precision in our knowledge.

There are myths on the ways of scientific discoveries and on the formation of new theories. Among them the myth on inductive method is frequently quoted and if not quoted, so it is implicitly hidden behind the activity of many institutions and scientists, especially in geosciences and in agronomy. Induction is a sort of heritage after positivism to which it is closely related. It is surviving even in neopositivism as a principal tool of scientific discovery. Inductive procedures are there melted with theoretical abstraction and this approach is a part of philosophical awareness of scientific community.

Inductive method starts at singular (or particular) statements and it passes from them to the universal statement. The procedure has no logical justification and it has been radically criticized by Popper (1934 and 1994,b). Inductive method assumes that the scientific progress is realized by the accumulation of new perceptual experience and that this collection of high number of observation is prerequisite condition to the birth of a new theory. It is this myth on methods of experimental physics which finds its theoretically justified extension in the theory on paradigm, introduced in this new context into philosophy and science by Kuhn (1970) and his school.

Paradigm is derived from the Greek paradeigma and its pre-Kuhnian meaning was example, pattern (Fowler and Fowler, 1964). Nowadays, paradigm is simply a predominant worldview in the realm of human thought and the Kuhnian school says e.g. that we live within an evolutionary paradigm, where the original paradigm of creation is replaced by evolution. The use of the term paradigm has rapidly spread out and all new theoretical approaches in specialized branches of sciences have been labeled by the term paradigm. It is not the term, but the Kuhnian procedure leading to a new paradigm, what has to be critically examined. The accumulation of experimental data could not lead to a new theory, to "a new paradigm".

There exists a principal logical argument against the theory on paradigm. The induction method realized by repetition is highly questionable since repetitions are always approximate. However, induction requires implicitly a full similarity in repeated tests. Similarity exists only in certain pre-defined aspects, but it can not be attained in all factors governing the experiments especially when we are dealing with complicated systems.

After Popper, the rational branch of recent philosophy rejects the inductive method, too, and the refusal is traceable e.g. in the earlier mentioned bootstrap method, where the data collection has no priority.

When we study procedures of meaningful experiments, we find that it is a planned action, where each step is directed by the theory, but never vice versa.

There are many practical examples on failure of the inductive method. Let me speak on soil physics. Heinrich has introduced a new method and a new term retention capacity, or field capacity in 1886. It denotes the amount of water held in the topsoil after a thorough wetting and subsequent drainage to the drier subsoil. The downward movement of water was assumed to stop within 2 to 3 days (Kutílek and Nielsen, 1994). Hundreds of papers published on the theme within about one century brought no new "paradigm" unless the theory on potential flow was applied. Some of the earlier "experimentation" was used only as a sort of verification of the theory applied to the defined problem.

The real application of physical theories to the solution of transport of water and solutes in soils, to soil compaction etc. was a difficult task. Because soil is a mechanically complicated medium neither rigid nor non-reactive to the transported matter, the application of physics was never a simple transcription of basic equations of hydrodynamics or diffusion. The basic equations have been considered as "old well corroborated theories". At a higher level new theories have been introduced fitting to the complicated soil medium and its porous system. However, new theories are not in contradiction to the old ones, but they include the old ones as a special case.

There is no indication that Buckingham's concept on soil water potential, Richards development of the partial differential equation of soil water flow, or Philip's solution of Richards equation would be a result of gradual accumulation of knowledge and data. Details and related quotations are in

Kutílek and Nielsen (1994). We find typical features of a deductive method in all instances and its principle is simply formulated: Theory first and its testing after that. Inductive method is not traceable. If we shift to the recent solutions of preferential flow, we find a typical example of the older well corroborated theory (Richards equation and its solutions for a rigid, homogeneous porous medium) which has been overcome by the theory of a higher level of universality. This new theory contains in itself the older theory. It is evident that the described procedure is not identical with Popper's falsification of a theory in a simple way. The extension of Popper's approach by Lakatos (1970) is well applicable in the quoted example. Research programs are composed of the hard core and of the envelope according to Lakatos. The hard core represents a set of corroborated statements. The envelope is either extending the hard core to new conditions, or it is completing the existing hard core by new factors related to new phenomena. Richards equation represents here the hard core and its modification to preferential flow stands for the envelope.

The principles of application of mechanics to soil tillage and to soil compaction are similar to transport processes.

In our earlier publication we have discovered cyclic features with periods lasting for about 20 years between the derivation and application of theories on soil hydrology (Kutílek and Rieu, 1998). We have explained main factors influencing this type of periodicity. The "empirical experimentation" of inductive character has been found among factors strongly acting against the formulation and use of new theories.

3. ON QUALITY AND TECHNOLOGY

It follows from the intrinsic meaning of the word that quality is related to technology. People are frequently confusing science and technology. Technology existed long before the birth of science. Science may or may not be related to technology at the outset of science. Science is an epistemological system, it is the curiosity about how and why the things happen in that complicated and fascinating world out there (Philip, 1991). Technology has not this type of aims. It has enabled our survival through invention of practical procedures and further on, it has brought tools for gradual improvement of man's environment.

Technology is a set of gradually modified tools and procedures mankind has used just from the origin of Homo sapiens sapiens. Those tools and procedures have been empirically developed and improved in the past by rules of thumb and by the method try and modify, accompanied by numerous repetitions of trials. Traditional methods were inherited and became the skeleton of individual cultures. Gradually, with the development of scientific knowledge, technology has translated the scientific discoveries into new products and into new rules of dealing with them. Technological advancement is now dominated by scientific knowledge, but induction is still its major procedure and many technological inventions are unthinkable without inductive procedures. Technology has flourished also without scientific advancement and even without science in the past. The combination of traditional methods with science is completed by a sort of "majority agreement" of the society and by "mutual understanding" between producers of technology and their users nowadays.

There are many further differences between science and technology. Even the ethos of technology is different from sciences. Technology serves the human community directly while for the development of science this type of criteria of serving could act even harmfully if the demand is requested pedantically (Kutílek and Novák, 1997).

Technology has not requirement on communality as the science has: Scientific knowledge must reside in the public domain and secrecy is the antithesis of this ethical norm on full and open communication in science. This ethic norm of science is often in conflict with commercial and military interests (Philip, 1991). This is not the case of technology. Patenting of new technological procedures and inventions are generally accepted by the public. New technological methods are often declared as secret in the frame of one institution or factory and their dissemination beyond the grounds of the institution is considered as a criminal act. The principal difference between science and technology is obvious in this aspect.

But nowadays science and advanced technology are inextricably linked, too and the recent technology is dominantly dependent on the findings of science and vice versa, technology puts even requirements upon science in short time scale. Technological products contribute to the technique of testing theories. Mutual bonds are distinct in some instances and hidden in some others. But in spite of all interrelations, we are dealing with two different categories.

There is another important feature separating science and technology: Determinism and predictability. A serious confusion of those two terms exists. Determinism denotes how nature behaves. Predictability is related to what we are able to observe, analyze and compute. Determinism can not be proved and tested unless we know all fundamental laws and unless we measure all initial conditions accurately. Any discussion of determinism outside the framework of the fundamental laws is useless. Prediction has stochastic features and we speak on a certain probability of our predictions.

As a practical example I am discussing here the scale related transport processes in small catchment hydrology (floods and droughts) and in environmental protection (pollution hazard of soils and of ground waters). While the theoretical solutions of soil hydrology and of transport of pollutants on the small scale, up to the pedon scale (10^0 m^2) have been more or less successfully formulated, the transition to higher scales is still not satisfactory. The situation is favorable for introduction of new approaches, among them is the requirement to apply the theory of chaos (Addiscott and Mirza, 1998 and quotations there). This theory describes complicated systems consisting of many subsystems linked together where the processes are nonlinear and influencing one the other by coupling coefficients. The parameters of each of the subsystem are variable in space and time. First of all, if we define two very close initial conditions and if we wait long enough, the two systems thus produced will be finally very different. Processes in soils at higher scale are however not influenced by different boundary conditions only. Generally speaking, the megasystem can not be fully described as the summation of all microsystems composing the megasystem. The behavior of soil in hydrologic catchment or on a higher taxon can not be explained only as sensitivity to spatial variation of initial conditions (in broader sense) including variation of soil physical parameters within the system. The whole "chaotism" does not shrink to our low ability to describe the initial conditions in all needed details. There are feed-back loops, increase of degrees of freedom etc. But this does not have any impact on determinism, since determinism does not mean full predictability in all instances in sciences (Bricmont, 1996).

I am skeptical to the effort to apply theory of chaos as a scientific tool explaining deterministically the process on the catchment in all details and for all existing situations. But I am considering the theory of chaos as applicable in technological sense, it is when epistemology is not considered as the aim. We come back to our requirement on an increased quality of our predictions on floods and droughts frequencies and on pollution hazard. Theory of chaos, similar to other theories dealing with upscaling looks as well suited to technological solutions of soil processes on large scales. Since we are working with technological procedures, we have to be aware of their limitations. They are similar to limits of all simulation modeling. A well performing model can offer reliable results on processes in systems slightly different only from the system for which the model has been tested. The same is valid for application of the theory of chaos to soil megasystems. In some instances, the technologists are using models not only for prediction, but they are replacing nature by models and we are then witnesses of exploitation of models for testing new hypothesis. The results of models substitute a direct observation of nature. It is obvious that this substitution of reality may lead to a blind alley not only in science, but in technology, too, if the employment of models is not very cautious.

My critical paragraph on application of the theory of chaos as epistemological tool in soil physics has nothing to do with a general assessment of the theory. Chaos theory has produced important new insights and it is a useful part of mathematics. But the mathematical issues are subtle and delicate.

4. CONCLUSIONS

Precision and quality are related to two different categories of human activity. The first is related to sciences, the second to technological applications. An increased precision may not lead to a higher quality since precision does not guarantee a higher level of technology and full predictability, especially if we deal with complicated systems. On the other hand side even if a high precision does not offer automatically high quality, it may contribute to the control of high quality in some instances. The requirement on high quality may stimulate the formulation of new theories. High quality of observation methods contributes to the corroboration of the earlier formulated theory. Thus, precision and quality are linked together by many ties but the increase of one of them does not mean automatically the increase of the other one.

REFERENCES

Addiscott, T.M., and Mirza, N.A., 1998, New paradigms for modelling mass transfer in soils. *Soil and Tillage Res.* **47**: 105-109.

Bricmont, J., 1996, Science of chaos or chaos in science? In *The Flight from Science and Reason* (P.R. Gross, N. Levitt and M.W. Lewis, eds.). New York Academy of Sciences, New York, pp.131-175.

Fajkus, B., 1997, *Současná filosofie a metodologie vědy* (In Czech, *Contemporary Philosophy and Methodology of Science*). Filosofia, Institute of Philosophy Publishers, Prague.

Fowler, H.W., and Fowler, F.G. (Eds.), 1964, *The Concise Oxford Dictionary*, 5th ed., Oxford, Clarendon Press.

Kuhn, T.S., 1970 *The Structure of Scientific Revolutions*, 2nd ed., University of Chicago Press.

Kutílek, M., and Nielsen, D.R., 1994, *Soil Hydrology*. Catena Verlag, Cremlingen-Destedt.

Kutílek, M., and Novák, V., 1997, Aspects of development of soil physics in Central and Eastern Europe and the impact of a totalitarian ideology upon it. In *History of Soil Science* (D.H. Yalon and S. Berkowicz, eds.), Advances in GeoEcology, Catena, pp.241-259.

Kutílek, M. and Rieu, M., 1998, Introduction to the symposium 1: New concepts and theories in soil physics. *Soil and Tillage Res.*, **47**: 1-4.

Lakatos, I., 1970, Falsification and methodology of scientific research programmes. In *Criticism and the Growth of Knowledge* (I. Lakatos and A. Musgrave, eds.) Cambridge Univ. Press, pp. 91-196.

Nichols, R.W., 1999, What drives discovery? *The Sciences*, **39**: 4.

Philip, J.R., 1991, Soils, natural science, and models. *Soil Sci.*, **151**: 91-98.

Popper, K.R., 1994a, *The Logic of Scientific Discovery*, 10th ed. Routledge, London and New York (1st German edition 1934, 1st English edition 1958).

Popper, K.R., 1994b, The Myth of Framework: In *Defense of Science and Rationality* (M.A. Notturno edt.), Routledge, London.

Schlick, M., 1938, *Positivismus und Realismus.* In *Gesammte Aufsätze, II.* Wien.

PART 1

SOIL AND ALLOCATION IN AGRICULTURE

Image Analysis and Microscopic Techniques to Characterize Soil Pore System

Soil Pore System Characterization

MARCELLO PAGLIAI and NADIA VIGNOZZI
Istituto Sperimentale per lo Studio e la Difesa del Suolo, Firenze, Italy

1. INTRODUCTION

To evaluate the impact of management practices on the soil environment it is necessary to quantify the changes which occur in the soil struct re. Soil structure is one of the most important properties affecting crop production because it determines the depth that roots can explore, the amount of water that can be stored in the soil and the movement of air, water and soil fauna. Soil quality is strictly related to soil structure and much of the environmental damage to intensively-farmed land such as erosion, compaction and desertification originate from soil structure degradation. To quantify soil structural changes following agricultural activities, besides traditional measurements such as aggregate stability and hydraulic conductivity, pore space measurements are being increasingly used. In fact, it is the size, shape and continuity of pores that affect most of the important processes in soils (Ringrose-Voase and Bullock, 1984). Detailed insight into the complexity of the pore system in soils can be obtained by using mercury intrusion porosimetry to quantify pores with equivalent pore diameter < 50 μm (micropores) within the soil aggregates (Fiès, 1992). Image analysis on thin sections prepared from undisturbed soil samples allows pores > 50 μm (macropores) to be quantified, which determine the type of soil structure (Pagliai et al., 1983b, 1984).

2. SOIL STRUCTURE

Soil structure may be defined either as "the shape, size and spatial arrangement of individual soil particles and clusters of particles (aggregates)" or as "the combination of different types of pores with solid particles (aggregates)". Soil structure has generally been defined in the former way and measured in terms of aggregate characteristics. These can be related to plant growth only empirically. In fact, it is the pore shape, the pore size distribution and the pore arrangement which affect many of the most important processes in soil that influence plant development such as storage and movement of water and gases, solute movements and ease of root growth. For this reason measurements of pore space are increasingly being used to characterize soil structure. In fact, between the particles arranged singly or in aggregates, there is an intricate system of pore spaces on which plant roots, micro-organisms and soil fauna depend for the storage and movement of water and air.

Soil porosity represents the liquid and gaseous soil phases. To characterize the pore system it is necessary, first of all, to determine the size distribution and shape of pores because the agronomic functions of pores depend on their size and shape.

3. SOIL THIN SECTION PREPARATION

For image analysis it is necessary to prepare thin sections of soil following the techniques of soil micromorphology developed by Kubiena (1970). The procedure consists in taking undisturbed soil samples using appropriate tools, containers and techniques taking care that the interior structure of the soil samples remains undisturbed. Then the soil samples, carefully packed, are transported to the laboratory, dried to avoid pronounced shrinkage phenomena, using appropriate methods, e.g. acetone replacement of the water (Murphy, 1986), and impregnated, under vacuum, with a polyester resin, which has the characteristic of polymerising slowly at room temperature without altering in any way the structure of the soil. Practically, this resin fills the pores of the soil. When the soil samples are hardened (generally after 4-6 weeks) they are made into vertically- or horizontally-oriented thin sections by using appropriate machines (Murphy, 1986). Their thickness is about 30 μm so that they can be analysed by the microscope in transmitted light. Thin sections can be examined using a polarising microscope at low magnification to observe soil structure. The size depends on the kind of machines available; for porosity measurement a size larger than 6x6 cm is recommended.

4. IMAGE ANALYSIS

The soil thin sections are analysed with image analysers (Murphy, 1977a, b; Pagliai et al., 1983b, 1984). With the technique of image analysis it is now possible to characterize soil structure by the quantification of soil porosity in all its aspects (pore shape, pore size distribution, irregularity, orientation, continuity, etc.) on thin sections, prepared from undisturbed soil samples (Bouma et al., 1977, 1982; Murphy et al., 1977a, b; Pagliai et al., 1983b, 1984; Pagliai, 1988). This morphometric technique has the advantage that the measurement and the characterization of pore space can be combined with a visual appreciation of the type and distribution of pores in soil at a particular moment in its dynamic evolution.

Technological and theoretical advances, regarding both sample preparation and image analysis, have improved the methods for direct quantification of soil pores. These methods allow the quantification of the effects of tillage practices on soil porosity and structure and in turn the definition of optimum tillage needs for sustainable agriculture (Mermut et al., 1992; Moran and McBratney, 1992).

Image analysers perform light-electronic analysis of the image obtained using a macroepidiascope or light or electronic microscope (Pagliai et al., 1983a ; Bruand et al., 1992; 1996). The image is filmed with a videocamera and shown on a monitor; the video signal passes through a processor that transforms the image into pixels and analyses them individually depending on the grey level in each. Pores are measured by programming the analyser with their corresponding grey level detected on a scale of 256 levels. Basic measurements of image analysis on pores include number, area, perimeter, diameters, projections, etc., and these are supplemented by derived quantities such as shape factors, size distribution, continuity, irregularity and orientation (Pagliai, 1988).

To characterise the microporosity (pores less than 50 μm) the thin sections must be covered by a layer of carbon, in order to allow conductivity to the electron beam, and analysed by a scanning electron microscope (SEM) with a module for backscattered electron scanning images (BESI) (Pagliai et al., 1983a). The thin sections can be examined by the SEM at several magnification. For example, if the analysis in the image analyser starts with BESI taken at x400 magnification, the size of the pixel will be 0.2 μm, therefore pores in the range of the storage pores 0.5-50 μm can be measured from the back-scattered electron scanning images (Pagliai and Vittori Antisari, 1993).

To characterise the macroporosity the thin sections can be analysed by a macroepidiascope and the image analyser being set to measure pores larger than 50 μm.

Two-dimensional images obtained can be transformed into data representing three-dimensional area percentages that are representative for three-dimensional volumes. Stereology techniques have been applied to achieve this objective (Ringrose-Voase and Bullock, 1984; Ringrose-Voase and Nortcliff, 1987; Mele et al., 1999).

Image analysis can be used not only on soil thin sections but also on polished faces of large soil blocks impregnated directly in the field with (fairly cheap) materials such as paraffin wax (Dexter, 1988), plaster of Paris (FitzPatrick et al., 1985), or resin (Moran et al., 1989).

5. SOIL POROSITY CHARACTERISATION

5.1 Pore Shape

The shape factors allow division of pores into different shape groups. For example, pore shape can be expressed by a shape factor [perimeter^2/(4π·area)] and pores divided into regular (more or less rounded) (shape factor 1-2), irregular (shape factor 2-5) and elongated (shape factor >5) (Pagliai et al. 1983b). These classes correspond approximately to those used by Bouma et al. (1977).

The regular pores are obviously those of a rounded shape and can be distinguished in two types according to their origin: the spherical pores formed by entrapped air during soil drying and the channels and chambers formed by biological activity (root growth and movement of soil fauna). Their distinction on soil thin sections is very evident because spherical pores (vesicles, according to Brewer, 1964) have very smooth walls, while channels, even though cut in a transversal way on thin section, present rough walls with deposits of insect escrements or root exudates. The presence of many spherical pores of the first type (vesicles) creates a vesicular structure typical of soils with evident problems of degradation.

The irregular pores are the common soil voids with irregular walls (vughs, according to the micromorphological terminology of Brewer, 1964) and can be isolated (packing voids) or interconnected. The dominant presence of these pores produce the typical vughy structure (Bullock et al.,

1985a). In cultivated soils these pores can be produced by the action of soil tillage implements.

Two types of elongated pores can be distinguished, i.e., cracks and thin fissures (planes). The former are typical of clay soils with a depleted soil organic matter content and they are visible at the surface when the soil is dry and has shrunk. The thin fissures are the most important, especially from an agronomic point of view, in fact, they are the typical transmission pores. An adequate proportion of this type of pore (over 10% of the total porosity) generally creates an angular to subangular blocky structure of good quality. Obviously for this to be true it is necessary for these pores to be homogeneously distributed in the soil matrix. In fact, for the characterization of these pores by image analysis, it is necessary to determine not only their shape and width, but also their length. With the same procedure of width determination it is also possible to determine the length of these elongated pores, which may reflect their continuity, and it is well known that the flow of water through soil depends on the continuity of large pores. Therefore the analysis of pore patterns allows the characterization and prediction of flow processes in soils.

For root growth and water movement not only the size and continuity of elongated pores are important but also their irregularity and orientation. The ratio convex perimeter/perimeter or convex area/area of elongated pores gives information about their irregularity, tortuosity and re-entrancy. As regards water movement, for example, the very regular and the moderately regular elongated pores play a different role. The very regular elongated pores are flat and smooth pores with accommodating faces, which tend to seal when the soil is wet, thus preventing water movement. In contrast, the moderately regular elongated pores have walls which do not accommodate each other. Therefore, these pores permit water movement even when the soil is wet and fully swollen (Pagliai et al., 1984). The ratio vertical/horizontal dimensions gives the orientation of elongated pores (Pagliai et al., 1984). It is easily understandable that many soil processes such as water movement, leaching, clay migration, etc., are strongly related to the orientation of pores in soil and these processes change radically depending on whether a vertical or horizontal pore orientation is dominant.

5.2 Pore Size Distribution

As already said, to characterize the pore system it is necessary, first of all, to determine the shape and size distribution of pores because the agronomic functions of pores depend not only on their shape bur also on their size.

The pore size distribution is a dynamic property. It is dependent upon the water content especially in fine textured soils subjected to swelling and drying (Kutilek and Nielsen, 1994). There are a lot of studies that show the changes of pore size distribution in Vertisols at different water content, i.e. when they are water saturated after the rain and the cracks are closed due to saturation and when they are dried and the cracks are open (Schweikle, 1982; Kutilek, 1983). The changes of pore size distribution are strictly correlated with the changes of hydraulic conductivity (Kutilek, 1996).

By means of image analysis pores of each shape group can be further subdivided into a selected number of size classes according to either the equivalent pore diameter for rounded and irregular pores or the width for elongated pores. The equivalent pore diameters are calculated from the area of regular and irregular pores, while the width of elongated pores is calculated from their area and perimeter data using a quadratic equation because it is assumed that elongated pores are long narrow rectangles (Pagliai et al., 1984).

Table 1 reports the most frequently used classification scheme of pore size distribution proposed by Brewer (1964). Detailed information about the applicability of hydraulic equations in the different classes of pores are reported in Kutilek and Nielsen (1994).

Table 1 - Morphologic pore size classification according to Brewer (1964).

Class	Subclass	Class limits (Equivalent diameter μm (10^{-6}m))
Macropores	Coarse	above 5000
	Medium	2000-5000
	Fine	1000-2000
	Very Fine	5-1000
Mesopores		30-75
Micropores		5-30
Ultramicropores		0.1-5
Cryptopores		less than 0.1

According to another of the widely used classifications, that of Greenland (1977) reported in Table 2, the very fine pores less than 0.005 μm, called "bonding pores", are critically important in terms of the forces holding domains and aggregates of primary particles together; pores of less than 0.5 μm are the "residual pores" for the chemical interactions at the molecular level; pores which have an equivalent pore diameter ranging from 0.5 to 50 μm are the "storage pores", i.e. the pores that store water for plants and for micro-organisms; and the pores ranging from 50 to 500 μm are those called "transmission pores" in which the movements of water are important for plants, and, moreover, they are the pores needed by feeding roots to grow into. The water content when pores larger than 50 μm have drained, corresponds to the field capacity of the soil. The wilting point commences when most pores larger than approximately 0.5 μm have emptied.

Table 2 - Classification of soil pores according to their size. Modified from Greenland (1977).

Equivalent diameter μm (10^{-6}m)	Water Potential (bar)	Name
<0.005	>-600	Bonding space
0.005 - 0.5	-600 / -6	Residual pores
0.5 - 50	-6 / -0.06	Storage pores
50 - 500	-0.06 / -0.006	Transmission pores
>500	<-0.006	Fissures

Pores larger than 500 μm can have some useful effects on root penetration and water movement (drainage), especially in fine-textured soils. However, a high percentage of this latter type of pore (above 70-80% of the total porosity) in soils is usually an index of poor soil structure, especially in relation to plant growth. This is because surface cracks, which develop after rainfall, when the stability of soil aggregates is poor, belong to this size class (Pagliai et al., 1981, 1983b). Until now the necessary proportion of large pores for air and water transmission and easy root growth has generally been inadequately defined. In fact, adequate storage pores (0.5-50 μm) as well as adequate transmission pores (50-500 μm) are necessary for plant growth (Greenland, 1981).

Characterization of the soil pore system by image analysis of thin sections can give detailed information about soil structural conditions, moreover if climate, agronomic and management data are known, an evaluation of soil physical vulnerability is possible. Hence, the soil pore system can be considered a good indicator of soil quality, nevertheless, as for other indicators, threshold values have to be known.

According to the micromorphometric method, a soil can be classified as follows where the total porosity represents the percentage of area occupied by pores larger than 50 μm per thin section (Pagliai, 1988):

Soil very compact	when total porosity is	< 5%
Soil compact	when total porosity is	5-10%
Soil moderately porous	when total porosity is	10-25%

Soil highly porous	when total porosity is	25-40%
Soil extremely porous	when total porosity is	>40%

A total macroporosity of 10% is considered to be the lower limit for good soil structural condition, anyway, only the complete evaluation, both quantitative and qualitative, of the soil pore system can produce exhaustive information on actual soil quality.

5.3 Types of Soil Structure

According to the definition of soil structure the different types of structure are originated by the combination and the spatial arrangement of different types of pores and aggregates. The definition of the several types of soil structure is fundamental for a first qualitative evaluation of soil physical fertility. The microscopic examination of soil thin section, by a polarizing microscope at low magnification, allows such a definition.

In general, three broad categories of soil structure can be recognized: single grained, aggregates and massive. According to Bullock et al. (1985a) the main types of soil structure can be summarized as follows:

Single grain structure. - Typical of sandy soils; the quartz (sand-sized) grains are completely loose and the fine materials in the intergranular spaces are very rare. The porosity is represented by packing voids delimitated by quartz grains. When these grains are of a more or less uniform shape and size a rather compact grain structure can be originated.

Bridged/pellicular grain structure. - The sand-sized grains are bridged or coated by fine materials, usually clay, which can cement some grains to each other, so originating microaggregates.

Vughy structure. - There are no separated aggregates and the mass is broken up by scattered but not interconnected irregular pores, "vughs", and occasional channels and chambers (Bullock et al., 1985a) (Fig. 1). Such a structure allows good soil aeration but the continuity of pores is limited. In the case of heavy rains pounding water can be present in the soil profile with consequent collapse of the soil structure. Such a type of structure is, therefore, unstable and transitory.

Vesicular structure. - There are no separated aggregates and the mass is broken up by rounded pores, "vesicles" (Bullock et al., 1985a), originated by entrapped air during drying processes (Fig. 2). This kind of structure is an indicator of degraded soils.

Crumb structure. - The soil aggregates are more or less rounded, often rugose, well separated from each other and rather compact inside. The porosity is represented by pore space (packing voids) which separates the aggregates. This type of structure is often originated by anthropogenic activities (soil tillage).

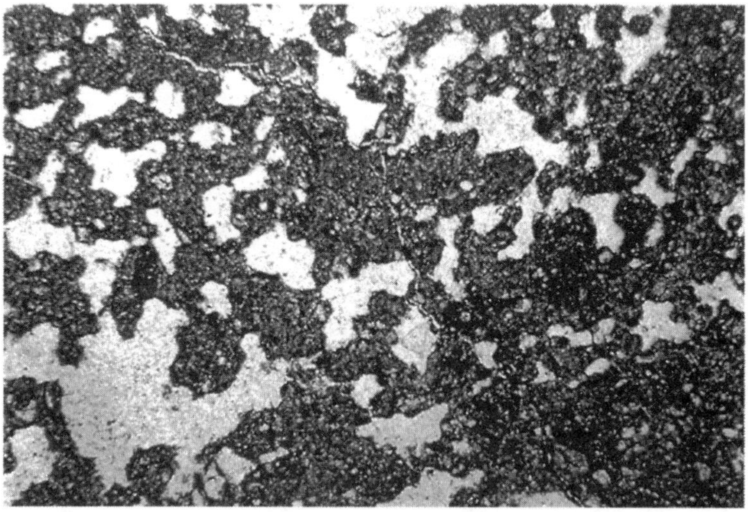

Figure 1 - Microphotograph of a vertically oriented soil thin section showing an example of vughy structure. The white areas represent the pores. Frame length 3×5 mm.

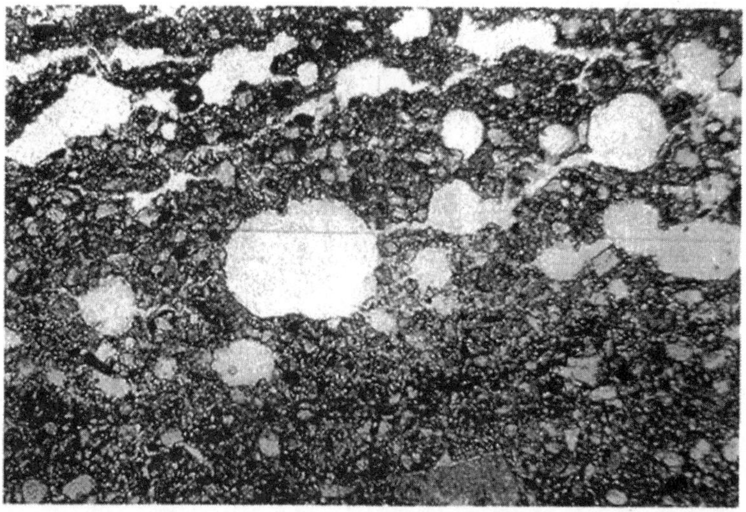

Figure 2 - Microphotograph of a vertically oriented soil thin section showing an example of vesicular structure. The white areas represent the pores. Frame length 3×5 mm.

Subangular blocky structure. - The soil aggregates are separated by elongated continuous pores (planes), are of different sizes and can be rather porous inside. Aggregate faces largely accommodate each other. From an agronomic point of view, this is the best type of soil structure because the continuity of elongated pores allows good water movement and facilitates root growth. Moreover, it is a rather stable soil structure.

Angular blocky structure. - The soil aggregates have angular edges and are separated by elongated pores of a regular shape. Aggregate faces normally accommodate each other. This is a typical structure of clay soils and it less stable than the subangular blocky structure because during the wetting process the soil swells and pores, which have accommodating walls, tend to seal up, thus, prevent water movement, resulting in soil degradation.

Platy structure. - The thin and flat soil aggregates are separated by elongated pores oriented parallel to the soil surface and, therefore, not continuous in a vertical sense. This leads, as a consequence, to a drastic reduction of water infiltration capacity. Soils with this type of structure are subject to water stagnation or runoff and erosion depending on their slope. The platy structure is typical of compact soils.

Prismatic structure. - The soil macroaggregates are divided into prisms separated by vertically oriented elongated pores with accommodating walls. Also this type of structure is typical of clay soils, especially in the B horizon, and is not very stable because the swelling of the soil when wet causes the closing of pores.

Massive structure. - The soil material is very compact, there are no visible separated aggregates. The porosity is very low and represented by small pores isolated in the soil matrix. This type of structure represents a bad "habitat" for plant development and is common in degraded soil with a low content of organic matter.

Complex structure. - The soil presents two or more types of soil structure.

6. RELATIONSHIPS BETWEEN SOIL POROSITY AND WATER MOVEMENT

The relationships between pore size distribution and soil water content are expressed by the capillary model, while the relationships between pore size distribution and water movement at specific water potentials have been developed in terms of several physical equations and models (Marshall, 1958; Childs, 1969).

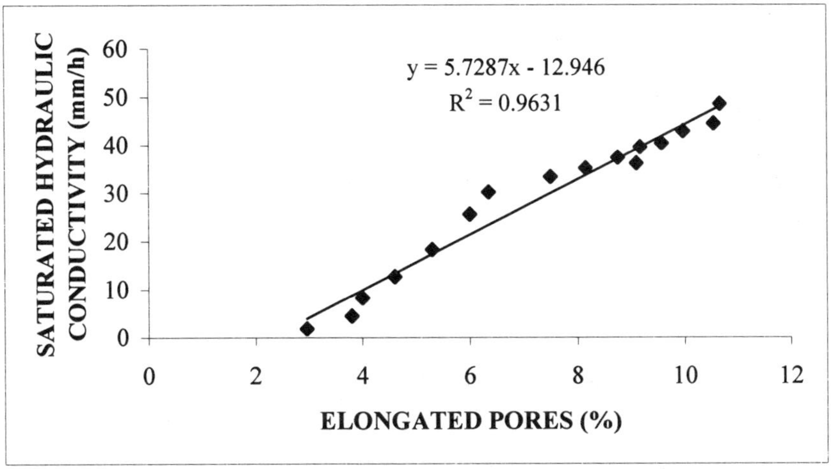

Figure 3. Correlation between soil porosity, formed by elongated pores, and saturated hydraulic conductivity in the surface layer (0-10 cm) of a loam soil cropped with maize.

The main limitation of these models is due to the assumption of the cylindrical shape of pores or the spherical shape of soil particles. The development of micromorphological techniques together with image analysis allow the improvement of such models. For example, Bouma et al. (1977) developed a method based on the preparation of undisturbed soil columns, saturated and then percolated with a 0.1% solution of methylene-blue that is adsorbed by the soil particles on the pore walls. Then vertical and horizontal thin sections are prepared. Pores are divided into three shape groups as already explained and then the pore size distribution is determined. For the planar elongated pores the total area, the area of the blue-stained pore walls, and their lengths, and the spatial distribution of the widths and lengths of the pores with blue-stained walls are determined. Particular attention should be paid to the measurement of the width of the necks of elongated pores because the hydraulic conductivity is determined by the necks in the flow system. Following this procedure the hydraulic conductivity (K_{sat}) can be calculated as proposed by Bouma et al. (1979). Further studies of Bouma (1992) confirmed that morphological information on the soil pore system is essential for the realization of water flux models.

The evolution of software for image analysis, that enables the acquisition of precise information about shape, size, continuity and arrangement of pores in soil, permit the simplification of the modelling approach. For example, Fig. 3 shows highly significant correlations between elongated pores and hydraulic conductivity.

Combined with image analysis, the use of fractal and fractal fragmentation models can help to characterize the geometry of a porous medium in relation to transport process (Kutilek and Nielsen, 1994). For example, the model of fractal fragmentation leads to a better understanding of relationships between aggregation, n-modal porosity and soil hydraulic properties.

7. EVALUATION OF THE IMPACT OF AGRICULTURAL MANAGEMENT PRACTICES ON SOIL STRUCTURE

Many anthropogenic activities affect soil structure. Parameters such as pore size distribution, pore shape, pore continuity and relative position of aggregates and pores are very important for evaluating induced modifications of soil structure, e.g., by different management practices such as soil tillage, application of manures and irrigation.

7.1 Soil Tillage

Long-term intensive arable cultivation has negative effects on soil physical properties, particularly on soil structure, with resulting effects on soil erodibility and crop yields. The need to check the degradation of soil structure has caused farmers to consider no-till and minimum tillage as an alternative to conventional tillage. Abandoning traditional farming rotations and adopting intensive monocultures, without applications of farmyard manure or organic materials to the soil, has decreased the soil organic matter content with evident degradation of soil structure. The resulting soil porosity conditions are often unfavourable for crop growth (Pagliai et al., 1983b; 1984; 1989; Shipitalo and Protz, 1987).

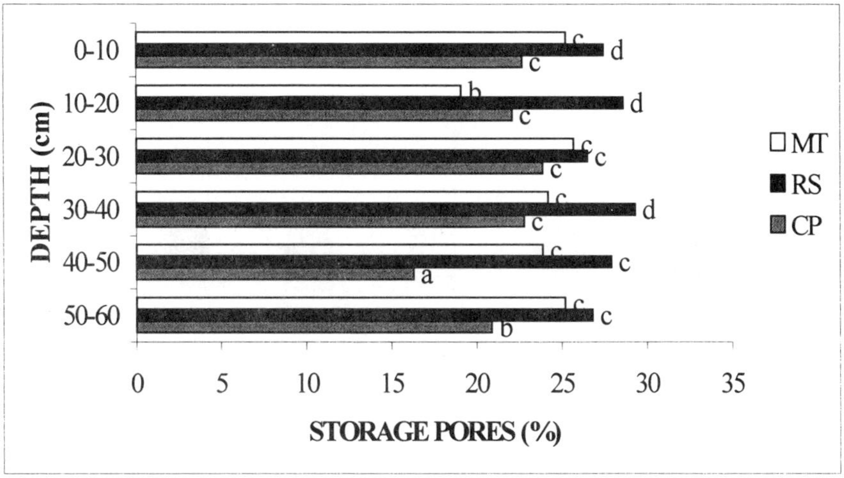

Figure 4. Effects of tillage systems on storage pores inside the aggregates measured by mercury intrusion porosimetry along soil profile (MT: minimum tillage; RS: ripper subsoiling; CP: conventional deep ploughing). Values differ significanly when followed by different letters at $P \leq 0.05$.

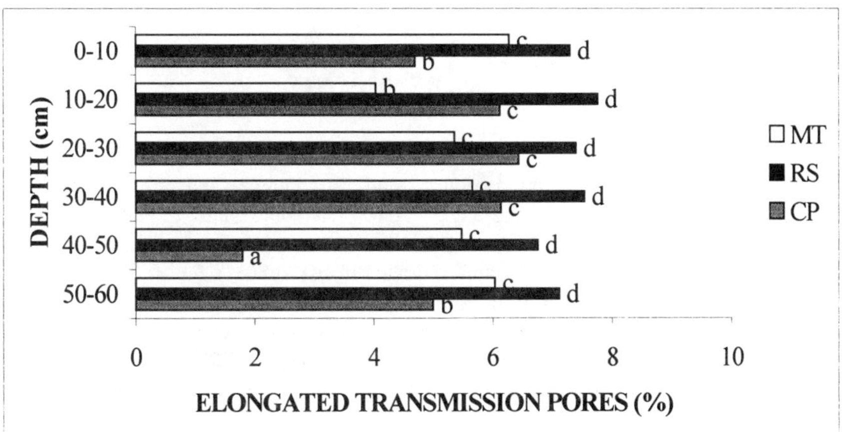

Figure 5. Effects of tillage systems on elongated transmission pore distribution along soil profile expressed as a percentage of total area occupied by pores ranging from 50-500 μm per thin section (MT, minimum tillage; RS, ripper subsoiling; CP: conventional deep ploughing). Elongated transmission pore values differ significantly when followed by different letters at $P \leq 0.05$ employing the Duncan's Multiple Range Test. (Modified from Pagliai et al., 2000).

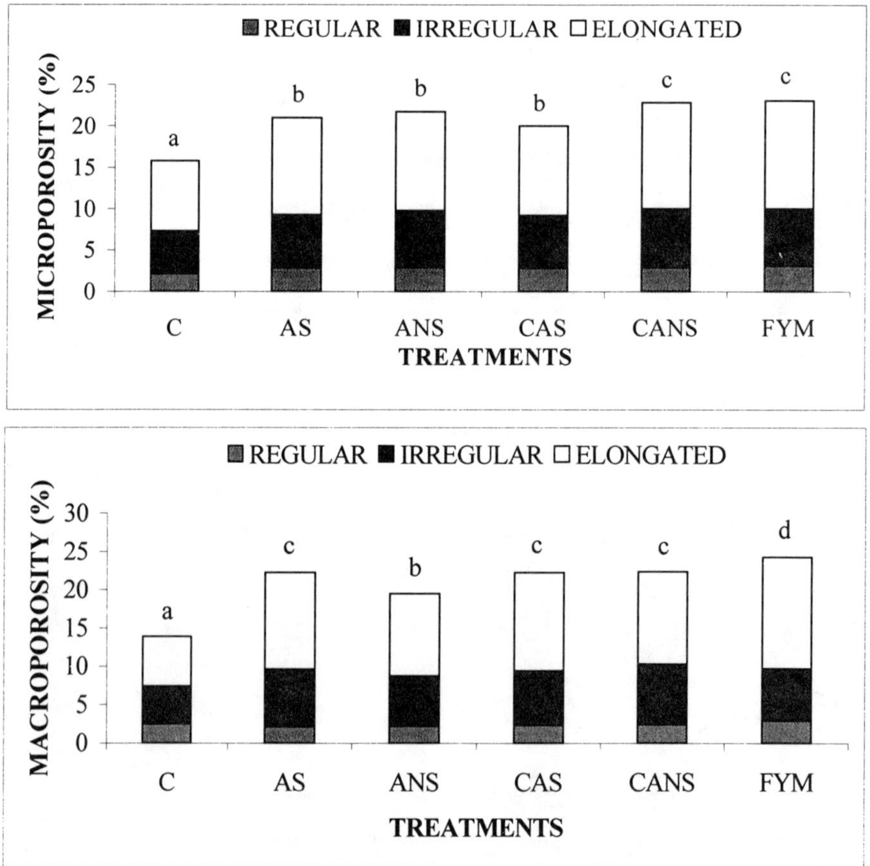

Figure 6. Effect of surface application of sludges, composts and farmyard manure (C: chemical fertilisation; AS: aerobic sludges; ANS: anaerobic sludges; CAS: compost of aerobic sludges and the organic fraction of urban refuse (40:60%); CANS: compost of anaerobic sludges and the organic fraction of urban refuse (20:80%); FYM: farmyard manure) on soil microporosity (0.5-50 μm) and macroporosity (>50 μm) of a sandy loam soil expressed as a percentage of total area of pores per thin section; mean of six replicates. Total porosity values followed by the same letter are not significantly different at the 0.05 level as determined by Duncan's multiple range test. (Modified by Pagliai and Vittori Antisari, 1993).

Long-term field experiments in different types of soils, representative of the typical pedological environments, have shown that alternative tillage systems, like minimum tillage, ripper subsoiling, etc., improve the soil pore system increasing the storage pores (Fig. 4) and the amount of the elongated transmission pores (Fig. 5) with respect to the continuous conventional ploughing. The resulting soil structure is more open and more homogeneous, thus allowing better water movement, as confirmed by the higher values of hydraulic conductivity measured in minimally tilled soils (Pagliai et al., 2000). Also the water content is generally higher in reduced tilled soils. The continuous conventional tillage, moreover, causes a decrease of soil organic matter content that is associated to a decrease of aggregate stability, leading, as a consequence, to the formation of surface crusts, with an increase of runoff and erosion risks, and compact structure.

Management practices can also affect biological activity and can therefore affect both the formation and preservation of biopores, important for water movement and root development (Kooistra, 1991; Pagliai and De Nobili, 1993).

The fractal parameters of pore surfaces as derived from micromorphological data can be useful to evaluate the impact of long-term management practices on soil structure (Pachepsky et al., 1996).

7.2 Applications of Manures

The application of organic materials to the soil has also been shown to enhance both soil porosity and pore size distribution. Fig. 6 summarizes results obtained in a long-term field experiment on a sandy loam soil in which annual application of sewage sludges and composts from urban refuse were compared to the applications of farmyard manure or chemical fertilizers (Pagliai et al., 1981; 1983a; Pagliai and Vittori Antisari, 1993).

Similar results were obtained with the application of pig slurry to a silty clay soil. In this experiment, a positive relationship between the rate of application and the increase of soil porosity was found. It also emerged that the time of application was important, with the best results being obtained for spring applications (Pagliai et al., 1983a; 1985). The increase of pore space and the modification of the pore system in soil treated with these organic materials was associated with an increase of aggregate stability (Pagliai et al., 1981). Thus, from these cited data there is clear evidence that sludges, composts and livestock effluents can improve the physical properties of soil in a similar way to manure.

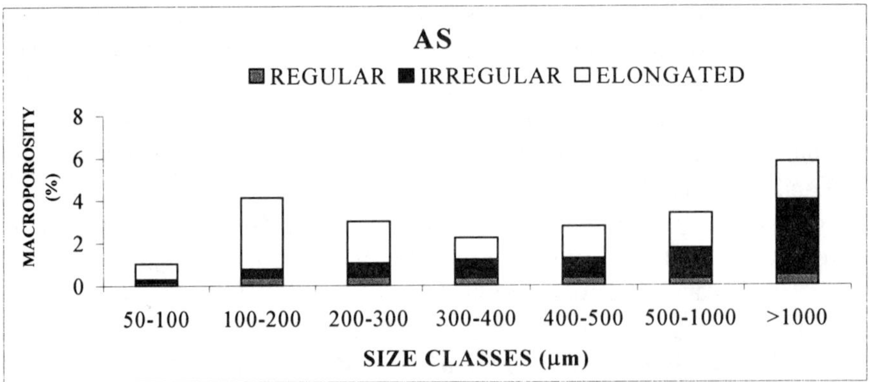

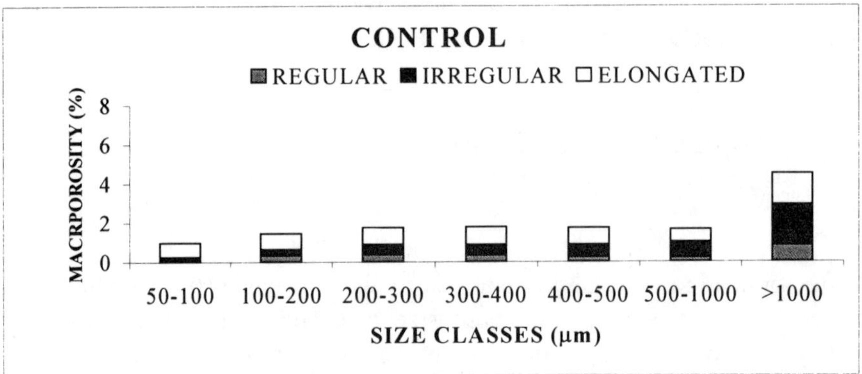

Figure 7. Pore size distribution of pores larger than 50 μm, according to the equivalent pore diameter for regular and irregular pores and width for elongated pores, in the control plots of a sandy loam soil and in those treated with aerobic sludges.

7.3 Irrigation

Intensive continuous cultivation is not the only cause of soil structure degradation; other management practices like irrigation are also important, especially in the longer term. Mathieu (1982) showed that after 15 years of impounding irrigation in clay soils, the structural conditions appeared greatly changed: massive structure, modification of the pore shape and pore size distribution, increased migration of clay particles from the ploughed horizon Ap to the B horizon. Pezzarossa et al. (1991) found that in a peach orchard near Verona (Italy) soil porosity of the surface layer decreased during the irrigation season and that the decrease was significantly greater when irrigation was by impounding rather than under sprinkler-irrigation, due to a reduction in elongated pores (Fig. 8). The latter was associated with a lower water content in the surface soil and reduced root density. The larger amount of water applied with this system caused progressive soil compaction, resulting in decreased porosity and structural degradation.

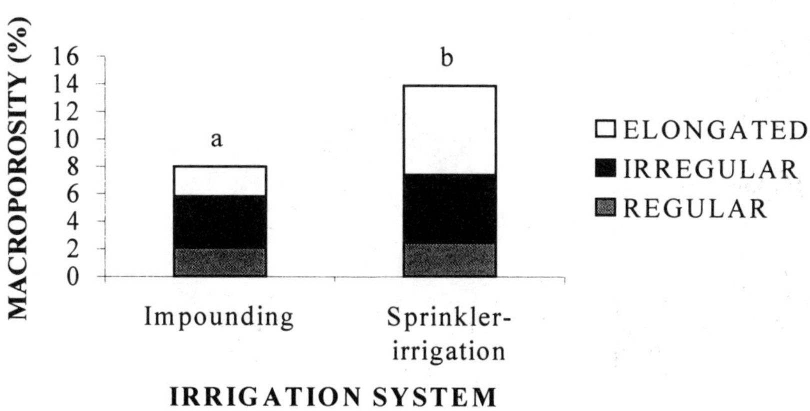

Figure 8. Effect of two irrigation systems on soil macroporosity (>50 μm) of a peach orchard sandy loam expressed as a percentage of total area of pores per thin section; mean of six replicates. Total porosity values followed by the same letter are not significantly different at the 0.05 level as determined by Duncan's multiple range test. (Modified by Pagliai, 1992).

Fig. 9 visually explains the results of Fig. 8: in fact, in the sprinkler-irrigated soil a subangular blocky structure can be observed, while in the soil irrigated by impounding the structure was more compact, rather massive, with drastic reduction of both the total amount and the size and continuity of elongated pores. This caused decreased water infiltration and, under these conditions, 30% of applied water was lost by run-off. It is clearly intuitive that run-off along the inter-row transported nutrients, particularly nitrogen and potassium. For example, results reported by Pagliai (1992) showed that the losses of potassium amounted to 25% of the quantity applied as fertilizer (150 kg/ha/year of K_2O).

Sprinkler irrigation consumes only half the amount of water needed for impounding and causes no evident damage to soil structure, so the loss of water by run-off is negligible but, unfortunately, this method is more expensive and complicated for the farmer. However, for the real conservation of environmental resources and over all to reduce risks of soil structure degradation, erosion and pollution, it is necessary to consider and rationalize all the management practices.

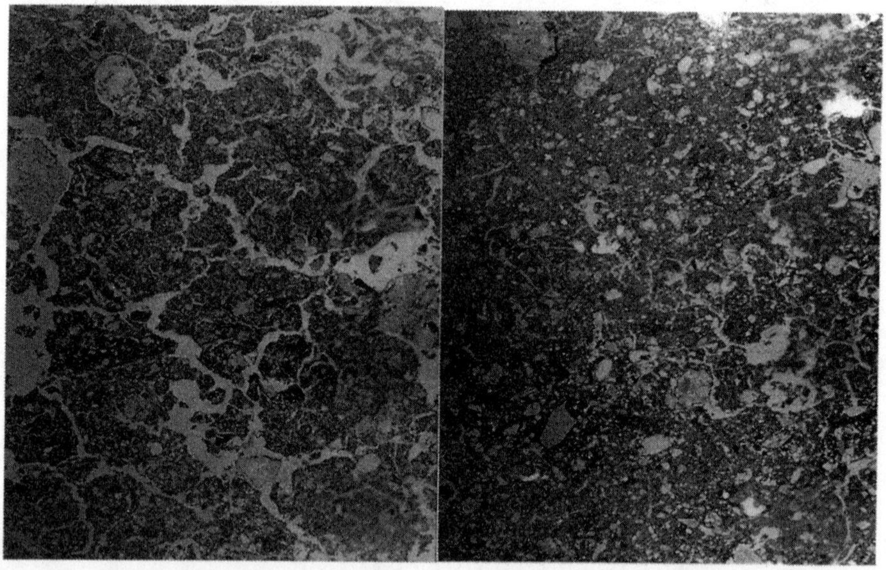

Figure 9. Macrophotographs of vertically oriented thin sections prepared from undisturbed samples from the surface layer (0-10 cm) of a peach orchard sandy loam under sprinkler irrigation (left) and irrigated by impounding (right). The transformation from a subangular blocky structure in sprinkler-irrigated soil to a rather massive structure in soil irrigated by impounding is very evident. Frame length 3×5 cm.

8. QUANTIFICATION OF SOME ASPECTS OF SOIL DEGRADATION

8.1 Soil Compaction

Soil compaction is another of the most important factors responsible for environmental degradation. It causes strong modifications to soil structure and reduces soil porosity.

Soil compaction is caused by a combination of natural forces, which generally act internally, and by man-made forces related to the consequences of soil management practices. The latter forces are mainly those related to vehicle wheel traffic and tillage implements and have a much greater compactive effect than natural forces such as raindrop impact, soil swelling and shrinking, and root enlargement. This is because trends in agricultural engineering over the last few decades have resulted in machines of a greater size and weight. Therefore, soil compaction has become one of the most significant aspects of soil degradation and problems of finding tyres, inflation pressures, etc., able to reduce soil compaction are far from being solved. It is therefore fundamental to evaluate the impact of wheel traffic on soil structure and porosity measurements can help to quantify the degradation effects of compaction.

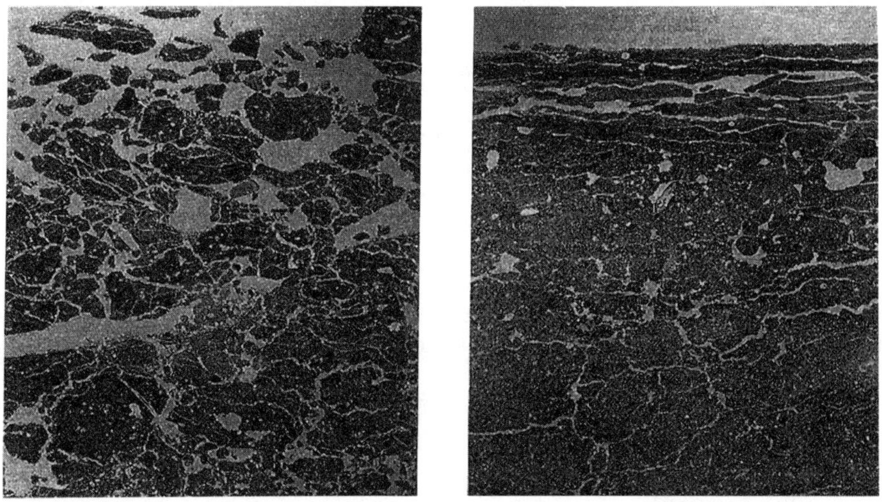

Figure 10. Macrophotographs of vertically oriented thin sections from the surface layer (0-5 cm) of the uncompacted (left) and compacted areas (right) of a loam soil. Plain polarized light. Pores appear white. The change of the subangular blocky structure of the uncompacted areas into a massive platy structure of the compacted areas is very evident. Frame length 3 cm.

Results showed that compaction, both in agricultural and forestry soils, not only reduces total soil porosity but also modifies the pore system. In fact, the proportion of elongated pores, useful for water movement and root growth is strongly reduced in compacted soil. The modifications to the pore system also changes the type of soil structure: the platy structure is a common feature in compacted cultivated soil (Fig. 10). Results also showed that the reduction of porosity and of elongated pores following compaction, is strictly related to the increase of penetration resistance and to the decrease of hydraulic conductivity (Fig. 11) and root growth (Marsili et al., 1998; Richard et al., 2001). Soil regeneration after compaction depends on the type of soil and on the degree of damage to the soil (Bullock et al., 1985b).

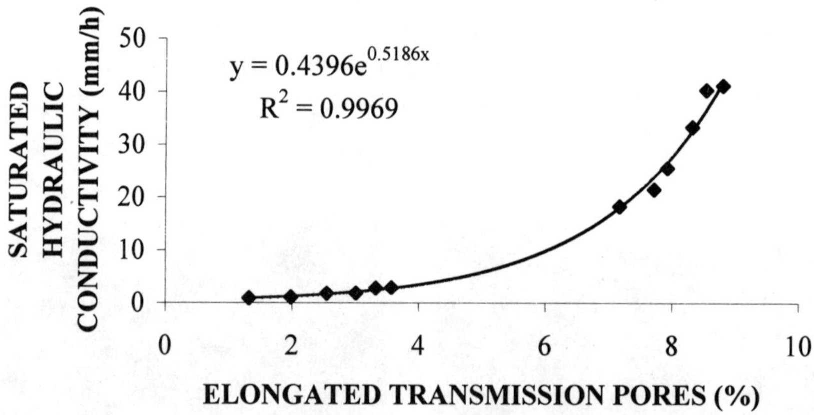

Figure 11. Exponential correlation between soil porosity formed by elongated pores and saturated hydraulic conductivity in the surface layer (0-10 cm) of the compacted (elongated pores less than 4%) and uncompacted areas (elongated pores greater than 7%).

8.2 The Formation of Compacted Layers along the Profile (Ploughpans)

Soil compaction is not only caused by wheel traffic but also to the shear strength of tillage implements, like the compact layer (ploughpan) formed at the lower limit of cultivation in continuous ploughed soils (Fig. 12). As already said, Fig. 5 shows the strong decrease of elongated transmission pores and over all of hydraulic conductivity, respectively in the 40-50 cm layer in conventionally tilled soil (Pagliai et al., 2000). Subsoil compaction is strongly under evaluated, even though such a ploughpan is largely widespread in the alluvial soils of the plains cultivated by monoculture and it is responsible of the frequent flooding of such plains in occasion of heavy rains concentrated in a short time (rainstorm), because the presence of this ploughpan strongly reduced drainage. Alternative tillage practices, like ripper subsoiling, are able to avoiding the formation of this compact layer.

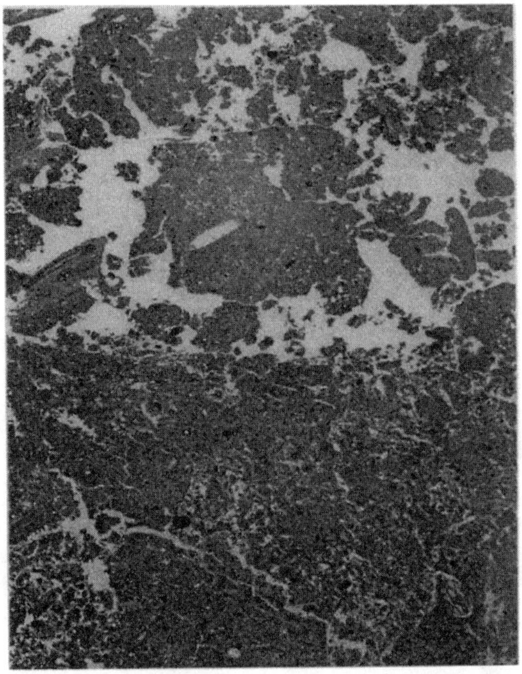

Figure 12. Macrophotographs of vertically oriented thin sections prepared from undisturbed samples from the 40-50 cm layer of a clay loam soil tilled by continuous deep ploughing. The lower limit of cultivation (ploughpan) is visible. Frame length 3×5 cm.

8.3 Soil Crusting

Surface crusts are another of the dangerous aspects of soil degradation, and are formed by raindrop impact, which causes the mechanical destruction of soil aggregates. The dispersed particles, that can be also translocated by runoff, in the following drying process their deposition causes the formation of a compact layer of horizontally oriented plate-like particles at the soil surface. This compact layer contains few, if any, large pores. Generally the surface crusts are formed by several thin layers of fine particles. These are intercalated by thin elongated pores parallel oriented to soil surface, not continuous in a vertical sense (Fig. 13), or by rounded pores (vesicles) formed by air trapped during drying. The presence of such pores in the topsoil formed a vesicular structure that can be regarded as an indicator of an unstable and transitory structure induced by poor stability of soil aggregates. Soil crusting reduces seedling emergence, soil-atmosphere gas exchange, water infiltration and increases surface runoff. Results showed that addition of organic materials to soil and reduced tillage practices are able to prevent crust formation. In intensively cultivated soils the surface aggregates are less rain-stable. The decrease of soil porosity and particularly of elongated transmission pores in conventional ploughed soils, as reported in Fig. 5, can be ascribed to the presence of the surface soil crusts that negatively affect the hydraulic conductivity (Pagliai et al., 2000). Such crusts were less developed in soil under minimum tillage and ripper subsoiling (Fig. 13).

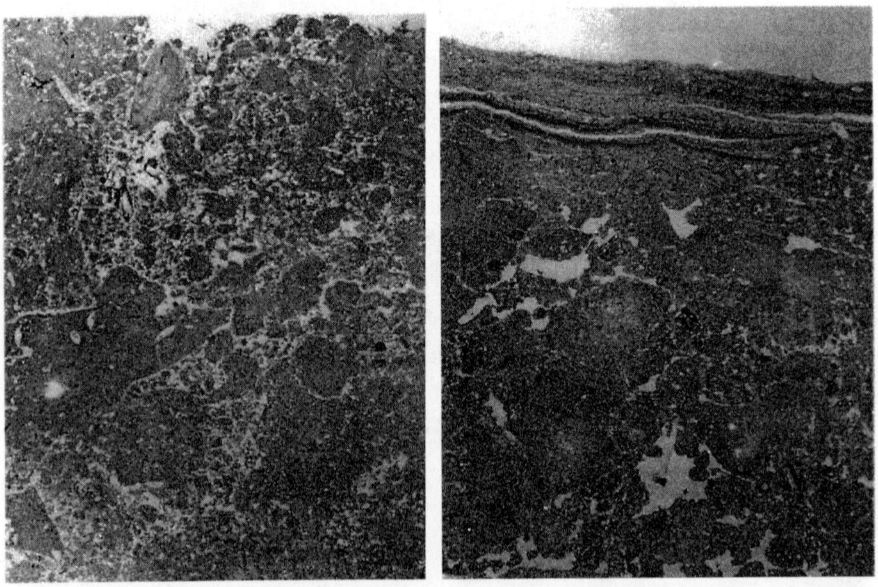

Figure 13. Macrophotographs of vertically oriented thin sections from the surface layer (0-5 cm) of the minimum tilled (left) and deep ploughed (right) soils. Plain polarized light. Pores appear white. The change of the subangular blocky structure of the MT soil into a platy structure of the DP soil, due to the surface crust, is very evident. Frame length 3 cm.

9. CONCLUSIONS

The characterisation of the soil pore system gives essential indications about soil quality and vulnerability in relation to degradation events mainly connected with human activity. Such a characterisation is especially useful in the study of the relationships between soil physical, chemical and biochemical properties and provides a realistic basis for understanding water retention and water movement in soil. In fact, the quantitative evaluation of water movement and solute transport along the macropores opens new horizons in the modelling of these phenomena. This is one of the new approaches in the study of soil, since up to now water movement in macropores has not been adequately considered. Some traditional concepts of soil physics need to be reconsidered or modified: for example, the concept of available water for plants should be associated with the concept of accessible water.

The characterisation of soil pore system, by means of image analysis on thin sections, can provide basic information for the study of soil. The major disadvantage of this technique is that the preparation of soil thin sections is both difficult and time consuming. However, many public and private laboratories are now equipped for the preparation of soil thin sections and the development of improved computer software has made the analysis of the images easy.

When the obstacle of the acquisition of soil thin sections is overcome, it will be possible to benefit from the full potential of this technique, most importantly to quantify the changes in soil structure following human activities. Therefore, on the basis of the acquired experience, it is possible to go deep into the analysis of soil thin sections in relation to aspects of water movement. The quantification of the size, continuity, orientation and irregularity of elongated pores allows the modelling of water movement and solute transport, or, at least, allows the prediction of the changes which can be expected following soil structural modifications, or following soil degradation due to compaction, formation of surface crusts, etc. The quantification of the damage caused by degradation processes also makes it possible to predict the risk of soil erosion.

REFERENCES

Bouma, J., 1992. Influence of soil macroporosity on environmental quality. *Advances in Agronomy*, **46**: 1-37.

Bouma, J., Jongerius, A., Boersma, O.H., Jager, A. and Schoonderbeek, D., 1977. The function of different types of macropores during saturated flow through four swelling soil horizons. *Soil Sci. Soc. Am. J.*, **41**: 945-950.

Bouma, J., Jongerius, A. and Schoonderbeek, D., 1979. Calculation of saturated hydraulic conductivity of some pedal clay soils using micromorphometric data. *Soil Sci. Soc. Am. J.*, **43**: 261-264.

Brewer, R., 1964. *Fabric and Mineral Analysis of Soils*. John Wiley, New York, 470 pp.

Bruand, A., Ould Mohamed, S., Pagliai, M. and Bégon, J.C. 1992. Modifications structurales (assemblage élémentaire) liées à l'activité racinaire au sein de sols limono-argileux. *C. R. Acad. Sci. Paris*, t. 315, Série II, p. 757-754.

Bruand, A., Cousin, I., Nicoullaud, B., Duval, O. and Bégon, J.C. 1996. Backscattered Electron Scanning Images of soil porosity for analyzing soil compaction around roots. *Soil Sci. Soc. Am. J.*, **60**: 895-901.

Bullock, P., Fedoroff, N., Jongerius, A., Stoops, G. and Tursina, T., 1985a. Handbook for soil thin section description. Waine Res. Pub., Wolverhampton.

Bullock, P., Newman, A.C.D., Thomasson, A.J., 1985b. Porosity aspects of the regeneration of soil structure after compaction. *Soil Tillage Res.* **5**, 325-341.

Childs, E.C., 1969. *An introduction to the physical basis of soil water phenomena.* London: Wiley-Interscience.

Dexter, A.R., 1988. Advances in characterization of soil structure. *Soil and Tillage Research*, **11**: 199-238.

Fiès, J.C., 1992. Analysis of soil textural porosity relative to skeleton particle size, using mercury porosimetry. *Soil Sci. Soc. Am. J.*, **56**: 1062-1067.

Fitz Patrick, E.A., Makie, L.A. and Mullins, C.E., 1985. The use of plaster of Paris in the study of soil structure. *Soil Use Management*, **1**: 70-72.

Greenland, D.J., 1977. Soil damage by intensive arable cultivation: temporary or permanent? *Phil. Trans. Royal Soc. London*, **281**: 193-208.

Greenland, D.J., 1981. Soil management and soil degradation. *Journal of Soil Science*, **32**: 301-322.

Kooistra, M.J., 1991. Micromorphological approach to the interactions between soil structure and soil biota. *Agriculture, ecosystems and environment*, **34**: 315-328.

Kubiena, W.L., 1970. *Micromorphological features of soil geography*. Rutgers Univ. Press, New Brunswick, 254 pp.

Kutilek, M., 1983. Soil physical properties of saline and alkali vertisols. In *Isotope and Radiation Technique in Soil Physics and Irrigation Studies*. IAEA, Vienna, pp. 179-190.

Kutilek, M., 1996. Water relation and water management of vertisols. In *Vertisols and technologies for their management* (N. Ahmad and A. Mermut Eds). Elsevier, pp. 201-230.

Kutilek, M. and Nielsen, D.R., 1994. *Soil Hydrology*. Geo-ecology textbook, Catena, Verlag.

Marshall, T.J., 1958. A relation between permeability and size distribution of pores. *Journal of Soil Science*, **9**: 1-8.

Marsili, A., Servadio, P., Pagliai, M. and Vignozzi, N., 1998. Changes of some physical properties of a clay soil following passage of rubber- and metal-tracked tractors. *Soil Tillage Res.*, **49**, 185-199.

Mathieu, C., 1982. Effects of irrigation on the structure of heavy clay soils in north east Marocco. *Soil Tillage Res.*, **2**: 311-329.

Mele G., Basile A., Leone A., Moreau E., Terribile F. and Velde B., 1999. The study of soil structure by coupling serial sections and 3D image analysis. In *Modelling of transport processes in soils at various scales in time and space.*. (Feyen and Wiyo Eds), Leuven (Belgium), 24-26 November 1999, pp 142-152.

Mermut, A.R., Grevers, M.C.J. and de Jong, E., 1992. Evaluation of pores under different management systems by image analysis of clay soils in Saskatchewan, Canada. *Geoderma*, **53**: 357-372.

Moran, C.J. and McBratney, A.B., 1992. Acquisition and analysis of three-component digital images of soil pore structure. II. Application to seed beds in a fallow management trial. *J. Soil. Sci.*, **43**: 551-566.

Moran, C.J., McBratney, A.B. and Koppi, A.J., 1989. A rapid method for analysis of soil macropore structure. I. Specimen preparation and digital binary image production. *Soil Sci. Soc. Am. J.*, **53**: 921-928.

Murphy, C.P., 1986. *Thin section preparation of soils and sediments.* AB Academic Publishers, Herts. pp. 149.

Murphy, C.P., Bullock, P. and Turner, R.H., 1977a. The measurement and characterization of voids in soil thin sections by image analysis: Part I. Principles and techniques. *Journal of Soil Science*, **28**: 498-508.

Murphy, C.P., Bullock, P. and Biswell, K.J., 1977b. The measurement and characterization of voids in soil thin sections by image analysis: Part II. Applications. *Journal of Soil Science*, **28**: 509-518.

Pachepsky, Ya., Yakovchenko, V., Rabenhorst, M.C., Pooley, C. and Sikora, L.J., 1996. Fractal Parameters of pore surfaces as derived from micromorphological data: effect of long term management practices. *Geoderma*, **74**: 305-319.

Pagliai, M., 1988. Soil porosity aspects. *Intern. Agrophysics*, **4**: 215-232.

Pagliai, M. 1992. *Losses of Potassium through run-off, erosion and "slash and burn" agriculture.* Proceedings 23rd IPI Colloquium. Prague, Czechoslovakia 12-16 October 1992. pp. 345-362.

Pagliai, M., Bisdom, E.B.A. and Ledin, S. 1983a. Changes in surface structure (crusting) after application of sewage sludges and pig slurry to cultivated agricultural soils in northern Italy. *Geoderma*, **30**: 35-53.

Pagliai, M. and De Nobili, M. 1993. Relationships between soil porosity, root development and soil enzyme activity in cultivated soils. *Geoderma*, **56**: 243-256.

Pagliai, M., Guidi, G., La Marca, M., Giachetti, M. and Lucamante, G., 1981. Effect of sewage sludges and composts on soil porosity and aggregation. *Journal Environmental Quality*, **10**: 556-561.

Pagliai, M., La Marca, M. and Lucamante, G., 1983b. Micromorphometric and micromorphological investigations of a clay loam soil in viticulture under zero and conventional tillage. *J. Soil Sci.*, **34**: 391-403.

Pagliai, M., La Marca, M. and Lucamante, G., 1985. Relationship between soil structure and time of landspreading of pig slurry. In *Long-term Effects of Sewage Sludge and Farm Slurries Applications* (J.H. Williams, G. Guidi and P. L'Hermite Eds.), Elsevier Applied Science Publishers, London, pp. 45-56.

Pagliai, M., La Marca, M., Lucamante, G. and Genovese, L., 1984. Effects of zero and conventional tillage on the length and irregularity of elongated pores in a clay loam soil under viticulture. *Soil Tillage Res.*, **4**: 433-444.

Pagliai, M., Pellegrini, S., Vignozzi, N., Rousseva, S., Grasselli, O. 2000. The quantification of the effect of subsoil compaction on soil porosity and related physical properties under conventional to reduced management practices. *Advances in GeoEcology*, **32**, 305-313.

Pagliai, M., Pezzarossa, B., Mazzoncini, M. and Bonari, E., 1989. Effects of tillage on porosity and microstructure of a loam soil. *Soil Technology*, **2**: 345-358.

Pagliai, M. and Vittori Antisari, L., 1993. Influence of waste organic matter on soil micro and macrostructure. *Bioresource Technology*, **43**: 205-213.

Pezzarossa, B., Pagliai, M. and Xiloyannis, C., 1991. Effetti di differenti sistemi di irrigazione su alcune caratteristiche fisiche di un terreno investito a pescheto. *Rivista di Frutticoltura*, **7-8**: 71-75.

Richard, G., Cousin, I., Sillon, J.F., Bruand., A. and Guerif, J., 2001. Effect of compaction on the porosity of a silty soil: influence on unsaturated hydraulic properties. *European Journal of Soil Science*, **52**: 49-58.

Ringrose-Voase, A.J. and Bullock, P., 1984. The automatic recognition and measurement of soil pore types by image analysis and computer programs. *Journal of Soil Science*, **35**: 673-684.

Ringrose-Voase, A.J. and Nortcliff, S., 1987. The application of stereology to the estimation of soil structural properties: a review. In *Soil Micromorphology* (N. Fedoroff, L.M. Bresson and M.A. Courty Eds.), Proc. VIIth Int. Working Meeting on Soil Micromorphology. Association Francaise pour l'Etude du Sol, Plasir, France, pp. 81-88.

Schweikle, V., 1982. *Gefügeeigenschaften von Tonböden.* Verlag Eugen Ulmer, Stuttgart.

Shipitalo, M.J. and Protz, R., 1987. Comparison of morphology and porosity of a soil under conventional and zero tillage. *Canadian Journal of Soil Science*, **67**: 445-456.

The Horizontal Hydraulic Conductivity of Vertical Interaggregate Capillary Cracks in Clay Soils

VICTOR.Y. CHERTKOV
Faculty of Agricultural Engineering, Technion, Haifa, Israel

1. INTRODUCTION

The contribution of an interaggregate-capillary crack network to the hydraulic conductivity of an aggregated clay soil, in different directions, is of significant interest for the three-dimensional preferential flow problem. Based on the crack network geometry and the cubic law for water flow in a slit Chertkov and Ravina (2001) considered the hydraulic conductivity of a capillary crack network and estimated the vertical hydraulic conductivity of a system of vertical interaggregate cracks in the clay of Beit Netofa, Israel, from Mualem's (1976) data. At sufficiently high water contents the contribution of interaggregate capillary cracks to the vertical hydraulic conductivity was more prevalent then that of the soil matrix. In the light of these results, we are also interested in the modeling of the horizontal hydraulic conductivity of vertical interaggregate capillary cracks in the frame of the same approach. The contribution of the cracks to the horizontal hydraulic conductivity is also expected to prevail. Validation of the model is based on the comparison between available data and the model prediction on the horizontal hydraulic conductivity of the vertical interaggregate cracks and on their specific volume. First, let us summarize the major concepts, results, and relationships of the basic model (Chertkov and Ravina, 2001).

The hydraulic conductivity, $K(\overline{n}, R_W)$ of a capillary-crack network with no rough walls, filled in part with water up to crack width, R_W, along a unit vector $\overline{n}$ is given by

$$K\left(\bar{n}, R_w\right) = \frac{g}{12\nu T\left(\bar{n}, R_w\right)} \int_0^{R_w} R^3 dL\left(\bar{n}, R\right) \tag{1}$$

where g is the acceleration of gravity, ν is the kinematic viscosity of water, $L(\bar{n}, R)$ is a specific length of traces of connected water filled cracks of width $<R$ at a soil cross-section normal to the unit vector $\bar{n}$, and $T(\bar{n}, R_W)$ is a spatial tortuosity of the water filled part of the crack network along the direction $\bar{n}$.

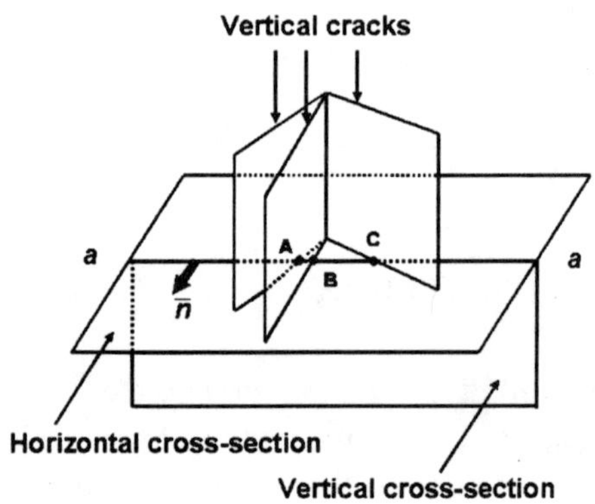

Figure 1. A system of vertical connected cracks with horizontal and vertical cross-sections

A system of the vertical interaggregate capillary cracks (Fig.1) is considered, with isotropic traces at a horizontal cross-section and width R distributed according to Poisson distribution. The length of the crack traces of width $< R$ at a horizontal cross-section, $L(R)$ is written then by

$$L(R) = L_t(1 - \exp(-R / \bar{R})), \qquad R < R_m \tag{2}$$

where R_m is the maximum width of the cracks, $\bar{R}$ is the mean crack width

$$\bar{R} \cong R_m / 4.6 \tag{3}$$

and $L_t \cong L(R_m)$ is the total specific length of crack traces at the horizontal cross-section (per unit area).

In the case of the capillary vertical-crack network, the vertical tortuosity of its water filled part, is equal to unity and does not depend on the maximum width of water filled cracks at a given water content, R_w.

Interaggregate capillary cracks are considered to be totally water filled at water content of the soil matrix in the range $\theta_r < \theta < \theta_s$ where θ_r and θ_s are the residual and saturated water content, respectively, i.e.,

$$R_w = R_m \tag{4}$$

Finally, in what follows, the expression for the specific volume of interaggregate vertical cracks, $V(\theta)$ will also be needed

$$V(\theta) = 2.05\frac{\overline{R}(\theta)}{l} \tag{5}$$

where l is the characteristic (mean) dimension of soil aggregates.

2. THEORY

Similar to Eq.(2) one can easily find that the specific number of crack traces of width < R intersecting a vertical cross-section, $N(R)$ (see points A, B, and C in Fig.1, per unit length of *aa* line) is given by

$$N(R) = N_t(1 - \exp(-R / \overline{R})) , \qquad R < R_m \tag{6}$$

with $N_t \cong N(R_m)$ denoting the total specific number of intersections between the crack traces and *aa* line (Fig.1).

According to Eq.(4) the tortuosity of the interaggregate (see Eq.(1)) crack network does not depend on R_w for any direction. It is also clear that the horizontal tortuosity of the vertical cracks with isotropic traces at a horizontal cross-section (Fig.1) does not depend on a direction $\overline{n}$ in the horizontal plane. Thus, in the case under study the tortuosity can be written as

$$T(\overline{n}, R_w) = T \tag{7}$$

with T denoting some constant.

By definition the spatial tortuosity of a crack network in a direction is equal to the ratio of the specific area of connected cracks (per unit volume) to the specific length of crack traces at a cross-section normal to the direction. Fig.1 shows that N_t (Eq.(6)) coincides with the total specific length of the crack traces at a vertical cross-section (per unit area), and L_t (Eq.(2)) coincides with the total specific area of vertical cracks (per unit volume). Thus, according to the above definition the constant horizontal tortuosity of the crack network, T is determined by the ratio

$$T = L_t / N_t \tag{8}$$

On the other hand two-dimensional tortuosity, T_2 of the isotropic connected-crack traces at the horizontal cross-section (Fig.1) is defined as (Chertkov and Ravina, 1999)

$$T_2 = L_t / N_t \tag{9}$$

(L_t and N_t values are identical to values of L_2 and L_1, respectively, used by Chertkov and Ravina (1999)). It follows from Eqs.(7)-(9) that in the case under consideration

$$T(\bar{n}, R_w) = T_2 \tag{10}$$

Finally, Fig.1 shows that $N(R)$ coincides with the specific length of crack traces (of width $<R$) at the vertical cross-section (per unit area). That is, in the case to be studied, in Eq.(1)

$$L(\bar{n}, R) = N(R) \tag{11}$$

In addition, from Eqs.(2), (6), and (9) it follows that

$$N(R) = L(R) / T_2 \tag{12}$$

Replacing in Eq.(1) $L(\bar{n},R)$ by $N(R)$ (Eq.(11)), $T(\bar{n},R_W)$ by T_2 (Eq.(10)), and R_W by $R_m(\theta)$ (Eq.(4)) one finds the horizontal hydraulic conductivity of the interaggregate-vertical-crack network (Fig.1) to be

$$K_h(\theta)=\frac{g}{12\nu}\cdot\frac{1}{T_2}\cdot\int_0^{R_m(\theta)} R^3 dN(R) \tag{13}$$

Replacing the $N(R)$ function in Eq.(13) from Eq.(12) and accounting for the fact that the vertical tortuosity of the crack network is equal to unity, $K_h(\theta)$ can be written as

$$K_h(\theta)=\frac{g}{12\nu}\cdot\frac{1}{T_2^{\,2}}\cdot\int_0^{R_m(\theta)} R^3 dL(R)=K_v(\theta)/T_2^{\,2} \tag{14}$$

with $K_v(\theta)$ denoting the vertical hydraulic conductivity function of the crack network. In the case under study $T_2=\pi/2$ (Chertkov and Ravina, 2001). Therefore Eq.(14) leads to the relation

$$K_v(\theta)/K_h(\theta)=(\pi/2)^2\cong 2.47 \tag{15}$$

Substituting for $K_v(\theta)$ in Eq.(14) the expression from Chertkov and Ravina's (2001) Eq.(18) (at $R_W=R_m=4.6\bar{R}$, see Eq.(3)) and accounting for $T_2=\pi/2$ one comes to the relation

$$K_h(\theta) = 3.56 \frac{g}{12\nu} \cdot \frac{\bar{R}(\theta)^3}{l} \tag{16}$$

where l is the characteristic (mean) dimension of soil aggregates.

Relations given by Eqs.(16), (5), and (15) should be tested. Available data (see the next Section) are only related to the saturated state.

3. DATA AND PROCEDURES TO BE USED

3.1 Data from McKay et al. (1993)

In the following we will take advantage of data from McKay et al. (1993). These researchers studied crack characteristics (width and specific volume) in the upper 1.5-5.5m of a weathered clay-rich till (Laidlaw site, 12 km south of Sarnia, Ontario, Canada) based on data of the saturated horizontal hydraulic conductivity. Tube piezometers or seepage collectors typically intersecting fewer then ten cracks were used for small-scale field measurements. The corresponding range of saturated horizontal hydraulic conductivity was 10^{-10} to 10^{-6} m/s. The hydraulic conductivity values from large-scale measurements were based on influx/discharge from trenches (depth 5.5m) intersecting thousands of cracks. The corresponding range of saturated horizontal hydraulic conductivity was 10^{-7} to $3 \cdot 10^{-7}$ m/s. In addition, the mean hydraulic conductivity of the clay matrix, K_m was determined to be $2 \cdot 10^{-10}$ m/s in the laboratory using an odometer (consolidation) test. For interpretation of the horizontal-hydraulic-conductivity data in terms of the crack characteristics McKay et al. (1993) used a model of two mutually orthogonal vertical crack systems with constant crack width and spacing as well as the cubic law for water flow in cracks with no rough walls. Estimates of the crack width and specific crack volume from McKay et al. (1993) were 1-43μm and $3 \cdot 10^{-5}$ to $2 \cdot 10^{-3}$, respectively.

McKay et al. (1993) also cited or hydraulically derived the crack characteristics for seven other clay-rich sites across North America (for depths up to 20m) based on data from D'Astous et al. (1989), Grisak (1975), Day (1977), Keller et al. (1986), Thompson (1990), Hendry et al. (1986), and Rudolph et al. (1991). The estimates of crack width from McKay et al. (1993) and the estimates of six sites (from the above seven) is emphasized to fall in the same remarkably narrow range. Data from D'Astous et al. (1989),

Grisak (1975), Day (1977), Keller et al. (1986), Hendry et al. (1986), and Rudolph et al. (1991) give upper values of the mean crack width 32, 4, 14, 11, 50, and 30μm, respectively. The hydraulically estimated mean width of interaggregate cracks ($\overline{R} \cong R_m / 4.6$) in Beit Netofa clay, Israel (Chertkov and Ravina, 2001; $\overline{R} = 1.3\mu\text{m}$) also fall in this narrow range. The same relates to Bronswijk's (1991) Dutch clay where the estimate based on Chertkov and Ravina's (2001) approach gives $\overline{R} = 4.6\mu\text{m}$. Accounting for this narrow range, in what follows, we take the mean value of crack width from the above six works and two our estimates ($\overline{R} = 1.3\mu\text{m}$ and $4.6\mu\text{m}$) as an upper boundary of the mean value of interaggregate crack width in a clay-rich soil. This upper boundary, $\overline{R}_{\text{up}}$ is

$$\overline{R}_{\text{up}} = (32 + 4 + 14 + 11 + 50 + 30 + 1.3 + 4.6) / 8 \cong 18\mu\text{m} \tag{17}$$

Note the estimates of McKay et al. (1993) of a crack width in the saturated state were not included because we used below their data on the saturated horizontal hydraulic conductivity from small- and large-scale measurements.

We assume from Fig.5 of McKay et al. (1993) that the maximum aggregate dimension for the Laidlaw site is $l_m \cong 0.1\text{m}$. Then, like Eq.(3) the mean aggregate dimension is

$$l \cong l_m / 4.6 \cong 0.0217\text{m} \tag{18}$$

The lower boundary of the mean crack width in the saturated state, $\overline{R}_{\text{low}}$ is determined from Eq.(16) after substitution for the value of l from Eq.(18), and for $K_h(\theta_s)$ the above mentioned hydraulic conductivity of the soil matrix ($K_m = 2 \cdot 10^{-10}$ m/s) from McKay et al. (1993) (standard values for $g = 9.81\text{m}/\text{s}^2$ and $\nu = 10^{-6}\text{m}^2/\text{s}$ at 20°C were used). Then

$$\overline{R}_{\text{low}} \cong 1\mu\text{m} \tag{19}$$

The values of $\overline{R}_{\text{low}}$ and $\overline{R}_{\text{up}}$ correspond with range boundaries of the saturated horizontal hydraulic conductivity (of the weathered clay of the Laidlaw site) at the small-scale measurements of McKay et al. (1993). We assume that the mean value

$$\overline{\overline{R}} = (\overline{R}_{\text{up}} + \overline{R}_{\text{low}})/2 \equiv 9.5\mu\text{m} \tag{20}$$

corresponds with the mean value of the saturated horizontal hydraulic conductivity at the large-scale measurements of McKay et al. (1993).

Since we are interested in this Subsection only in the saturated horizontal hydraulic conductivity, to simplify notation hereafter in this Subsection and in Table 1 (see Results and Discussion) we will denote the saturated horizontal hydraulic conductivity from small-scale measurements and large-scale ones by K and $\boldsymbol{K}$, respectively. Based on the data given by Eq.(17)-(20) one can make a number of model predictions of the Laidlaw site clay. Equations (16), (17), and (18) estimate the upper boundary of the saturated horizontal hydraulic conductivity, K_{up} from the small-scale measurements. Equations (16), (20), and (18) give an estimate of the mean value of the saturated horizontal hydraulic conductivity, $\overline{\boldsymbol{K}}$ from the large-scale measurements. Equations (5), (19), and (18) enable one to estimate the lower boundary of the specific crack volume in the saturated state, V_{low}. Similarly, Eqs. (5), (17), and (18) estimate the upper boundary of the specific crack volume in the saturated state, V_{up}.

Based on K_{up} and $\overline{\boldsymbol{K}}$ values one can estimate the total range of the saturated horizontal hydraulic conductivity from the large-scale measurements (for the Laidlaw site clay) in the following way. Standard deviation of the saturated horizontal hydraulic conductivity from the small-scale measurements, δ can be estimated by

$$\delta \cong (K_{\text{up}} - K_{\text{low}})/2 \tag{21}$$

with $K_{\text{low}} \cong K_m$ denoting the lower boundary of the saturated horizontal hydraulic conductivity from the small-scale measurements (K_m is the mean hydraulic conductivity of the clay matrix). The large-scale measurements embrace ~1000 cracks, and the small-scale ones embrace ~10 cracks (McKay et al., 1993). Hence, the standard deviation of the saturated horizontal hydraulic conductivity from the large-scale measurements, Δ can be estimated by

$$\Delta \cong \delta / \sqrt{1000/10} = 0.1\delta \tag{22}$$

Thus, the lower ($\boldsymbol{K}_{\text{low}}$) and upper ($\boldsymbol{K}_{\text{up}}$) boundaries of the range of the saturated horizontal hydraulic conductivity from the large-scale measurements are estimated from Eqs.(21), (22) and $\overline{\boldsymbol{K}}$ value as

$$\boldsymbol{K}_{\text{low}} \cong \overline{\boldsymbol{K}} - \Delta, \qquad \boldsymbol{K}_{\text{up}} \cong \overline{\boldsymbol{K}} + \Delta \tag{23}$$

3.2 Data from Bouma and Dekker (1981)

For validation of the relation given by Eq.(15) (for a vertical-interaggregate-crack system with isotropic orientations of crack traces at a horizontal cross-section and crack width and spacing distributed according to Poisson's distributions) we took advantage of data from Bouma and Dekker (1981). The data are related to the saturated vertical (K_v) and horizontal (K_h) hydraulic conductivity of four types of clay soils. Four separate measurements were conducted in every soil. We are interested in the data on K_v and K_h values for soil type 3 only because, unlike other soil types, this soil (from the province of Gelderland, Holland) did not contain vertical worm channels with a diameter of ~4mm. Hence, the vertical and horizontal hydraulic conductivity in this soil were only stipulated by the vertical cracks (Bouma and Dekker, 1981). The four K_v values were K_v^i = 495, 23, 115, and 47 cm/day (the superscript i = 1, 2, 3, 4), and corresponding K_h values were K_h^i = 131, 43, 42, and 26 cm/day (the superscript i = 1, 2, 3, 4). Using these data one can estimate the mean value, $\overline{K_v / K_h}$ of the ratio, K_v / K_h as

$$\overline{K_v / K_h} = \Big(\sum_{i=1}^{4} K_v^i / K_h^i \Big) / 4, \tag{24}$$

and then a standard deviation of the mean value, $\delta(\overline{K_v / K_h})$. These estimates will be compared with the model prediction (Eq.(15)).

4. RESULTS AND DISCUSSION

Table 1 shows that the model predictions of K_{up}, $\boldsymbol{K}_{low}$, $\boldsymbol{K}_{up}$, V_{low}, and V_{up} found as stated above. Accounting for the model approximations and experimental errors they do not contradict the data (of K_{up}, $\boldsymbol{K}_{low}$, and $\boldsymbol{K}_{up}$) and estimates (of V_{low}, and V_{up}) from McKay et al. (1993) (Table 1).

It is worth noting that the model under study considers a vertical crack system isotropic at a horizontal cross-section unlike the two orthogonal crack systems of McKay et al. (1993). For this reason the number of horizontal ways for water in the model to be considered is more than in the case of McKay et al. (1993) (per unit area of the horizontal cross-section). Therefore, it is not surprising that the upper boundary of the mean crack width (Eq.(17)) is less than 43 μm from McKay et al. (1993).

Table 1. Data and estimates (McKay et al., 1993) against model prediction of saturated horizontal hydraulic conductivity and specific volume in the saturated state of vertical interaggregate capillary cracks

	Saturated horizontal hydraulic conductivity			Specific volume in the saturated state	
	Small-scale measurements	Large-scale measurements			
Type	Upper boundary K_{up}	Lower boundary $\boldsymbol{K}_{low}$	Upper boundary $\boldsymbol{K}_{up}$	Lower boundary V_{low}	Upper boundary V_{up}
	----------m/s----------				
Model	$7.8 \cdot 10^{-7}$	$7.6 \cdot 10^{-8}$	$1.5 \cdot 10^{-7}$	$9.4 \cdot 10^{-5}$	$1.7 \cdot 10^{-3}$
Data	10^{-6}	10^{-7}	$3 \cdot 10^{-7}$		
Estimates				$3 \cdot 10^{-5}$	$2 \cdot 10^{-3}$

The model prediction of the ratio $K_v / K_h = 2.47$ (Eq.(15)) does not contradict the experimental estimate of the mean value, $\overline{K_v / K_h}$ of the ratio, and its standard deviation, $\delta(\overline{K_v / K_h})$ based on Bouma and Dekker's (1981) data (see Eq.(24) and the end of the previous Section) as

$$K_v / K_h = \overline{K_v / K_h} \pm \delta(\overline{K_v / K_h}) = 2.2 \pm 0.8 \qquad (25)$$

Possible reasons for the differences between this experimental estimate and the model prediction (Eq.(15)) are the following. Some contribution to the horizontal hydraulic conductivity in Bouma and Dekker's (1981) experiments (for the soil type 3) can be connected to the vertical interblock and horizontal interaggregate cracks, besides the vertical interaggregate cracks. Furthermore, the distribution of the vertical interaggregate cracks at the horizontal cross-section in the experiments can deflect from isotropic.

It is worth noting that knowing the mean value of the interaggregate-crack width at the saturated condition, $\overline{R}$ (for instance, $\overline{R}_{\mathrm{up}}$ from Eq.(17)) and following Chertkov and Ravina's (2001) approach, one can estimate not only the saturated vertical hydraulic conductivity of an interaggregate-crack system, $K_{\mathrm{v}}(\theta_s)$, but also the vertical hydraulic conductivity function of the system, $K_{\mathrm{v}}(\theta)$ and then the horizontal one, $K_h(\theta)$ from Eq.(15).

Finally, it is to be emphasized that in the frame of Chertkov and Ravina's (2001) approach to be applied in this work, in estimating the horizontal hydraulic conductivity, a vertical crack system is considered to be isotropic at a horizontal cross-section, and, in addition, crack width and spacing are characterized by Poisson distributions. In this relation the model under study is more realistic compared to that of McKay et al. (1993). As for the assumption of smooth crack walls (McKay et al., 1993; Chertkov and Ravina, 2001) one should account for the fact that corresponding works (Tsang, 1984; Brown, 1987; Renshaw, 1995; Ge, 1997) relate to rocks. It is natural to consider that the wall roughness of water filled capillary cracks in swelling clay soils is considerably lower than in rocks because the crack walls in the soils are much more deformable and movable.

5. CONCLUSION

The results obtained show that the basic model (Chertkov and Ravina, 2001) can be applied for prediction of the horizontal hydraulic conductivity of vertical interaggregate capillary cracks in swelling soils. It is also interesting to consider, based on the model, the vertical and horizontal hydraulic conductivity of the system consisting of interconnected vertical and horizontal cracks.

ACKNOWLEDGEMENTS

The research was supported in part by a Grant from the G.I.F., the German-Israeli Foundation for Scientific Research and Development.

REFERENCES

Bouma, J., and L.W. Dekker. 1981. A method for measuring the vertical and horizontal K_{sat} of clay soils with macropores. *Soil Sci. Soc. Am. J.* **45**:662-663.

Bronswijk, J. J. B. 1991. Drying, cracking, and subsidence of a clay soil in a lysimeter. *Soil Sci.* **152** (2):92-99.

Brown, S.R. 1987. Fluid flow through rock joints: the effect of surface roughness. *J. Geophys. Res.* **92**(B2):1337-1347.

Chertkov, V.Y., and I. Ravina. 1999. Tortuosity of crack networks in swelling clay soils. *Soil Sci. Soc. Am. J.* **63**:1523-1530.

Chertkov, V.Y., and I. Ravina. 2001. The effect of interaggregate capillary cracks on the hydraulic conductivity of swelling clay soils. *Water Resour. Res.* **37**:1245-1256.

D'Astous, A.Y., W.W. Ruland, J.R.G. Bruce, J.A. Cherry, and R.W. Gillham. 1989. Fracture effects in the shallow groundwater zone in weathered Sarnia-area clay. *Can. Geotech. J.* **26**:43-56.

Day, M.J. 1977. *Analysis of movement and hydrochemistry of ground-water in the fractured clay and till deposits of the Winnipeg area*, Manitoba. M.Sc. thesis, Univ. of Waterloo, Waterloo, Ont., Canada.

Ge, S. 1997. A governing equation for fluid flow in rough fractures. *Water Resour. Res.* **33**:53-61.

Grisak, G.E. 1975. The fracture porosity of glacial till. *Can. J. Earth Sci.* **12**(3):513-515.

Hendry, M.J., J.A. Cherry, and E.I. Wallick. 1986. Origin and distribution of sulfate in a fractured till in southern Alberta, Canada. *Water Resour. Res.* **22**(1):45-61.

Keller, C.K., G. van der Kamp, and J.A. Cherry. 1986. Fracture permeability and ground water flow in clayey till near Saskatoon, Sakatchewan. *Can. Geotech. J.* **23**:229-240.

McKay, L.D., J.A. Cherry, and R.W. Gillham. 1993. Field experiments in a fractured clay till. 1.Hydraulic conductivity and fracture aperture. *Water Resour. Res.* **29**:1149-1162.

Mualem, Y. 1976. *A catalogue of the hydraulic properties of unsaturated soils.* Research Project no. 442. Technion-Israel Institute of Technology. Haifa. Israel.

Renshaw, C.E. 1995. On the relationship between mechanical and hydraulic apertures in rough-walled fractures. *J. Geophys. Res.* **100**(12):24629-24636.

Rudolph, D.L., J.A.Cherry, and R.N. Farvolden. 1991. Groundwater flow and solute transport in fractured lacustrine clay near Mexico City. *Water Resour. Res.* **27**(9):2187-2202.

Thompson, D. 1990. *Hydraulic evidence of Wisconsinan-aged open fractures in a deep clayey till.* M.Sc. thesis, Univ. of Waterloo, Waterloo, Ont., Canada.

Tsang, Y.W. 1984. The effect of tortuosity on fluid flow through a single fracture. *Water Resour. Res.* **20**(9):1209-1215.

Estimation Error in Statistical Distribution:
Soil Hydraulic Parameters

MIROSLAV KUTÍLEK[1*], MIROSLAV KREJČA[2], and JANA KUPCOVÁ-VLAŠIMSKÁ[3]
[1]*Emeritus Professor, Czech Technical University, Prague,*
[2]*School of Technological Education, Písek,*
[3]*Konstruktiva, Prague, Czech Republic (all)*

1. INTRODUCTION

Let us denote all materials we are dealing with in agriculture as agricultural materials without distinction between their nature and origin. When we use this term, we suppress at this level whether they form a part of ecological environment, or that they are produced by natural processes in the field, or if they are products of food processing. The solution of transport and transformation processes in agricultural materials belongs to the main tasks of the recent research. For the solution of relevant equations we have to know the physical characteristics of the materials, as e.g. diffusion coefficient, thermal conductivity, hydraulic conductivity etc. It is rarely possible to determine them exactly, i.e. in experiments where the boundary conditions are exactly defined and kept during the experiment and where the studied material is stable and homogeneous in time and in the domain of the experiment. The procedure of the experiment has to be formulated in accordance with physical laws governing the process. Since all conditions are rarely met, various types of approximate steps are used for the estimation of the required characteristics. In the estimates, two types of errors are combined: Error due to approximate character of the applied method, and error due to the low precision of our instruments.

Heterogeneity and space and time variability are typical for majority of agricultural materials. Estimates of physical parameters are usually related to a smaller scale then is the technological scale and statistical procedures are then applied for the evaluation of the estimated parameters over a larger scale. One of the principle questions is the extent of the role of the estimation error upon probability density function of a parameter. We are presenting results of our study performed on one important part of agricultural materials, on soils and we are assuming that our conclusions are applicable to many other heterogeneous agricultural materials.

Soils are spatially variable natural bodies due to their genetic development, actual vegetation cover, recent type of land use and methods of soil cultivation. All soil properties vary at relatively short distance with various coefficients of variation CV ranging from less than 15% to over 100% (Wilding, 1985, Jury, 1989) on one soil taxon within a landscape unit of a few ha. Soil physical properties are grouped into two classes: 1. Capacity parameters which have a static character with usually small CV. 2. Transport characteristics including hydrodynamic parameters and fluxes which are dynamic, their CV is relatively high (Kutílek and Nielsen, 1994). For modeling of transport processes and for prognosis on their results, the knowledge of hydrodynamic parameters is required. Using the term transport processes, we understand e.g. fluxes of water in soil-plant-atmosphere system, recharging of groundwaters, fluxes of solutes in soil profile and consequent potential pollution of groundwaters. Owing to the spatial variability of hydrodynamic parameters their statistical evaluation is applied and for a proper understanding of the variation, the probability density function PDF is estimated.

For the determination of soil hydrodynamic parameters in this study we use field methods based upon certain approximations (Angulo-Jaramillo et all., 2000) and the procedure of inverse solution is applied. We are searching the value of the soil parameter P, however our observation is $(P + e)$, where e is the estimation error related to the applied method.

The aim of this research is to study the role of the estimation error e upon PDF of soil hydrodynamic parameters. Since we do not know *a priory* the value of P, we are restricted in this study to the relative comparative procedure. Thus we have evaluated a population of infiltration tests by five different infiltration equations and the PDF of resulting sorptivity and saturated hydraulic conductivity have been determined.

2. MATERIALS AND METHODS

2.1 Site Description

Experiments have been performed on two locations:

1. Arenic Chernozem of the carbonate variety (FAO) in Central Bohemia (Czech republic). The soil has been developed on a Quaternary fluvial terrace of gravel sand overlaid by sandy-loamy topsoil of variable thickness. The soil is typical by its low structural stability in A horizon. Infiltration tests were performed on a rectangular net with spacing 7.5 m. The net consisted of 10 rows and 7 columns, total number of infiltration tests was 70.
2. Deep Ferralsol in Nigeria, with loamy texture and high stability aggregation. Infiltration tests were performed on four transects. Two parallel transects (A,B) were at the mutual distance 38 m and the distance of two parallel transects (C,D) located perpendicularly to A,B was 8 m. Spacing of infiltration tests in the transects was 2 m. Total number of infiltration tests was 128.

2.2 Infiltration Tests

Double ring infiltrometers were used with ponding depth (= positive pressure head on the soil surface) in ranges 2 to 2.5 cm, the diameter of the inner ring was 37.5 cm, of the outer ring 60 cm. We have assumed that the experimental conditions were very close to the conditions of one-dimensional infiltration with Dirichlet boundary conditions on the top boundary. The mean initial soil water content of the topsoil on location 1 was $\theta_i = 0.222$ at the depth 10-15 cm. The value of θ_i was not determined on location 2. Since there was no rain during the time of infiltration measurement and the tests were performed after a dry rainless period, we have assumed that θ_i of the topsoil did not change practically within the time of the whole set of measurements. In the tests, we have measured cumulative infiltration I [L] in time t up to 120 min. The time t_c of the quasi constant infiltration rate, $q_c = dI/dt \approx$ const. was reached earlier, $q[LT^{-1}]$. This was due to our limited accuracy measuring technique where q was oscillating at $t > t_c$ around the mean q which was taken as quasi q_c. Infiltration data on location 2 were kindly offered to us by Dr. O.A. Folorunso.

2.3 Evaluation of Infiltration Tests

The measured data on cumulative infiltration in time $I(t)$ were fitted to the following approximate infiltration equations derived from the theoretical solution of the process:

Green-Ampt's (1911) equation is exact for Dirac delta soil (Philip, 1957b, 1973):

$$I(t) = Kt + \lambda \ln(1 + \frac{I(t)}{\lambda}) \tag{1}$$

where K is the saturated hydraulic conductivity [LT^{-1}], and the parameter $\lambda = (\theta_S - \theta_i)(h_0 - h_f)$, with θ_S saturated soil water content assumed to exist above the step-like wetting front, h_0 the pressure head on the soil surface [L], h_f the pressure head on the wetting front. The equation is an approximation for all other soils than Dirac delta soil. It means that the equation is approximate for our field soils.

In Philip's (1957b) algebraic equation the first two terms are kept from the theoretically derived time series equation (Philip, 1957a). The truncation error transforms the equation to the approximate form:

$$I(t) = St^{1/2} + At \tag{2}$$

with S sorptivity [$LT^{-1/2}$], which is an estimate of the real sorptivity of infinite series solution (Kutílek et al., 1988) and parameter A [LT^{-1}] which is related to K by $A = mK$. Theoretically derived and most frequently used m is close to $m = 0.361$ (Philip, 1987).

Three parameters equation was derived from the Philip's (1957a) series solution (Kutílek and Krejča, 1987, English version in Kutílek and Nielsen, 1994). The truncation error was reduced, when compared to eq. (2), but the approximate type of the equation is still kept:

$$I(t) = c_3 t^{1/2} + c_4 t + c_5 t^{3/2} \tag{3}$$

where parameters c_3 [$LT^{-1/2}$], c_4 [LT^{-1}] and c_5 [$LT^{-3/2}$] are related to the Philip's series solution. The estimate of sorptivity is $S = c_3$, and approximation of $K = (3c_3 c_5)^{1/2} + c_4$.

Swartzendruber's (1987) infinite series solution is in a way derived from the Philip's series solution of infiltration and three parameters are kept in the approximate equation:

$$I(t) = \frac{S}{c_6}\left[1 - \exp(-c_6 t^{1/2})\right] + Kt \tag{4}$$

where S is the estimate of sorptivity, K is the estimate of saturated hydraulic conductivity, and c_6 [$T^{-1/2}$] is an empirical parameter.

Brutsaert's (1977) three parameters approximate equation is

$$I(t) = Kt + \frac{S^2}{BK}\left\{1 - \frac{1}{1 + (BKt^{1/2})/S}\right\} \tag{5}$$

where B is an empirical coefficient, dimensionless, S and K are estimates of sorptivity and saturated hydraulic conductivity.

The estimates of sorptivity S and saturated hydraulic conductivity K behave as fitting parameters during the fitting procedure (Kutílek et al., 1988). Their physical interpretation corresponds to the studied parameter P plus the estimation error e.

We have used the chi-square goodness-of-fit test and Kolmogorov-Smirnov test on the fit (STATGRAPHICS, 1985) for the determination of the probability density function of sorptivity estimates and of saturated hydraulic conductivity estimates, both obtained by fitting experimental data to equations (1), (2), (3), (4), and (5). Further on, for comparative reasons, PDF of directly measured experimental data q_1 and q_{120} were evaluated, too, since q_1 is closely related to sorptivity and q_{120} to saturated hydraulic conductivity. Both are free of estimation error, but their relations to S and to K are not constant for each of the experimental plot.

3. RESULTS AND DISCUSSION

In many instances of our data sets, the evaluation of PDF according to Kolmogorov-Smirnov test and according to the chi-square goodness-of-fit test have offered the same results. The second procedure was sensitive to the number of classes and due to this requirement the PDF differed in some instances from the first procedure, as it could be expected theoretically. In Tables 1 and 2, there are results obtained by Kolmogorov-Smirnov test only.

On experimental site 1 with Arenic Chernozems, the infiltration data were not applicable to fitting procedure according to equation (1) of Green and Ampt. Therefore, this equation was excluded from PDF testing on site 1. Philip's two parameters equation offered non realistic values of hydraulic conductivity K with either $K = 0$ or even $K < 0$ in about 30% of tests. Brutsaerts equation (5) led on site 1 to unrealistic values of parameter $B < 0$ or to $B = 0$ in 16 % of instances. These tests were identical with tests belonging to 30% of non-realistic output in K according to eq. (2). We have therefore formed a shortened set of 49 infiltration tests, where we have omitted the infiltration experiments not applicable for fitting according to eq. (2). We have determined PDF of parameters obtained from equations (2), (3), (4), and (5) in this short set. Full set with all 70 infiltration experiments was kept for PDF determination of parameters obtained by eq. (3) and (4).

On site 2 with Ferralsols all equations were applicable and no fitting problems occurred there. In parallel transects A, B, the number of infiltration tests was 64 and the same number of tests was performed in parallel transects C, D. In all transects A, B, C, D we have worked with a set containing 128 data.

3.1 PDF of Sorptivity

Evaluation of distribution functions of sorptivity *S* estimates is in Table 1.

Table 1. First two best fitting distributions of S[$LT^{-1/2}$] sorptivity, and q_1 [LT^{-1}],infiltration rate after the first minute of infiltration

Equation	Site 1				Site 2				Site 2	
	Short set		Full set		Transects A, B		Transects C, D		Transects A, B, C, D	
	S	q_1	S	q_1	S	q_1	S	q_1	S	q_1
-	-	E	-	B	-	E	-	E	-	N
		G		W		G		G		G
(2)	LG	-	W	-	G	-	N	-	N	-
	E		G		E		W		G	
(3)	LG	-	E	-	G	-	N	-	E	-
	G		G		E		W		G	
(4)	LG	-	E	-	G	-	H	-	N	-
	E		G		LG		G		G	
(5)	LG	-	N	-	LG	-	G	-	G	-
	G		G		E		E		E	

Type of probability density function: N - normal, LG - lognormal, W - Weibull, G - gamma, B - beta, E – Erlang

The PDF of *S* are supplemented by evaluation of infiltration rate after the first minute, q_1 which is theoretically closely related to sorptivity. However, PDF of q_1 differ in majority of instances from PDF of *S*. The only exception is in site 2, all transects A, B, C, D together, where *S* obtained by equations (2) and (4) show the same PDF as q_1. There is distinct curtosis and skewness of *S* and q_1 distributions indicating that there is not normal distribution in majority of instances. The exception is in 4 cases, where normal PDF was evaluated. Lognormal PDF was equally frequent as gamma PDF and both were more frequent than other tested PDF. For q_1 we have found normal PDF in one case, while Erlang PDF was prevailing in remaining sets. We are concluding that the PDF of *S* estimates depends not only upon distribution of theoretical *S* which is unknown to us, but upon the nature of approximation in the applied equation. A not unique type of PDF in directly measured q_1 is probably due to a lack of accuracy even in the measuring technique with double ring infiltrometers and with not strictly constant pressure head on the surface. In addition to it, the variation in soil heterogeneity on vertical may cause variation of divergence of flow paths and a not unique deviation from theoretically assumed 1-dimensional flow. Low population in short sets may play a role in discussed differences, too, especially due to soil heterogeneity. For population of 128 experiments (all transects A, B, C, D) there was an increased identity in PDF between q_1 and S estimates.

Table 2. First two best fitting distributions of K [LT^{-1}], saturated hydraulic conductivity, and q_{120} [LT^{-1}], infiltration rate after 120 minutes of infiltration

Equation	Site 1				Site 2				Site 2	
	Short set		Full set		Transects A, B		Transects C, D		Transects A, B, C, D	
	K	q_{120}	K	q_{120}	K	q_{120}	K	q_{120}	K	q_{120}
-	-	B E	-	W B	-	LG E	-	G B	-	LG G
(1)	-	-	-	-	LG E	-	G E	-	E G	-
(2)	LG G	-	-	-	LG E	-	E G	-	LG G	-
(3)	W N	-	B W	-	LG G	-	E G	-	LG G	-
(4)	B W	-	W E	-	LG G	-	E B	-	LG E	-
(5)	W B	-	-	-	LG E	-	LG N	-	LG G	-

Type of probability density function: N - normal, LG - lognormal, W - Weibull, G - gamma, B - beta, E – Erlang

3.2 PDF of Saturated Hydraulic Conductivity

In Table 2 there are the results of distribution evaluation of estimates of saturated hydraulic conductivity K together with evaluation of quasi steady infiltration rate after 120 minutes, q_{120}. The PDF of q_{120} and PDF of K estimates was more frequently identical than was the case of S and q_1. However, a complete identity did not exist. A distinct curtosis and skewness of K and q_{120} distributions was found indicating that there is not normal distribution in all instances. The type of prevailing PDF was different on site 1 and on site 2. On site 1 was the Weibull PDF more frequent than other distributions. On site 2, lognormal PDF was distinctly prevailing and the next most frequent was Erlang PDF. This was dominant in transects C, D. There is a tendency showing that the increase of population in the studied sets leads to a better agreement between PDF of directly measured q_{120} and PDF of the *K* estimates. A greater variation of PDF types in site 1, when compared to site 2 is probably due to the instability of soil structure on site 1. Owing to it *K* was probably time dependent.

4. CONCLUSIONS

The comparison of PDF for the estimates of parameters *S* and *K* shows that their PDF is deformed by the error *e*, i.e. we are determining PDF of an estimate ($P + e$), where *P* is the studied soil parameter. We assume, that the error *e* is a summation of two components: 1. Error due to the approximate character of the equation applied for fitting the experimental data. 2. Error due to differences between the field reality and the conditions of the theoretical development of equations, where soil physical homogeneity with depth and stability in time during infiltration is assumed. Thus PDF of our observation ($P + e$) does not offer a reliable information on PDF of the studied parameter.

ACKNOWLEDGEMENTS

The authors thank Dr. O.A. Folorunso for offering us his infiltration data.

The publication of the paper was supported by the research grant project No 103/02/0971 of the Grant Agency of the Czech Republic.

REFERENCES

Angulo-Jaramillo, R., Vandarvaere, J.-P., Roulier, S., Thony, J.-L., Gaudet, J.P., and Vauclin, M., 2000, Field measurement of soil surface hydraulic properties by disc and ring infiltrometers. A review and recent developments. *Soil Till.Res.* **55**:1-29.

Brutsaert, W., 1977, Vertical infiltration in dry soil. *Water Resour.Res.* **13**:363-368.

Green, W.H., and Ampt, G.A., 1911, Studies on soil physics: I. Flow of air and water through soils. *J. Agric. Sci.* **4**:1-24.

Jury, W.A., 1989, Spatial variability of soil properties. In: *Vadose Zone Modelling of Organic Pollutants* (S.C. Hern, and S.M. Melancon, eds.) Lewis Publishers, pp.245-269.

Kutílek, M., Krejča, M., 1987, Tříparametrická rovnice infiltrace Philipova typu (in Czech, A three-parameters infiltration equation of the Philip´s type solution). *Vodohosp.čas.*, **35**:52-61.

Kutílek, M., Krejča, M., Haverkamp, R., Rdendón, L.L., Parlange, J.Y., 1988, On extrapolation of algebraic infiltration equations. *Soil Technol.* **1**:47-61.

Kutílek, M., and Nielsen, D.R., 1994, *Soil Hydrology.* Catena Verlag, Cremlingen-Destedt, 370 pp.

Philip, J.R., 1957a, The theory of infiltration: 1. The infiltration equation and its solution. *Soil Sci.* **83**:345-357.

Philip, J.R., 1957b, The theory of infiltration: 4. Sorptivity and algebraic infiltration equations. *Soil Sci.* **84**:329-339.

Philip, J.R., 1987, The infiltration joining problem. *Water Resour.Res.,* **23**:2239-2245.

Swartzendruber, D., 1987, A quasi-solution of Richards′ equation for the downward infiltration of water into soil. *Water Resour.Res.* **23**:809-817.

STATGRAPHICS, 1985. The Braegen Group Inc., Toronto, Ontario.

Wilding, L.P., 1985, Spatial variability: its documentation, accommodation and implication to soil surveys. In: *Soil Spatial Variability* (D.R. Nielsen, and J. Bouma, eds.), Pudoc, Wageningen, pp.125-178.

The Influence of Tillage Methods on the Infiltration in Soil

Changes of the saturated hydraulic conductivity K of topsoil

SVATOPLUK MATULA
Czech University of Agriculture in Prague, Prague 6 – Suchdol, Czech Republic

1. INTRODUCTION

Over the centuries, conventional agricultural practice around the developed countries encouraged extensive soil tillage. Tillage can have both favourable and unfavourable effects on different soil physical properties of the treated topsoil. If the topsoil water content is not too high or too low when the tillage is performed the short-term effect of tillage is generally favourable; the top layer cohesive strength is decreased, the total porosity is increased. Tillage implements break up the large clods, incorporate organic matter into the topsoil, reduce the weeds and prepare a seedbed. Over longer periods, tillage treatments have detrimental effects on surface soil structure such as hastening of the oxidation processes of soil organic matter due to mixing and stirring the soil and reducing the aggregating effect of the soil. When the tillage treatments are carried out especially on the wet soil, the stable soil aggregates are crushed or smeared, macro porosity is decreased and puddled soil is created. This action exposes organic matter that had been protected inside the aggregates, hastening its loss by decomposition (Brady, Weil, 1999). Some studies have shown that aggregation and the associated desirable soil physical properties (bulk density, porosity, degree of aeration, capillary water capacity and others) and hydro-physical properties related to infiltration rate decline after long periods of tilled row-crop cultivation. Le Bissonnais and Arrouays (1997) published, that the infiltration rate under rain agrees well with aggregate stability measurements and the soil structure stability declined as organic carbon declined.

The classical type of tillage, applied since Middle Ages is mouldboard plough followed often by secondary tillage operation for example harrowing. Repeated traffic activities related to mouldboard ploughing and using heavy machinery cause soil compaction. In recent years, conservation tillage practices have been largely developed. Those systems reduce greatly tillage treatment, and in some cases no tillage treatment at all is rather successfully applied. The reviews of conservation tillage treatments were published by Blevins and Frye (1992) and Carter (1994). The Conservation Technology Information Center, West Lafayette, IN, USA divided the conservation tillage systems into 5 types (Brady, Weil, 1999) (Table 1.1).

Table 1.1 General Classifications of Different Conservation Tillage Systems

No-till	Soil undisturbed prior to planting, which occurs in narrow seedbed, 2.5 to 7.5 cm wide. Weed control primarily by herbicides.
Ridge till (till, plant)	Soil undisturbed prior to planting, which is done on ridges 10 to 15 cm higher then row middles. Residues moved aside or incorporated on about one-third of soil surface. Herbicides and cultivation to control weeds.
Strip till	Soil undisturbed prior to planting. Narrow and shallow tillage in row using rotary tiller, in row chisel, and so on. Up to on- third of soil surface is tilled at planting time. Herbicides and cultivation to control weeds.
Mulch till	Soil surface disturbed by tillage prior to planting, but at least 30% of residues left on or near soil surface. Tools such as chisels, field cultivators, disks and sweeps are used (e.g. stubble mulch). Herbicides and cultivation to control weeds.
Reduced till	Any other tillage and planting system that keeps at least 30% of residues on the surface.

Klute (1982) presented a review of the tillage effects on the hydraulic properties of soil. Infiltration of water through soil profile and surface runoff on the agricultural soils are strongly dependent on a type of tillage treatment of a topsoil layer. One of the dominant soil hydraulic properties related to those processes is saturated hydraulic conductivity (K) of the topsoil.

This study is focused on the influence of a repeated tillage treatment of the same area to saturated hydraulic conductivity (K) on a relatively homogeneous soil profile.

The transient and steady state infiltration of water into the vadose zone of soil reflects the pore system resulting from the juxtaposition of soil particles and peds. The pore system is characterised by soil hydraulic properties that quantify the size, distribution and continuity of the system. The most important properties are saturated hydraulic conductivity, the matric and tension flux potentials, and the sorptivity (Elrick and Reynolds, 1992). These properties can be measured in a laboratory on core samples, however the field measurements using infiltrometers and permeameters are far more reliable in results and have been extensively used around the world.

Elrick and Reynolds (1992) basically divided existing infiltrometers into:

- *Tension infiltrometers*
- *Pressure infiltrometers*

Angulo-Jaramillo et al (2000) published a review and recent developments of using infiltrometers *in situ.* The existing techniques to infer hydraulic properties from measurements of either transient flow rates or steady ones, that emanate from a disc infiltrometer are discussed here. Similarly the flow of water from a ring infiltrometer is described mathematically and analysed. The description of water and solute flow from tension disc infiltrometer using one-dimensional two region mobile-immobile water model and solute exchange equation are reviewed in detail.

Angulo-Jaramillo et al (2000) divided the infiltrometers into:

- *Tension disc infiltrometers*
- *Pressure ring infiltrometers*

In the presented study the laboratory-made pressure ring infiltrometer (Matula and Kozakova, 1997) was applied in situ.

2. MATERIALS AND METHODS

The pressure ring infiltrometer (Matula and Kozakova, 1997) consists of a Mariotte bottle, mounted on the top of a single iron infiltration ring of 0.15 m inner diameter, driven to a short distance (0.10 m) into the soil top layer. A wide range of steady water pressure can be applied to the soil surface inside the infiltration ring using a moveable air tube within the Mariotte bottle. The schematic diagram of the pressure ring infiltrometer (Matula and Kozakova, 1997) is shown on Fig. 2.1.

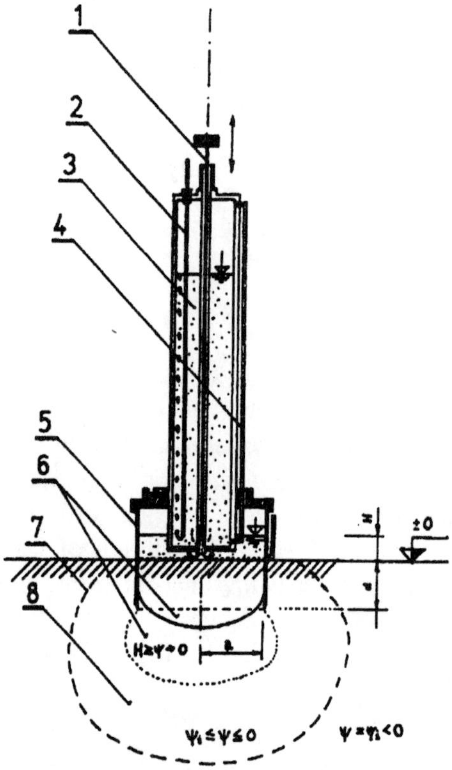

Figure 2.1 Pressure ring infiltrometer (Matula and Kozakova, 1997): 1. piston valve to open /close the water outlet, 2. moveable air tube of Mariotte bottle to set the water level (*H*) inside the infiltrometer ring, 3. Mariotte bottle, 5. infiltrometer ring, 6. bulb full of field saturated soil, 7. wetting front, 8. wetting zone

The field experimental work was conducted in 1997 and repeated in 2000 again at the experimental field of the Research Institute of Plant Production (RIPP), Praha - Ruzyně on a Hapludalfs (US Classification) / Orthic luvisol (FAO). The whole experimental site was separated into 4 tillage treatment areas (TTA4, TTA1, TTA2, TTA3) in 1997-1999 and 4 TTA in 1999 – 2000 (TTA01, TTA02, TTA03, TTA04), that where maintained using different tillage treatments (see Fig. 2.2).

The TTA were segregated mechanically (paths) and marked from each other with an application of chemical spray (Gramoxon). Land slope has been about 4 % within each area. The tillage practises maintained on these separated TTA are presented in Table 2.1.

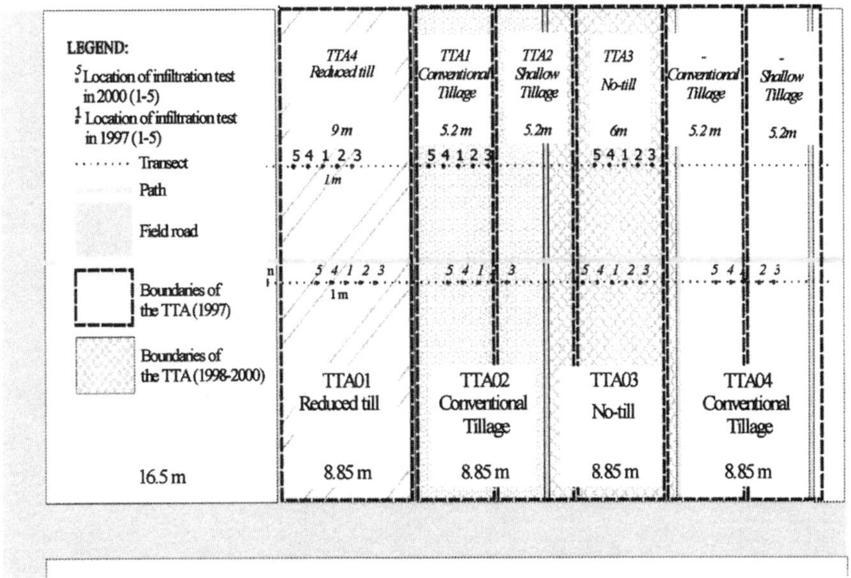

Figure 2.2 The location of tillage treatment areas in 1997 and from 1998 to 2000 and the location of transects.

The TTA were segregated mechanically (paths) and marked from each other with an application of chemical spray (Gramoxon). Land slope has been about 4 % within each area. The tillage practises maintained on these separated TTA are presented in Table 2.1.

Table 2.1 Tillage treatments, crop and particle densities for 3 selected tillage treatment areas and different depths

TTA	Soil treatment	Crop:	Depth of soil (m)	Particle density ρ_d (kg m^{-3})
1, 02, 04	Conventional	Wheat	0.0 – 0.1	2630
4, 01	Reduced till	Wheat	0.1 – 0.2	2600
3, 03	No-till	Wheat	0.0 – 0.1	2580
			0.1 – 0.2	2590
All	No treatment - subsoil	-	0.2 – 0.3	2610 - 2640

The particle size distribution curves of the topsoil in TTA and subsoil of the experimental field are presented in figures Figs. 2.3 - 2.7. Particle densities of the topsoil and subsoil, measured in 2000 are also in Table 2.1.

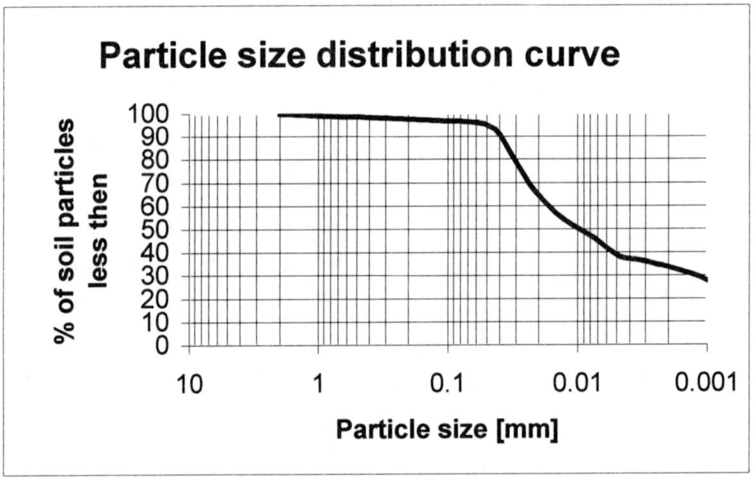

Figure 2.3 Particle size distribution curve – topsoil, conventional tillage (TTA02, TTA04, TTA1), depth 0 – 0.1 m

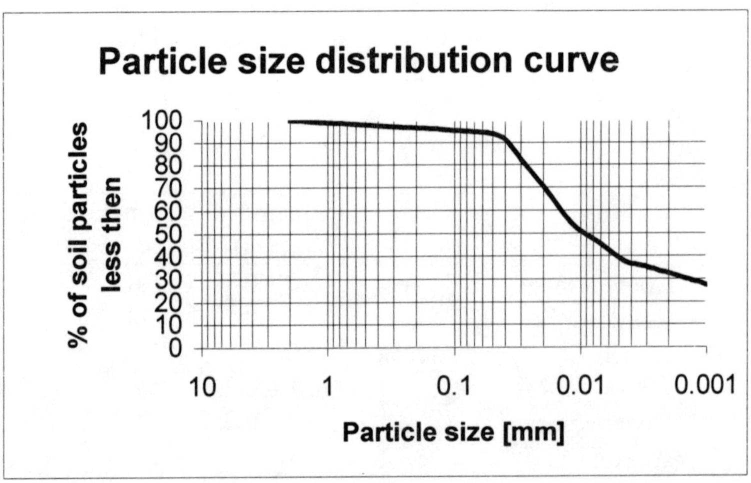

Figure 2.4 Particle size distribution curve – topsoil, reduced till (TTA01, TTA4), depth 0.1 – 0.2 m

Transects were marked across the whole site perpendicular to the land slope, rows and/or machinery movement (Fig. 2.2).

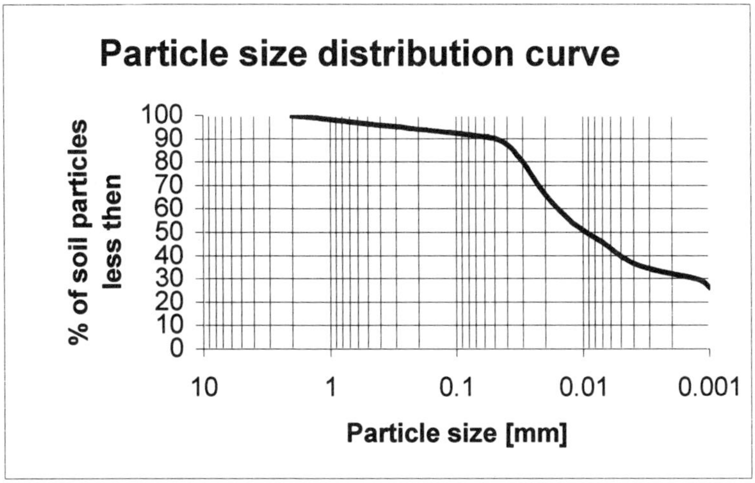

Figure 2.5 Particle size distribution curve – topsoil, no-till (TTA03, TTA3), depth 0 – 0.1 m

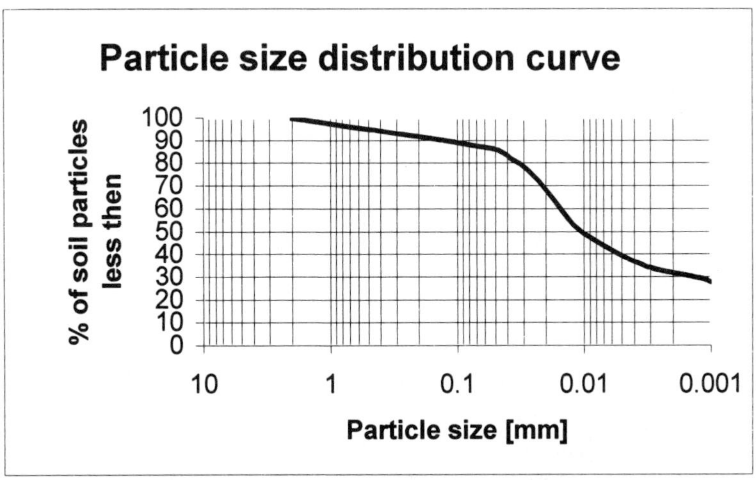

Figure 2.6 Particle size distribution curve – topsoil, no-till (TTA03, TTA3), depth 0.1 – 0.2 m

A series of 5 infiltration tests were carried on each TTA and in both experimental years (Fig. 2.2). The first run of infiltration tests was conducted in 1997 (Matula and Krivohlavy, 1998a,b), the second one in 2000. An equidistant space 1 m between the test locations was used. The pressure ring infiltrometer (Matula and Kozakova, 1997) with iron ring 0.15 m inner diameter was inserted 0.08 m into the soil. The results of the infiltration experiments are presented as cumulative infiltration (*I*) versus time (*t*) in selected time steps. The infiltration rate *v (t)* was calculated from this data. The results are presented on the Figs 3.1, 3.2, and 3.3.

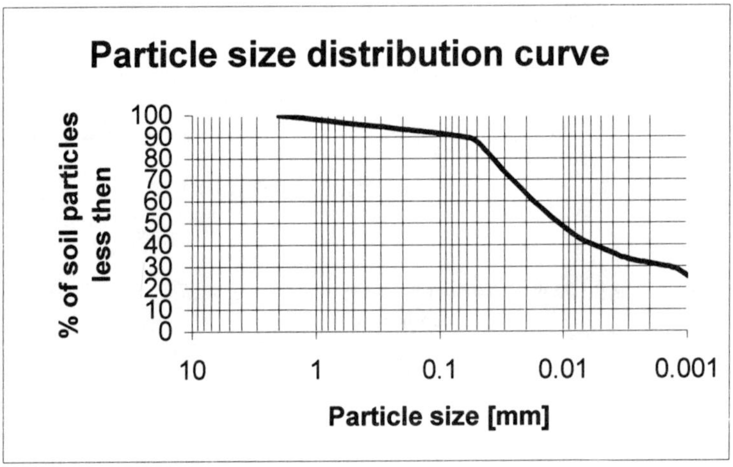

Figure 2.7 Particle size distribution curve – subsoil, a typical one (for all TTA), depth 0.2 – 0.3 m

The flow into unsaturated soil from the pressure ring infiltrometer occurs at two stages. Initially the flow goes through an initial transient phase and then approaches steady state. During the transient period both the field saturated bulb and the surrounding wetting zone increase in size by migrating downwards and outwards from the infiltration surface. After reaching the steady state, the field-saturated bulb remains essentially constant in size and shape but the wetting zone continues to increase in size by the outward migration of the wetting front. The final size of the steady state field saturated bulb is determined by the positive pressure head H applied by the infiltrometer, the dimensions of the infiltration surface and soil properties, in particular by texture, structure, initial water content and/or initial (background) water potential (ψ_i) (Elrick and Reynolds, 1992).

The transient flow can be classically described by the Philip's (1957) equation. Philip introduced a series solution to describe cumulative infiltration under ponded conditions at time t with $t = 0$ at the beginning of the infiltration process as:

$$I = S\, t^{1/2} + A_1\, t + A_2\, t^{3/2} + A_3\, t^2 + \ldots \tag{1}$$

where:

I = cumulative infiltration (m)
S = sorptivity (m $s^{-1/2}$)
t = time (s)
A_1, A_2, A_3 = constants depending upon the soil hydraulic properties and boundary conditions

Equation (1) can be truncated after the first two terms because the series converges rapidly for a relatively short time. The three-dimensional flow can be neglected only for short periods at the beginning of the infiltration process and the equation can be rewritten as:

$$I = S\, t^{1/2} + A\, t \tag{2}$$

where:
A= constant represents gravity term (m s^{-1})

Talsma (1969) found a linear relationship between the cumulative infiltration (I) and the square root of time during this early stage (the initial minutes) with sorptivity as the slope of the line:

$$I = S\, t^{1/2} \tag{3}$$

Equation (2) can be rewritten for the pressure infiltrometer with the applied positive pressure head (H) as:

$$I = S_H\, t^{1/2} + A\, t,\ H > 0,\ \text{constant} \tag{4}$$

where:
S_H = sorptivity for the ponded conditions (m $s^{-1/2}$)
Note that the term $A\, t$ can be neglected for an early stage of infiltration.

Sorptivity in equation (4) incorporates the effects of the positive pressure head (H) and the capillarity of unsaturated soil.

The empiric relationship between unsaturated hydraulic conductivity (k) and water potential (ψ) has been presented by Gardner (1958) :

$$k(\psi) = K\exp(\alpha\psi)\ ,\quad \psi \le 0 \tag{5}$$

where:
k = unsaturated hydraulic conductivity (m s-1)
K= saturated hydraulic conductivity (m s-1)
α= Alpha parameter, the slope of ln (k) versus ψ
ψ= water potential (m)
Gardner (1958) defined the flux potential as:

$$\phi(\psi) = \int_{\psi_i}^{\psi} k(\psi)\, d\psi,\quad \psi_i \le \psi \tag{6}$$

where:
ϕ = flux potential ($m^2\ s^{-1}$)
ψ_i = initial (or background) water potential (m)
ψ = water potential as a variable (m)
Several specific values for fixed upper limits of the integration can be defined:

$$\phi_m = \int_{\psi_i}^{0} k(\psi)\, d\psi \tag{7}$$

in which ϕ_m is matric flux potential ($m^2\ s^{-1}$) and:

$$\phi_v = \int_{0}^{H} k(\psi)\, d\psi = KH \tag{8}$$

in which ϕ_v is velocity potential ($m^2\ s^{-1}$)

Note, that ϕ_m represents the unsaturated area under the $k(\psi)$ curve and ϕ_v the saturated area.

The driving flux potential (ϕ_d) is defined in this case as:

$$\phi_d = \phi_m + \phi_v \tag{9}$$

Substituing (5) into (7) and integrating gives the following for the pressure infiltrometer:

$$\alpha (\psi) = (K - k_i (\psi)) / \phi_m (\psi) \tag{10}$$

where:

α = sorptive number or alpha parameter (Philip, 1985, 1987) (m^{-1})

$k_i = k (\psi)$ for $\psi = \psi_i$

When k_i is much smaller relative to K (Scotter el al.,1982), equation (10) can be simplified to:

$$\alpha (\psi) \cong K / \phi_m \equiv \alpha^*, K >> k_i \tag{11}$$

The relation between α and α^* depends upon ψ_i. If ψ_i is at or below field capacity value, then $K >> k_i$ and $\alpha^* = \alpha$. If ψ_i is closed to zero (saturation conditions of the soil), then α^* can be larger than α (Kachanoski et al., 1989).

Elrick and Reynolds (1992), based on Philip (1958), determined from equation (4) the sorptivity S_m for the ponded situation with water potential ψ = 0:

$$S_m = S_H / (1 + \alpha^* H)^{1/2} \tag{12}$$

Raats (1976) described the Alpha parameter (α^*) as the indicator of the relative magnitude of the capillarity and gravitational components of water flow through unsaturated soil. The smaller the value of α^* the greater the relative contribution of soil capillarity to the flow (Elrick and Reynolds, 1992). Equations (4) and (12) can be used for sorptivity determinations from the pressure ring infiltrometer field experiments.

The infiltration during the steady state flow (for a long period of infiltration) may be described by the equation formulated by Philip (1985), Reynolds and Elrick (1990, 1991), Reynolds et al. (1985) and Elrick and Reynolds (1992):

$$Q = \pi a^2 k (\Psi) + G \phi (\Psi) \tag{13}$$

where:

Q = the steady flow rate of water into the soil out of the pressure ring infiltrometer (m^3 s^{-1})

a = the radius of the infiltration ring (m)

G = a "geometry function" which accounts for the geometry of the infiltration surface

The first term on the right hand side of the equation (13) represents the flow due to gravity and the second one the influence of H and capillarity of unsaturated soil.

Using equations (6), (7), (8) and (9) equation (13) can be rewritten for the pressure ring infiltrometer as:

$$Q = \pi a^2 \; K + (a / G) \; (K H \; + \phi_m) \qquad (14)$$

or

$$Q = (\pi a^2 + (a \; / G) \; (H + (1 / \alpha^*))) \; K \qquad (15)$$

The "geometry function" G is a dimensionless shape parameter determined by numerical solution from Richard's equation. Reynolds and Elrick (1990) found that G is essentially independent of H, K and ϕ_m for a constant radius a and depth of insertion d of the infiltrometer ring. Elrick and Reynolds (1992) published a detailed description of the theory of flow from permeameters and infiltrometers including the equations for the saturated hydraulic conductivity (K). The final equation for K determined from the pressure ring infiltrometer test is calculated from equation (15) as:

$$K = Q \, G / (a \, H + a^2 \, G \, \pi + a / \alpha^*) \qquad (16)$$

where:

K = saturated hydraulic conductivity (m s^{-1})

Q = steady flow rate of water into the soil out of the pressure ring infiltrometer ($m^3 \, s^{-1}$)

G = the geometry function, defined by Reynolds, Elrick, 1990 for the pressure ring infiltrometer as:

$$G = 0.316 \, (d / a) + 0.184 \qquad (17)$$

d = depth of infiltration ring insertion (m)
a = the radius of the infiltration ring (m)
H = a positive pressure head (m)
α^*= Alpha parameter (m^{-1})

Equation (16) employs parameter α^*, which has to be chosen. White and Sully (1987) suggested setting the value to 10 m^{-1} for most soils. Elrick and Reynolds (1992) in their Table 1 - 1 use soil texture/structure categories for site estimation of α^*. It ranges from $\alpha^* = 1$ m^{-1} for compacted, structure less, clayey materials to $\alpha^* = 36$ m^{-1} for coarse, gravely sands or soils with large macro-pores. For most soils they recommended a value of $\alpha^* = 12$ m^{-1}

(Elrick and Reynolds, 1992). Matula and Kozakova, 1997, found the values of α^* for their average measured K values around $\alpha^* = 12.55$ and 13 m^{-1}, which are very close to the value recommended by Elrick and Reynolds. The value $\alpha^* = 12\ m^{-1}$ was also chosen for the present calculations.

3. RESULTS AND DISCUSSION

The infiltration tests were carried on each TTA. Fifteen infiltration tests were taken (5 on each area) in total in 1997 and 20 tests (again 5 on each area) in 2000.

The initial water content was measured before the beginning of the tests using the TDR TRIME FM3; the results of measurements for the 3 types of tillage treatment are in Table 3.4.

The measured *I (t)* was then evaluated, the *v (t)* was calculated for each test. The results of cumulative infiltration *I (t)* and infiltration rate *v (t)* for selected three tillage treatments are plotted in Figs. 3.1, 3.2, and 3.3.

The calculated values of saturated hydraulic conductivity (*K*) for the same tillage treatments and some basic statistics are in Tables 3.1, 3.2, and 3.3.

Table 3.1 *K* values and basic statistics related to *K* for TTA4 and TTA01 in 1997 and 2000.

TTA4 1997		**Reduced till (restricted tillage)**			
Infiltration ring no.:	1	2	3	4	5
K (cm s^{-1})	0.00175	0.00180	0.00352	0.00242	0.00178
Mean of *K* values		0.00225		cm s^{-1}	
Median of *K* values		0.00180		cm s^{-1}	
TTA01 2000					
K (cm s^{-1})	0.02487 *)	0.00079	0.00091	0.00060	0.00059
Mean of *K* values		0.00555 *)	(0.00072)	cm s^{-1}	
Median of *K* values		0.00079		cm s^{-1}	

**) Note: An underground passage produced by a field mouse was found connected after digging the infiltration ring out of the soil layer when the test was over. This event made the K value, representing ring no. 1 significantly higher. The calculations excluding ring no.1 are in brackets.*

Equations (1) and (2) and α^* based on Elrick et al. (1989) were applied to the steady state flow rate of water into the soil, when *I* is linearly related to *t* and the *K* values were calculated.

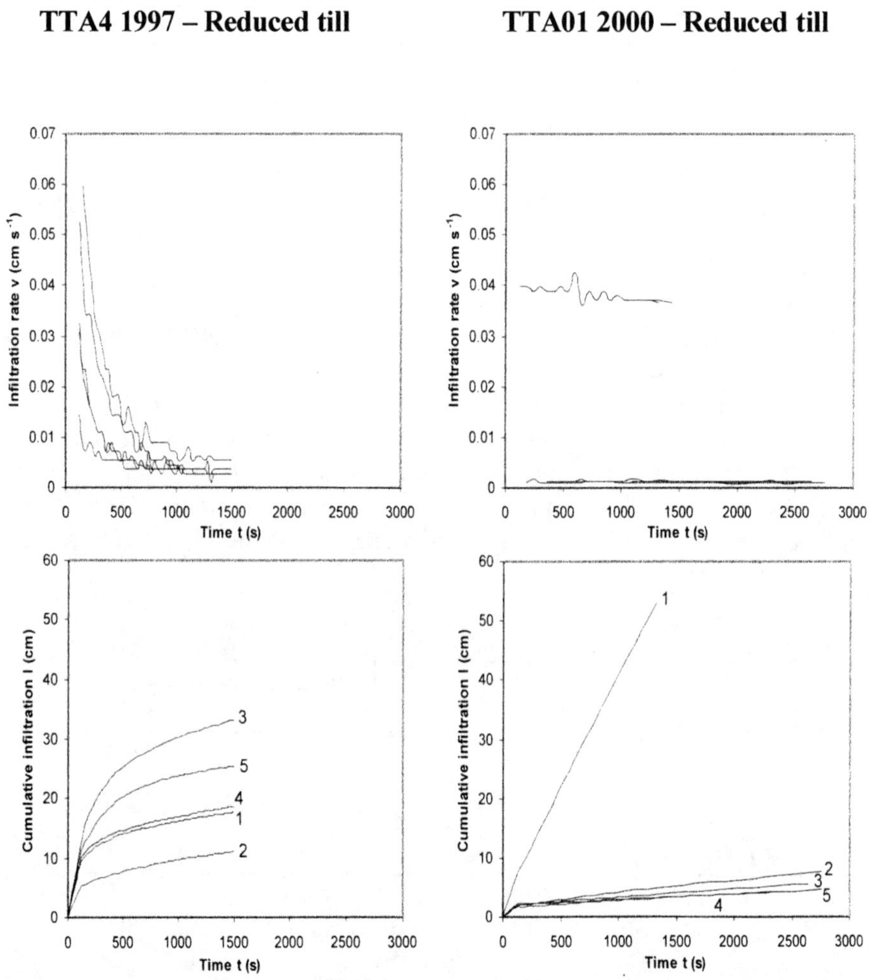

Figure 3.1 Infiltration rate *v (t)* and cumulative infiltration *I (t)* for TTA4 and TTA01 in 1997 and 2000 (Numbers added to cumulative infiltration curves represent the transect test numbers. The infiltration rate curves are not marked due to available space, however the highest I(t) relates the highest v(t) and vice versa).

TTA3 1997 – No-till **TTA03 2000 – No-till**

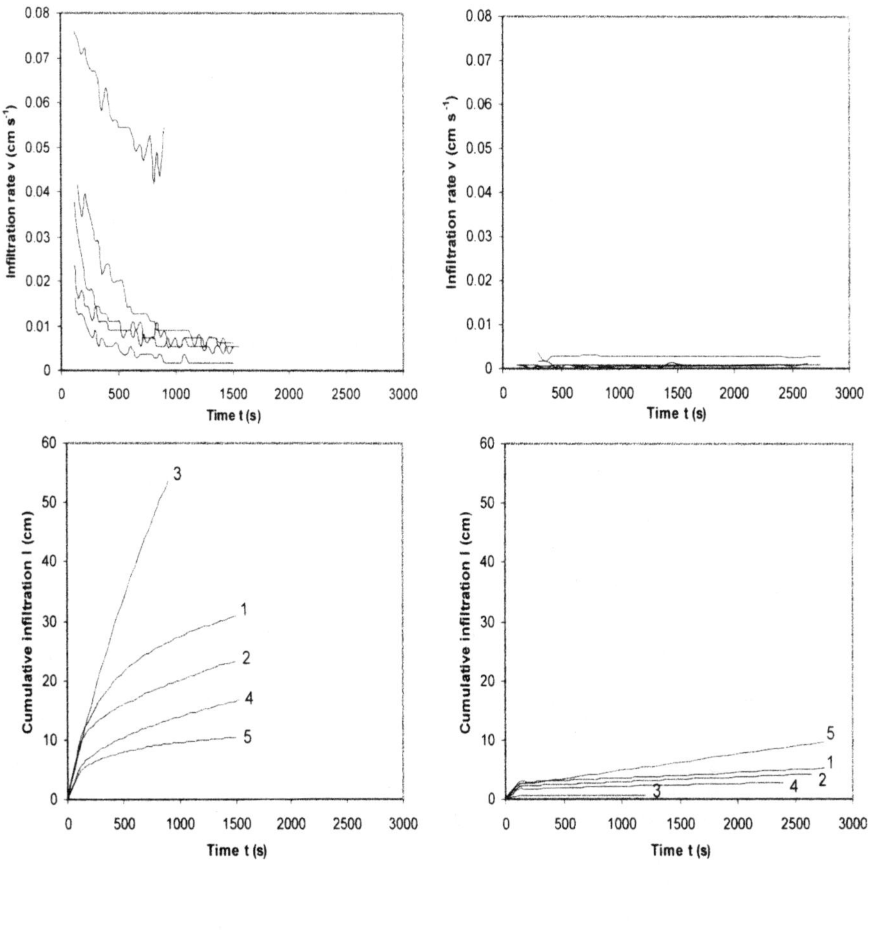

Figure 3.2 Infiltration rate *v (t)* and cumulative infiltration *I (t)* for TTA3 and TTA03 in 1997 and 2000. (Numbers added to cumulative infiltration curves represent the transect test numbers. The infiltration rate curves are not marked due to available space, however the highest I(t) relates the highest v(t) and vice versa).

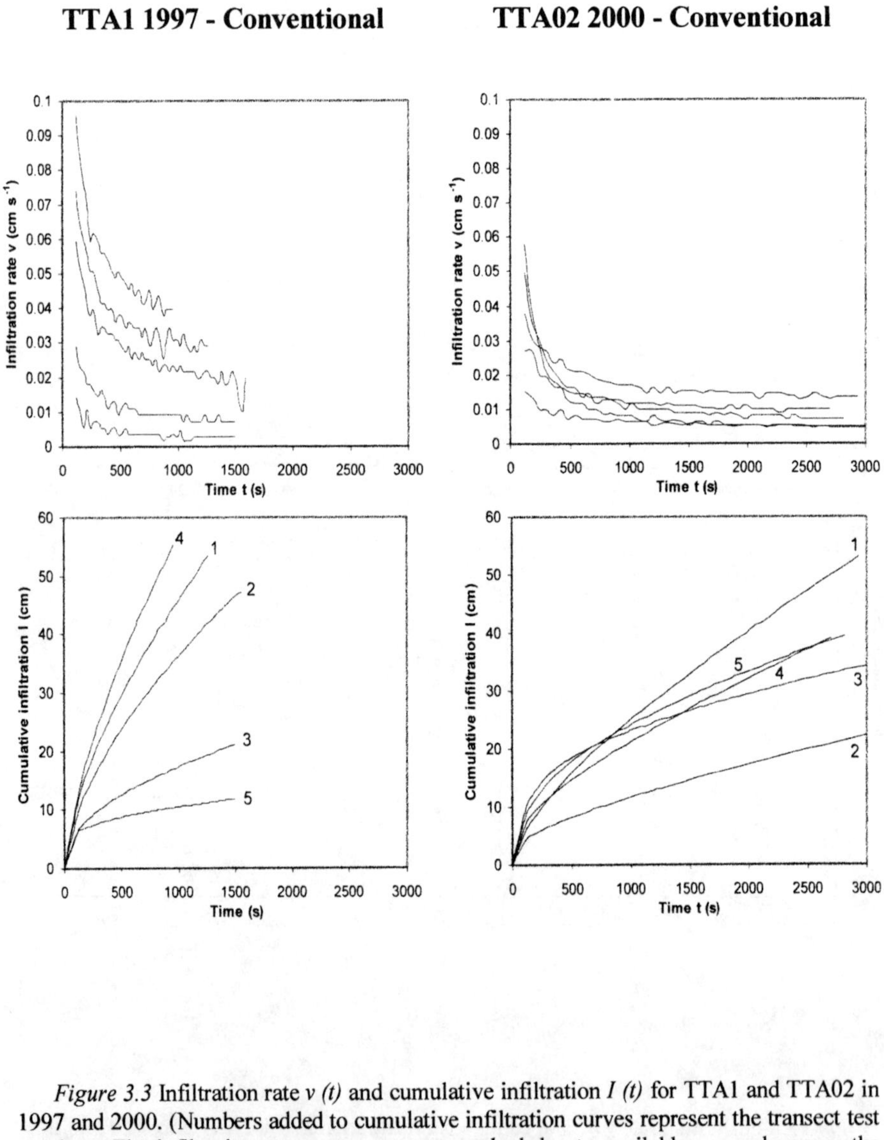

Figure 3.3 Infiltration rate *v (t)* and cumulative infiltration *I (t)* for TTA1 and TTA02 in 1997 and 2000. (Numbers added to cumulative infiltration curves represent the transect test numbers. The infiltration rate curves are not marked due to available space, however the highest I(t) relates the highest v(t) and vice versa).

Table 3.2 *K* values and basic statistics related to *K* for TTA3 and TTA03 in 1997 and 2000.

TTA3 1977			**No-till**		
Infiltration ring no.:	1	2	3	4	5
K (cm s^{-1})	0.00372	0.00411	0.03530 *)	0.00380	0.00123
Mean of *K* values		0.00963 *)	(0.00410)	cm s^{-1}	
Median of *K* values		0.00380		cm s^{-1}	
TTA03 2000					
K (cm s^{-1})	0.00062	0.00061	0.00008	0.00030	0.00177
Mean of *K* values		0.00068		cm s^{-1}	
Median of *K* values		0.00030		cm s^{-1}	

*) *Note: A field mouse nest was found close to the ring after digging the infiltration ring out of the soil layer when the test was over. This event again makes the K value, representing ring no.3 significantly higher. The calculations excluding ring no. 3 are in brackets.*

Table 3.3 *K* values and basic statistics related to K for TTA1 and TTA02 in 1997 and 2000.

TTA1 1997			**Conventional tillage**		
Infiltration ring no.:	1	2	3	4	5
K (cm s^{-1})	0.02092	0.01458	0.00475	0.03298	0.00175
Mean of *K* values		0.01500		cm s^{-1}	
Median of *K* values		0.01458		cm s^{-1}	
TTA02 2000					
K (cm s^{-1})	0.01124	0.00154	0.00902	0.00124	0.00264
Mean of *K* values		0.00514		cm s^{-1}	
Median of *K* values		0.00264		cm s^{-1}	

Note: Conventional tillage creates large heterogeneity of the top layer

Table 3.4 Initial water content (θ_i) of the topsoil for selected TTA in 1997 and 2000, measured by the TDR.

TTA No.:	Location of the infiltration tests (see *Figure 2.2*) and measured initial water content (θ_i) (% by volume)					Year	Tillage treatment
	1	2	3	4	5		
	θ_i	θ_i	θ_i	θ_i	θ_i		
4	19.0	21.8	21.7	21.7	21.0	1997	Reduced
01	28.6	27.3	22.6	22.9	24.5	2000	Reduced
3	24,2	25,6	25,5	25,8	25.6	1997	No till
03	34.4	31.60	32.7	27.5	30.7	2000	No till
1	15.3	14.9	16.1	13.2	21.7	1997	Conventional
02	25.7	29.6	17.6	31.0	25.5	2000	Conventional

The comparison of K values for each location of the infiltration tests and years 1997 and 2000 for reduced till, no-till, and conventional tillage are presented in Fig. 3.4.

The technological reasons (applied technology of treatment) resulted unfortunately in the changes of the size of TTA in 1999 in the fields with a span equal to 8.85 m. However the type of treatment has not been significantly changed.

4. CONCLUSIONS

After 3 years of very similar tillage treatment applied on the TTA, the repeated infiltration tests were carried out. The fixed equidistant space 1 m between infiltration rings was kept strictly, thus the same locations were always selected for the tests. Unfortunately, the 3-year data set collected until now is not enough for a detailed geo-statistical evaluation. The experimental tests will continue during the coming years. However some conclusions based on the experiments carried out till now can be made. The initial water content was measured before the beginning of the tests using the TDR TRIME FM3 in all cases. As can be seen from the Table 3.4 the initial water content (θ_i) was higher in 2000 than in 1997 in all locations of the infiltration tests. The initial value of water content (θ_i) has influence on the infiltration rate of an early and medium stage of ponded infiltration (Kutilek and Nielsen, 1994). The infiltration rate (v) supposes to decrease as θ_i increase. This can be seen in all presented cases as is demonstrated in Figs. 3.1, 3.2, and 3.3.

A ponded infiltration has a linear relationship between I and the square root of time during the early stage (the initial minutes) with S_H as the slope of the line, see Eqs. (3) and (4). In this stage S_H plays a dominant role; an influence of A term (A relates to K) is negligible. The value of S_H depends on diffusivity and θ_i. However, for the calculation of the K values, the infiltration rate for larger t (I and thus v are constant with t) and an influence of S_H in this time of infiltration is negligible.

The fluctuation of the measured data of the infiltration and also K values (Fig. 3.4) in the case of conventional tillage was created by the quality of ploughing and of course location of measurement is important within a particular TTA. It was found that the heterogeneity of the top layer after conventional tillage was enormous and it is possible to detect by the tests the resultant troughs of ploughing. However the conventional ploughing gave not significant changes in K values after 3 years. This work is more focused on the influence of repeated reduced till and no-till on infiltration and K values.

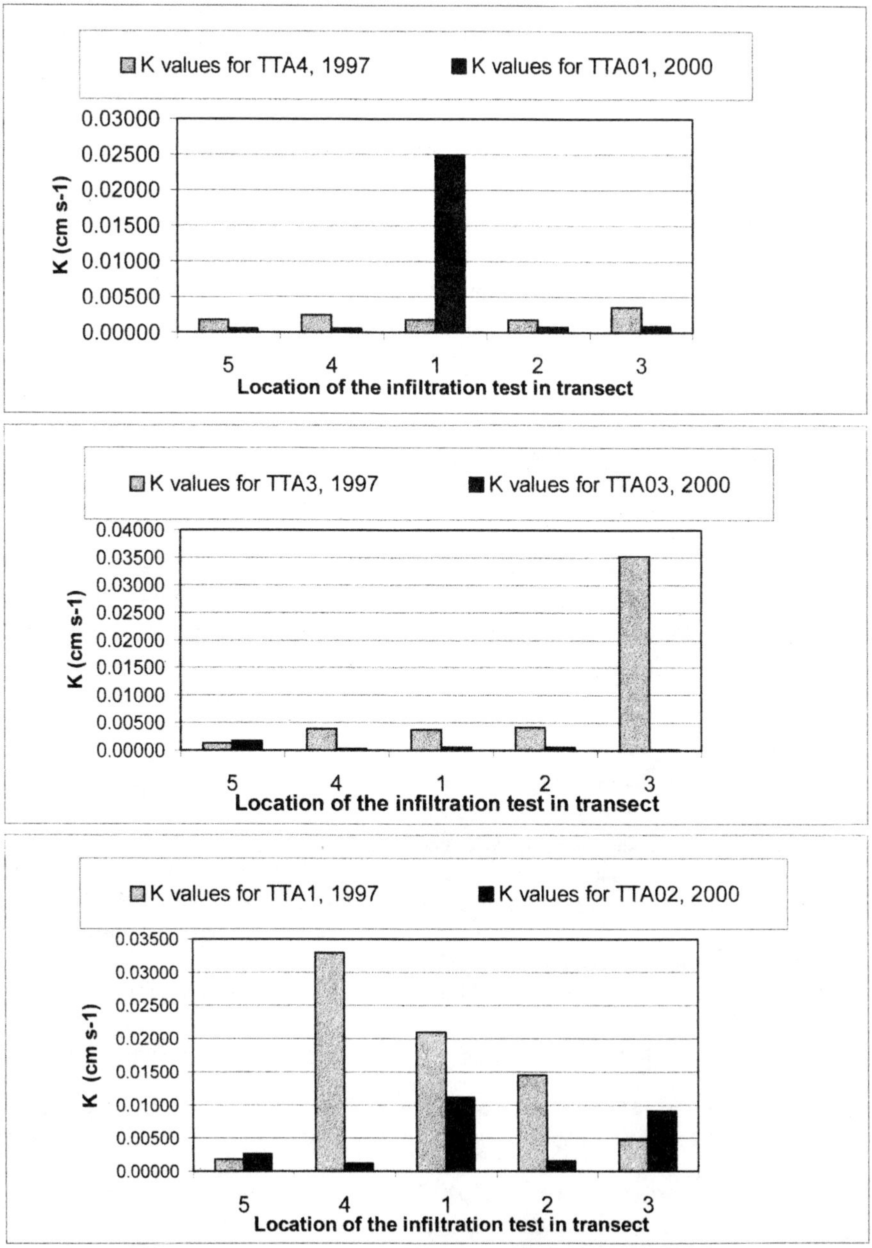

Figure 3.4 Calculated values of K (cm s^{-1}) for reduced till, no-till and conventional tillage for the transect tests in 1997 and 2000

Reduced till treatment and no-till treatment show a significant decrease of *v (t)* values after three years compared to the situation in 1997 (Figs. 3.1, 3.2, and 3.3). The *K* decreased approximately 3 times for reduced till and 6 times for no-till treatment (see Tables 3.1, and 3.2 and Fig. 3.4).

The decrease on this type of soil may cause several negative results from a surface soil hydrology and agricultural points of view as:

- Surface runoff increase
- Decrease of water storage in the soil profile
- Decrease quality and yield of the plant production during dry years
- Possibly increase of the soil compaction of surface layer
- Possibly increased soil erosion in hilly areas

ACKNOWLEDGEMENT

This work was partly supported by the Project no. MSM 412100004. I thank to Dr. Pavel Ruzek from the RIPP for the use of his experimental fields and his co-operation and Pavel Krivohlavy, MSc. for the co-operation in the matter of fieldwork and data processing. I thank the referee for helpful corrections and comments of an earlier draft of the paper.

REFERENCES

Angulo-Jaramillo, R., Vandervaere, J-P., Roulier, S., Thony, J-L., Gaudet, J-P., Vauclin, M., 2000, Field measurement of soil surface hydraulic properties by disc and ring infiltrometers. A review and recent developments. *Soil Tillage and Research* **55:** 1–29.

Blevins, R.L., and Frye, W.W., 1992, Conservation tillage: An ecological approach to soil management. *Advances in Agronomy,* **51:** 33-78.

Brady, N.C., and Weil, R.R., 1999, *The Nature and Properties of Soils,* Twelfth Edition, Prentice Hall Inc., New Jersey, USA.

Carter, M.R., 1994, *Conservation Tillage in Temperate Agroecosystems*, Boca Raton, Fla, Lewis Publishers, USA.

Elrick, D.E., Reynolds, W.D. and Tan, K.A., 1989, Hydraulic conductivity measurements in the unsaturated zone using improved well analyses. *Ground Water Monitoring Revue* **9**: 184-193.

Elrick, D.E. and REYNOLDS W.D., 1992, Methods for analysing constant-head well permeameter data. *Soil Science Society of America Journal* **56:** 309– 312.

Gardner, W. R., 1958, Some steady-state solutions of the unsaturated moisture flow equation with application to evaporation from water table. *Soil Science* **85**, 228-232.

Kachanoski, R.G., Elrick, D.E., and O'Neil, B., 1989, *Effect of management on soil hydraulic properties: method development.* Dept. of Land Resources Sciences, University of Guelph, Ontario, Canada.

Klute, A., 1982, Tillage effects on the hydraulic properties of the soil: A review. In: P. W. Unger and D. M. van Dooren (Editors), Predicting Tillage Effects on Soil Physical Properties and Processes. *ASA Spec. Publ.* **44**: ASA, Madison, WI, USA, 29–42.

Kutilek, M., and Nielsen, D.R., 1994, *Soil Hydrology*, Catena Verlag, Cremlingen-Dested, Germany.

Le Bissonnais, Y., and Arrouays D., 1997, Aggregate stability and assessment of soil crustability and erodibility: II. Application to humic loamy soils with various organic carbon contents. *European Journal of Soil Science* **48:** 39-48.

Matula, S., Kozakova, H., 1997, A Simple Pressure Infiltrometer for Determination of Soil Hydraulic Properties by in situ Infiltration Measurements. *Rostlinna vyroba/Plant Production Journal* **43**: (9), 405- 413.

Matula, S., Krivohlavy, P. 1998a, Vliv povrchoveho zpracovani pudy na jeji pudne-hydraulicke vlastnosti. In: Vyuziti ruznych zpusobu zpracovani pudy pri pestovani rostlin. *Sborník odborneho semináře VURV, VUZT, VUP a CZ BIOM* : 61-72.

Matula, S., Krivohlavy, P., 1998b, Tillage effects on the soil-hydraulic properties in arable land. *Annual meeting of DVWK (Deutscher Verband für Wasserwirtschaft und Kulturbau),* Potsdam, Germany.

Philip, J.R., 1957, The theory of infiltration: 4. Sorptivity and algebraic infiltration equations. *Soil Science* **84**: 257-267.

Philip, J.R., 1985, Aproximate analysis of the borehole permeameter in unsaturated soils. *Water Resources Research* **21**: 1025-1033.

Philip, J.R., 1987, The quasilinear analysis, the scattering analog, and other aspects of infiltration and seepage. In: *Y. S. Fok (ed.) Infiltration, development and application.* Water Resources Research Center, Honolulu, USA, 1-27.

Raats, P.A.C., 1976, Analytical solutions of simplified flow equation. *Transactions of ASAE* **19**: 683 – 689.

Reynolds, W. D. and Elrick, D.E., 1990, Ponded infiltration from a single ring: I. analysis of steady flow. *Soil Science Society of America Journal* **54**: 1233- 1241.

Reynolds, W. D. and Elrick, D.E., 1991, Determination of hydraulic conductivity using a tension infiltrometer. *Soil Science Society of America Journal* **55**: 633-639.

Scotter, D.R., Clothier, B.E. and Harper E.R., 1982, Measuring saturated hydraulic conductivity and sorptivity using twin rings. *Australian Journal of Soil Research* **20:** 295-304.

Talsma, T., 1969, In situ measurement of sorptivity. *Australian Journal of Soil Research,* 7, 269-276.

White, I., and Sully, M.J., 1987, Macroscopic and microscopic capillary length and time scales from field infiltration. *Water Resources Research,* **23:** 1514-1522.

Bifacial PV Modules in Solar Trackers and Concentrators

A New Approach to Supplying Power in Agriculture

MARTIN LIBRA[1] and VLADISLAV POULEK[2]

[1] *Czech University of Agriculture in Prague, Prague*

[2] *Poulek Solar Co. Ltd, Brandys nad Labem, Czech Republic (both)*

1. FUNDAMENTAL ASPECTS OF THE ENERGY PRODUCTION

Permanently sustainable development has been a well discussed theme during the last years. Some experts believe in its feasibility, others are sceptical. Renewable energy usage is a real problem considering limited reserves of classical energy sources and ecological aspects. In addition to the widely discussed arguments, we would like to emphasise that exhaust heat, dust and fly ash, sulphur-, carbon- and nitrogen-oxides are not produced in the case of energy production in solar power stations, hydro-electric power plants and wind-power plants. The mentioned oxides cause greenhouse effect. Exhaust heat is usually two times higher than the electric energy produced in thermal power stations and nuclear power plants. Atmosphere heating and climate changes are well known problems. From the point of view of the permanently sustainable development, the energy production from the renewable energy sources is the only way to sustain the economic development.

Fig.1 shows development of the energy production in the last century and the forecast for next two centuries. The curve 1 corresponds to the total energy production, curve 2 corresponds to the electric energy production and the area 3 shows the forecast of the future development. The permanently sustainable development will be possible only in the case of the limiting curve with limited value $E = 10^{16}$ kW.h.year^{-1}.

This chapter deals with solar energy. The direct photovoltaic (PV) energy transformation is sufficiently effective to be used for example to supply portable solar pumping systems in agriculture. Especially in outlying places, the long electric supply lines are expensive. Similarly, petrol for combustion engine driving electric generators is costly. Fig.2 shows one example of the power supply in a house. It is possible to use DC voltage of 12/24= V or change the voltage to 120 or 220 V AC by an inverter. The battery is used to level the discontinuities in solar radiation over time.

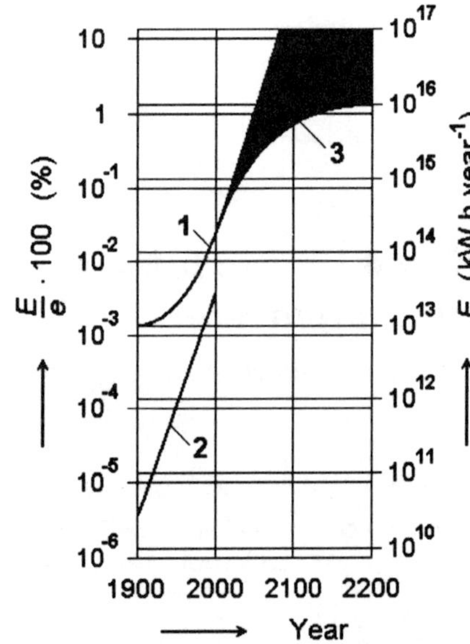

Figure 1. The development of the energy production (1 - total energy production, 2 - electric energy production, 3 - forecast of the development, e =7.2x10^{17}kW.h.year^{-1} - solar energy falling on the Earth)

It is possible to use more simple arrangement for solar pumping systems in agriculture with direct connection between solar PV panels and a water pump without incorporating a battery. At night, it is not possible to pump in this case. In most applications the night pumping is not necessary.

The solar constant $I_s = 1{,}367$ W/m^2 is the average radiation intensity over the earth atmosphere in the average Earth distance from the Sun. The maximum radiation intensity is up to $I_{max} = 1{,}100$ W/m^2 on the Earth surface (below the atmosphere) during the sunny day. The efficiency of the best semiconductor photovoltaic solar panels based on the monocrystalline silicon is over 20% and the efficiency of the best solar panels based on the polycrystalline silicon is up to 15%. We can calculate, that it is possible to obtain the maximum power $P = 160 - 200$ W from the area of $S = 1$ m^2 in this case. The other solar panels based for example on GaAs, CdTe, CdS, CdSe, InP and so on have some advantages, but the main disadvantage is the much higher price. That is the reason, why these panels are used in special applications, where the price isn't the limiting parameter. Especially in space applications in power supply for the satellites the limiting parameter is the highest efficiency and not the price. For example GaAs has higher resistance to the hard radiation in addition to the efficiency over 25% and the efficiency decrease with the increasing temperature is very small.

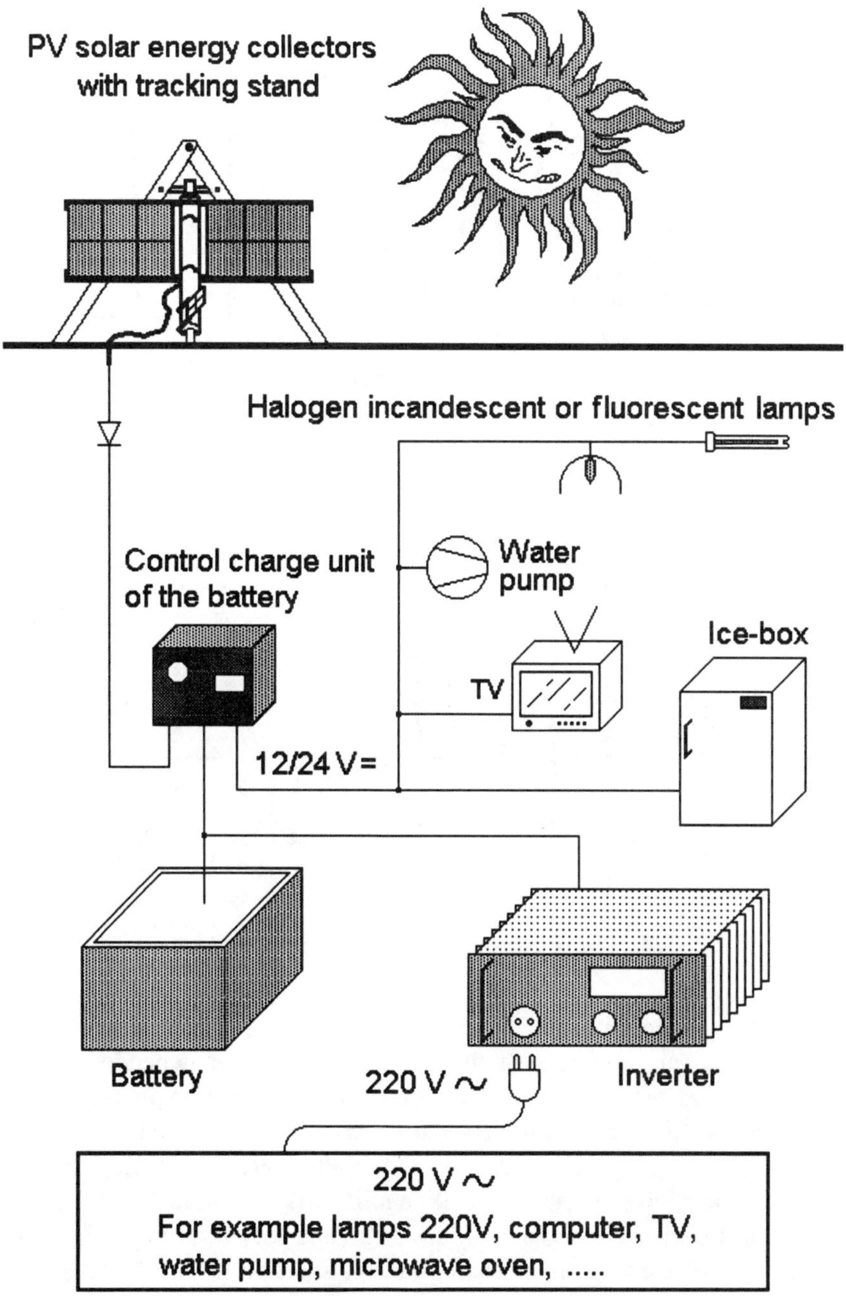

Figure 2. Example of the solar power supply in a house

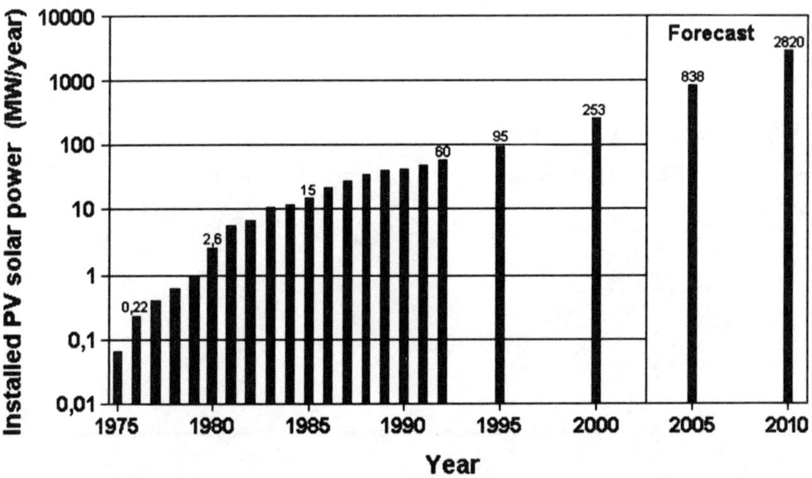

Figure 3. Production, sale and installation of photovoltaic solar panels

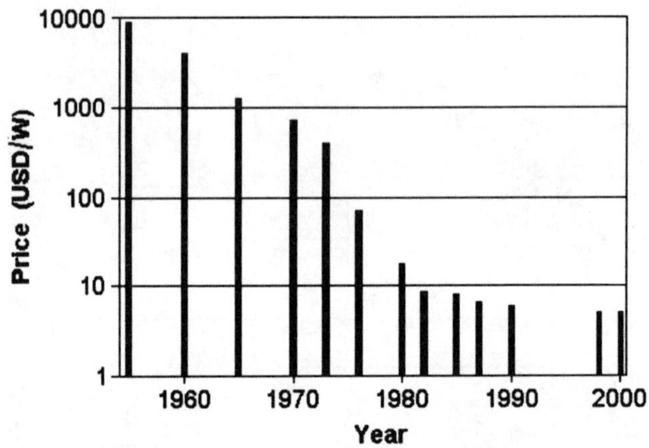

Figure 4. Photovoltaic solar panels prices

Fig.3 shows progress in solar PV panels production and sales during the last 25 years. We can approximate the last years by linear dependence in the semilogarithmic scale and it really corresponds with the exponential. Fig.4 shows the development of solar PV panels prices. We can see the substantial price decrease during the period of 1970-1980. Today the average price of the solar panels based on the crystalline silicon is approximately 5 USD per watt and has been nearly constant during the last ten years. Fig.5 shows that most of PV solar energy is produced in Japan and in the USA. This fact is logical because these countries have high level of industry and low level of classical energy sources and they must import especially crude petroleum.

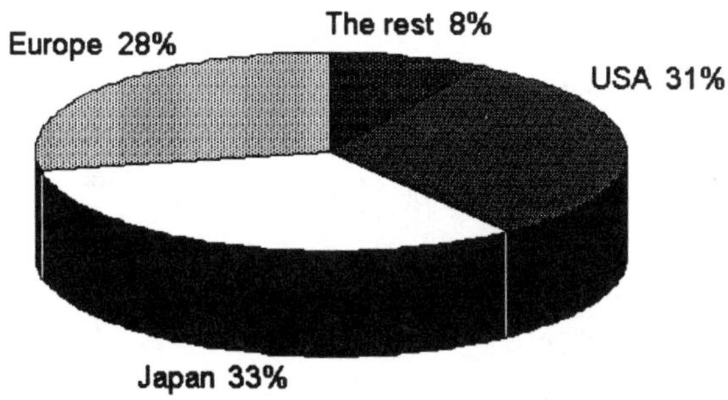

Figure 5. Electric energy production by photovoltaic solar panels

They are existing solar power stations with maximum power higher then 1 MW. But we can't expect that the solar power stations will replace classical power stations with the same maximum power. The utilisation of the thermal power station and nuclear power plant is approximately 85-90%, but utilisation of the solar power plant is only 30% in desert places with the best solar conditions. Nevertheless 1 km^2 of the solar panels based on polycrystalline silicon with tracking stands mentioned below can yield energy 4.3×10^8 kW.h.year^{-1}. The area does not correspond with 1 km^2 of the power station area because distances between panels and areas for inverters must be considered. The energy production in the whole world was 10^{14} kW.h.year^{-1} in the year 1997 (UNO, 1997). Comparison of these two values shows, that it would be theoretically possible to produce the energy only in the solar power stations. There are desert areas large enough on the Earth surface for example in north Africa, south USA, Mexico, Chile, Mongolia, west Australia, Kazakhstan and so on. The energy would be transported in the form of electrical energy by classical wires or by superconductive wires or liquid hydrogen would be produced and the energy transported in this form. We must incorporate some energy loss during the transport. Production of 1 kg liquid hydrogen needs energy approximately 50 kW.h (water electrolysis and liquefaction) but in this case we must include some energy for water transport and liquid hydrogen transport.

2. THE TRACKING STRATEGY

The solar radiation consist of three parts. The direct solar radiation includes most of energy. It is coming only from the sun circle in a defined direction. Smaller amount of energy is in diffuse circumsolar radiation which is diffused by atmosphere but it keeps the direction of the major angle. Only low energy is in isotropic diffuse radiation which does not depend on the direction. The isotropic diffuse radiation dominates in the case of cloudy weather. Fig.6 shows the iso-curves of radiation intensity in dependence on the direction from the cloudless sky. The dotted lines correspond with diffuse circumsolar radiation, the full lines correspond with isotropic diffuse radiation. The sun is 30° over horizon and we can see, that the radiation

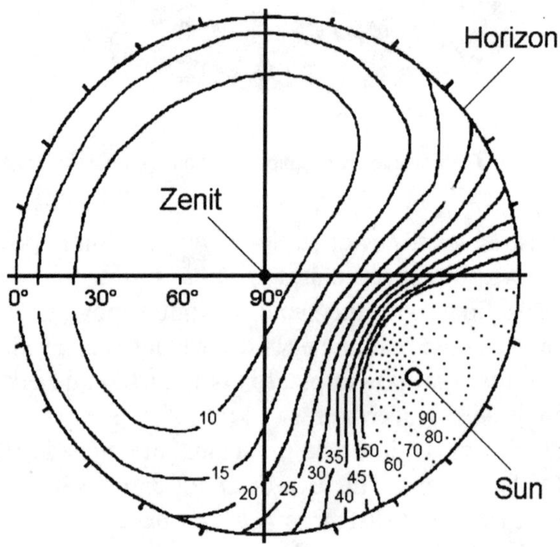

Figure 6. Iso-curves of radiation intensity in dependence on the direction

intensity from the opposite side is constant for the most part of the sky. The increase near the opposite horizon causes the reflection.

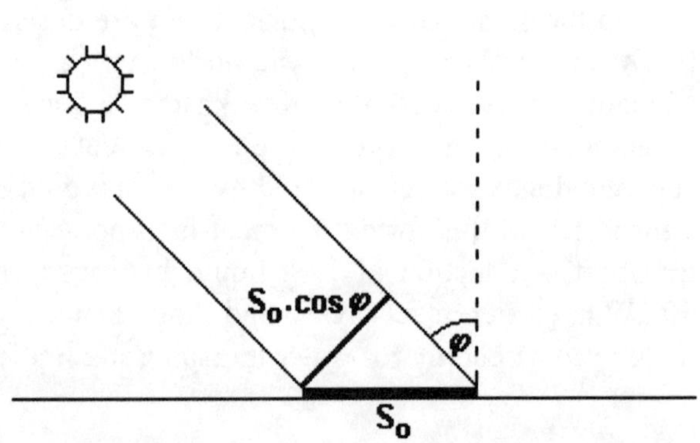

Figure 7. Angle of incidence φ of the solar radiation

Sun is moving across the sky during the day. In the case of fixed solar collectors the projection of the collector area into the plane, which is perpendicular to the radiation direction, is given by the function of cosine of the angle of incidence (Fig.7). The higher is the angle of incidence φ, the lower is the power. The solar tracker, a device that keeps photovoltaic or photothermal panels in an optimum position perpendicularly to the solar radiation during daylight hours, can increase the collected energy by up to 40%. This increase is raised to 80% when solar energy is used for water pumping because during the day the pump is longer time over the limit power value. Theoretical calculation of the energy surplus in the case of tracking collectors is as follows: We assume the maximum radiation intensity $I = 1{,}100\ \mathrm{W.m^{-2}}$ is falling on the area which is oriented perpendicularly to the direction of radiation. We assume the day length $t = 12\ \mathrm{h} = 43{,}200\ \mathrm{s}$ as well as the night length and we compare the tracking collector which is all the time optimally oriented to the sun and the fixed collector which is oriented perpendicularly to the direction of radiation only at noon. We mark the collector area S_o.

a) *For fixed collector we calculate*: The projection of this area on the area which is oriented perpendicularly to the radiation direction is equal $S = S_o \,.\cos\varphi$ and the angle φ is changing in the interval $\varphi \ni \left\langle \frac{-\pi}{2}; \frac{+\pi}{2} \right\rangle$ during the day. The angular velocity of the sun moving across the sky is equal to $\omega = 2\pi / T = 7.27\,.10^{-5}\ \mathrm{s^{-1}}$ and the differential of the acting energy is equal to $\mathrm{d}W = I\,S\,\mathrm{d}t$. When we do not consider the atmosphere influence, we can calculate the energy, which has fallen on the collector area $S_o = 1\ \mathrm{m}^2$ during one day as

$$W = \int_{-21600}^{+21600} I\,S_o \cos\omega t\,\mathrm{d}t = I\,S_o \left[\frac{\sin\omega t}{\omega}\right]_{-21600}^{+21600} =$$

$$= \frac{2\,I\,S_o}{\omega} = 3.03\,.10^7\ \mathrm{W.s} = 8.41\ \mathrm{kW.h}\,. \qquad (1)$$

b) *For the tracking collector which is all the time optimally oriented with respect to the sun,* and when the atmospheric influence is not considered, we can calculate the energy, which has fallen on the collector area $S_o = 1\ \mathrm{m}^2$ during one day as

$$W = I\,S_o\,t = 4.75\,.10^7\ \mathrm{W.s} = 13.2\ \mathrm{kW.h}\,. \qquad (2)$$

Comparison between Eq.(1) and Eq.(2) shows the energy surplus of 57% when we do not consider the atmosphere influence. We would really obtain this surplus for example on the Moon surface. On the Earth surface the sun is shining through a thick atmosphere layer after sunrise and before sunset. In the morning and in the evening, the radiation intensity acting on the area which is oriented perpendicularly to the radiation direction is much lower

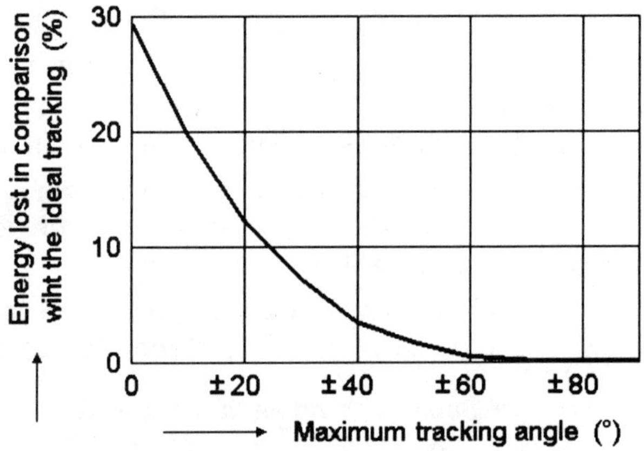

Figure 8. Energy lost in dependence of the maximum tracking angle in comparison with the ideal tracking

than at noon. On the other hand, the day can be longer than 12 h at higher latitude. That is the reason, that the energy surplus can be really as much as 40% on the Earth surface. The addition of the below-mentioned soft concentrator can increase the energy surplus up to 100%.

Fig.8 shows the dependence of the energy loss on the maximum tracking angle in comparison with the ideal tracking. The value $\pm 0^{o}$ corresponds with the fixed stand. We can see, that the tracking angle more than $\pm 60^{o}$ does not yield energy surplus. That is why our below-mentioned solar tracker TRAXLE™ was constructed for tracking angle $\pm 60^{o}$.

3. SOLAR TRACKERS

Although solar trackers can boost the energy gain of PV arrays by up to 40% and pumping capacity by up to 80%, their application was not frequent at present. Why? Relatively high prices and medium reliability prevented widespread use of solar trackers. Reliability and cost/performance ratio will decide about the future trends of solar trackers.

Active trackers: Electrooptical solar trackers are usually composed of at least one pair of antiparallel connected photoresistors or photovoltaic solar cells which are, by equal intensity of illumination of both elements, electrically balanced so that there is either no or negligible control signal on a driving motor. By differential illumination of electrooptical sensors a differential control signal occurs which is used to drive the motor to orient the apparatus in such a direction where illumination of the electrooptical sensors is equal and balance is restored. Such high accuracy trackers are intended mainly for concentrator solar systems. These trackers are complex and, therefore, expensive and unreliable. Active solar trackers based on clockworks or combining both principles exist.

Passive trackers: Passive solar trackers are based on thermal expansion of matter (freon) or on shape memory alloys. They are usually composed of a couple of actuators working against each other which are balanced by equal illumination. By differential illumination of actuators, the resulting unbalance of forces is used for orientation of the apparatus in such a direction where equal illumination of actuators and balance of forces is restored. Passive solar trackers, compared to active trackers, are less complex but they are working with lower efficiency and at low temperatures are not working at all.

4. NEW SOLAR TRACKER

The construction of the new tracking stand TRAXLE™ for solar devices is original and has recently been patented (Poulek at all, 1998, 2000a, 2001). The apparatus enables backtracking within 5 minutes. It is an important advantage in comparison with popular passive trackers where backtracking times as long as one hour are usual. The tracker works also at low temperatures down to –40°C with average tracking accuracy $\pm 5^{\circ}$. The low cost solar tracker is based on a new arrangement of solar cells connected directly to a reversible DC motor. Solar cells both sense and provide energy for tracking. Costly and unreliable electronics has been completely eliminated.

Fig.9 shows the principle of the tracker. Two antiparallel sensing/driving solar cells are connected to reversible DC motor, the transmission is self-locking. When solar collectors are oriented eastwards and the sun moves from the east to the west (Fig.9b), the angle β between solar radiation and sensing/driving cells increases until the power of the driving DC motor is high enough to move solar collectors. Then the collectors start to move and the angle β starts to decrease until the power of DC motor is lower than what is necessary to move solar collectors. Step by step the collectors moves westwards during the day. Additional antiparallel solar cell enables backtracking of the tracker in the morning. In Fig.9b it is seen that the sun can move the collectors from any position when it is cloudy weather during part of the day and the sun start shining for example in the afternoon. Backtracking is possible from any position. The plane of the sensing/driving

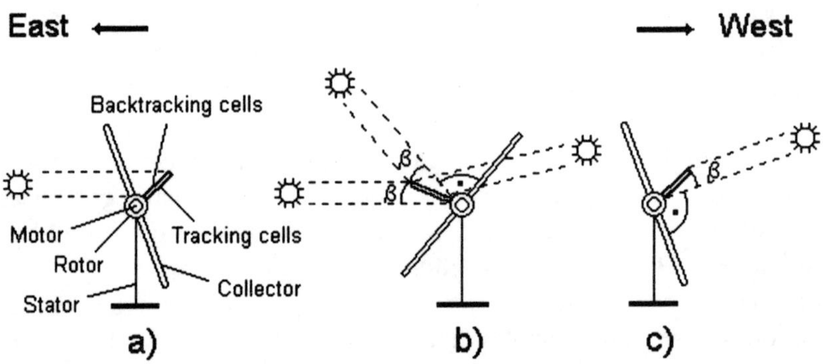

Figure 9. The principle of the new solar tracker

cells is inclined from the plane parallel to the axle and perpendicular to the collectors approximately by 20° eastwards. This angle is important due to prompt morning backtracking and accuracy of tracking. Fig.10 shows the complete wiring diagram.

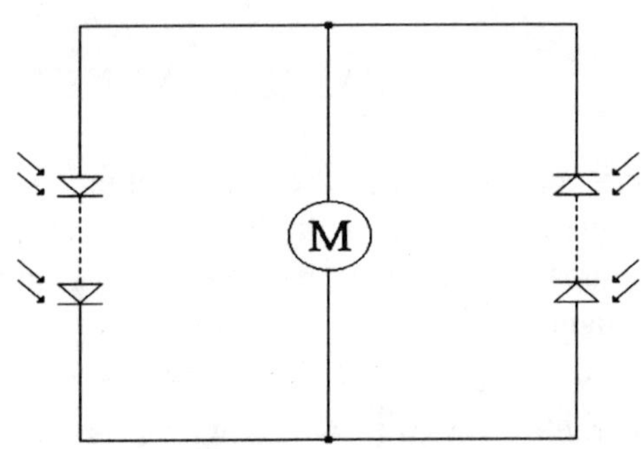

Figure 10. Complete wiring diagram of the tracker

Backtracking time of the new tracker is only a few minutes even in cold climate (-40°C) due to the higher efficiency of the semiconductor solar cells. Backtracking time of the passive trackers (either freon and shape memory) is more than 1 hour (Rauth at all, 1995) at the same temperature. The area of the auxiliary solar panel with sensing/driving cells is about 1% of the area of the moved solar collectors. Good balance is necessary. The above mentioned tracking accuracy of $\pm 5^{\circ}$ is sufficient, because the values of the cosine function are approximately equal to 1 within this interval ($0.996 \div 1$, see Fig.7).

Recently we developed solar tracker with 360° tracking angle and with vertical tracking axle. Unlike standard trackers with 120° tracking angle, it can work in higher latitude in regions above polar circle, for example in Russia, Canada, Alaska, Scandinavia, Antarctica and so on, where sun can shine 24 hours a day.

5. THE TRACKING CONCENTRATOR

Our new tracking soft ridge concentrator (C = 1.6 -1.7) (Poulek at all, 2000b) can double photovoltaic energy harvest in comparison with fixed panels and substantially reduce the price of PV energy. The new system combines the simple low-cost tracker (Poulek at all, 1998) with flat booster mirrors. Unlike V-trough concentrator (Klotz, 1995, Nann, 1991), the "outer" mirror has been eliminated (Fig.11). On single axis trackers the mirrors have to be extended beyond the PV panels to ensure uniform illumination of panels at seasonally variable elevation of the sun. On polar axis trackers with seasonally adjustable slope of the axle, the extended mirror is not needed.

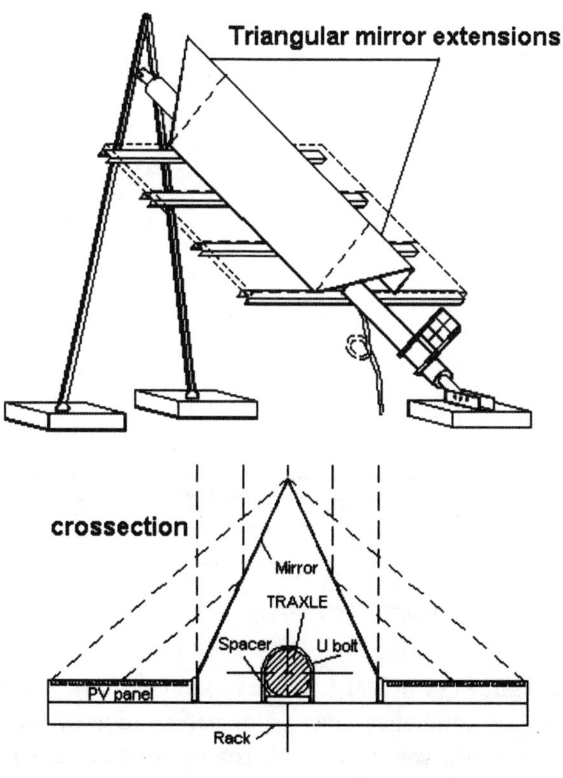

Figure 11. The tracking ridge concentrator.

It is advantageous that soft concentrators for photovoltaics do not need highly specular expensive mirrors. Weather resistant (at least 10 years) mirrors with high total reflectance are needed. The mirror can be made of rolled stainless steel sheet with special surface finish (Nostell at all, 1998), of rolled aluminium alloy sheet (plated with pure aluminium) protected by a weather resistant polymer (PVF) film (Nostell at all, 1998, Perers at all 1994), of silver coated polymer (acrylic) film (Schissel at all, 1994) or sheet, of aluminium coated polymer (acrylic) film (Nostell at all, 1998) or sheet, or silver coated hardened glass.

The new tracking soft concentrator is very compact, simple and reliable. Proven existing tracker hardware is used. Unlike V-trough concentrators no additional mirror supporting structures on the contrary the ridge concentrator acts as a load carrying structure. Wind induced torque is strongly reduced. Application of the ridge concentrator on different tracking systems is also very easy. It can be used on polar or horizontal single axis trackers, two axis trackers as well as on down mentioned 360° trackers for space and terrestrial applications.

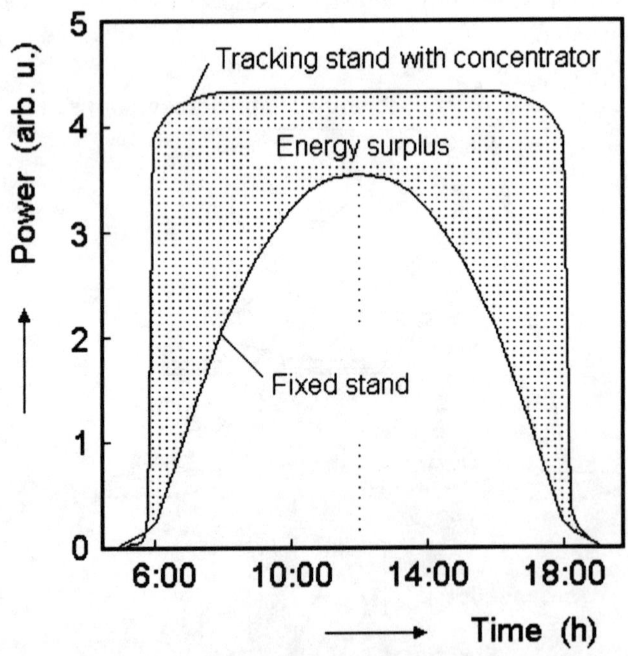

Figure 12. Idealised dependence of the power on the time

Concentration ratio (1.6-1.7) reduces temperature of PV panels, consequently increases efficiency and avoids degradation of the encapsulant. The new design also improves (against V-trough) air flow around PV panels (improved cooling). Concentration ratio (2-2.4) of standard V-trough concentrators frequently caused browning of the EVA encapsulant while elevated temperature reduces efficiency of the semiconductor PV panels.

The new tracking ridge concentrator can double annual energy harvest (100% energy surplus) in comparison with fixed panels and pumping capacity surplus can be as high as 150%, because during the day the pump is longer time over the limit power value (see Fig.12). Fig.12 shows the idealised dependence of the power on the time during the day for collectors with fixed and tracking stand. Produced energy corresponds with the area below the curve, as given by Eq. (3)

$$E = \int_{\Delta t} P.dt \,, \tag{3}$$

where P is the power and t is the time.

Two years comparison of energy production between fixed tilt PV panels and PV panels mounted on polar axis tracking ridge concentrator was started in May 1999 (Prague region, 50° N). The results show that e.g. on a clear day in June (6.8 kWh/m^2), the energy surplus of 107% was observed.

6. BIFACIAL SOLAR PV MODULES

Bifacial PV modules are recently delivered by several manufacturers. Because there is either no or low price difference in the price of such bifacial and monofacial modules it is reasonable to use bifacial modules which could produce 5-20% more energy (in comparison with monofacial PV module with the same nominal output power).

Polar axis solar tracker and/or tracking concentrator is always mounted on high support structures (to avoid contact of the rotating PV array with the ground). It improves back side energy collection in comparison with fixed PV arrays or even roof integrated PV modules. The energy gain can be in the range 10-15% (for typical albedo 0.3).

Additionally solar trackers/concentrators are usually oriented to the west, before backtracking, in the morning. The bifacial modules enable to collect direct back side solar radiation before backtracking. According to local climate the resulting energy gain could be 2-5%.

It is also very advantageous that bifacial PV modules (fixed), transparent for infrared radiation, has lower operating temperature against monofacial ones (about 5-9°C). It is especially advantageous by solar trackers and tracking soft (C = 1.6) concentrators where PV modules are exposed to higher solar radiation than on fixed racks. As indicated above solar trackers/concentrators are always mounted on high support structures which improve cooling of PV modules by air flow. Measured temperature of bifacial c-Si PV modules on the tracker was usually lower (by 5-8°C) than that of roof integrated monofacial c-Si modules.

Even in soft concentrators (C = 1.6) with bifacial PV modules there is very low temperature difference against roof integrated monofacial modules (lower than 5°C). The reduced temperature of bifacial modules can also increase the energy gain by 2-5%.

The reduced temperature is also very important for lifetime of PV modules in soft (C = 1.6) concentrators. It should help to avoid degradation of polymer encapsulants of modules caused by higher temperatures (of monofacial modules).

The new bifacial PV modules with reduced temperature sensitivity (HIT) can further increase the system energy gain by up 10%. A synergic combination of all above effects can boost energy gain by 15-25% in comparison with the same tracker/concentrator with monofacial modules.

The polar axis solar tracker with c-Si bifacial PV modules will therefore deliver by about 50% more energy than fixed c-Si monofacial PV array with the same rated output power. The tracking bifacial soft concentrator will even double the energy gain against fixed monofacial PV array (Fig.13). Concerning the PV pumping systems there is 100% and 150% pumping capacity surplus for tracks and tracking concentrators respectively.

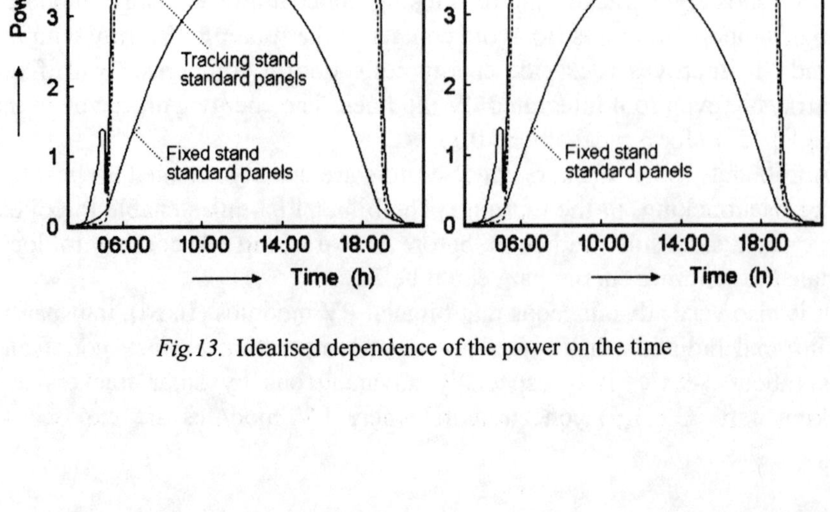

Fig.13. Idealised dependence of the power on the time

The customer e.g. in Spain will therefore get the support 0.4 EUR per kWh of the 5 kWh systems. The 5 kW tracking bifacial system will deliver energy equivalent to 7.5 kWh fixed system, but for 7.5 kW system the support would be only 0.2 EUR per kWh. There is 5 kW limit for the 0.4 EUR/kWh support in Spain, for bigger systems the support is 0.2 EUR/kWh.

The same facts are valid also for tracking ridge concentrators where the energy gain is about 100% (i.e. 5 kWh tracking bifacial concentrator gives the same energy output like the 10kWh fixed monofacial system).

Another calculation shows that 5 kW tracking concentrator is by about 2,700 EUR more expensive than 5 kW fixed mono stand but the double energy gain will pay back the price within less than 6 months (5 kW x 0.4 EUR x 10hrs. x 180 days = 3,600 EUR) in sunny arid climate.

6.1 Operating Temperature Measurement of Bifacial Modules

Bifacial as well as monofacial reference PV modules were delivered by the same manufacturer. The design of both modules has been very similar. The only difference is back surface grid of bifacial cells and back surface glass encapsulant of the bifacial module. Cell temperatures were calculated using the temperature dependence of the open circuit voltage.

Our measurements of both bifacial glass/c-Si/glass and monofacial glass/c-Si/foil flat plate PV modules indicate that the operating temperature $T_{op} = 41°C$ of bifacial modules was by 12°C lower than that of monofacial ones $T_{op} = 53°C$ at AM 1.5 solar radiation, wind speed below 1 $m.s^{-1}$ and ambient temperature $T_{amb} = 22°C$.

Another measurement indicate that bifacial PV modules at soft concentrators ($C_{geometrical} = 1.5$, $C_{optical} = 1.35$) are still by about 3°C less hot $T_{op} = 50°C$ than monofacial modules at one sun radiation.

The study shows that main reasons are as follows:

1. Bifacial Si PV cells are transparent for the infrared part of the solar radiation. As the infrared radiation above 1,100 nm represents more than 20% of the solar energy a proportionally lower energy is absorbed in the bifacial module in comparison with non transparent monofacial one.
2. Opaque back surface protection foil also contributes to the higher temperature of monofacial modules. Typical packing density of c-Si PV modules is about 0.85. It means that 15% of the module area, not covered by PV cells, can absorb substantial quantity (about 50%) of the incident solar radiation while there is negligible absorption in the bifacial modules.

The experimental results are in good agreement with our calculation as well as with soft concentrator (C = 1.6 and 2.2) measurements (Stacey at all, 1984) of PV modules (both glass/Si/glass and glass/Si/foil) with monofacial cells only.

In conclusion using soft concentrators ($C_{geometrical}$ = 1.5-1.7 , $C_{optical}$ = 1.35-1.45) with bifacial PV modules the over heating of modules is eliminated.

Additionally module mounted on high pedestal solar trackers/concentrators can be well cooled by air while fixed modules are frequently roof integrated so the natural air flow is substantially restricted.

7. SOLAR PUMPING SYSTEMS

Solar pumping systems offer the main usage for the solar trackers in agriculture. Pumping is the most effective application. As written above, the surplus of the pumped water can be up to 80% in the case of solar tracking system without concentrator and up to 150% in the case of solar tracking system with above mentioned soft ridge concentrator. Fig.14 shows an example of a simple arrangement of the solar pumping system without batteries, the solar collectors are connected directly to the water pump.

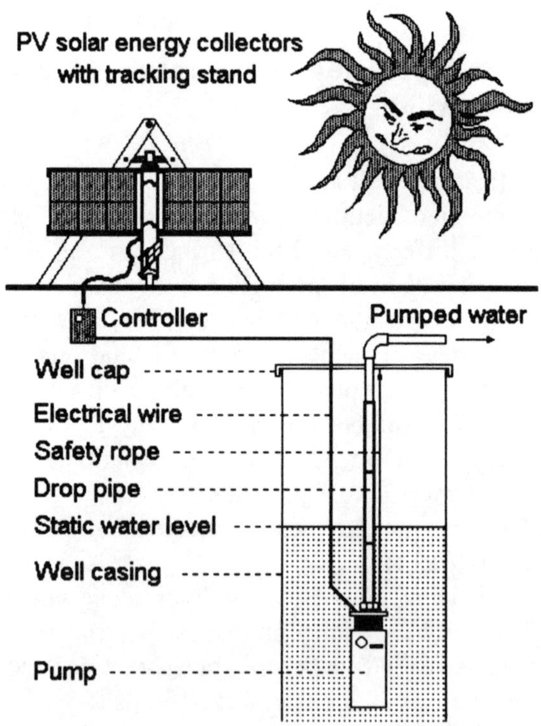

Figure 14. Simple arrangement of the solar pumping system

We used this arrangement in our experiments with water pump SOLARJACK™ SDS-D-228 (Photocomm, Inc., USA), with solar tracking stand PST 301 (Poulek Solar, Ltd., Czech rep.) and with two 55W solar panels T15 (Tetom, Czech Rep.). We pumped the water to the height of approximately 5 meters. Our tests validate the theoretical calculations. In clear day in June near Prague (50° N) we measured the following values. In the case of a fixed stand, the amount of pumped water was 1,745 l/day, in the case of a tracking stand, the amount of pumped water was 2,975 l/day (surplus 71%). In the case of a tracking stand with soft ridge concentrator, the amount of pumped water was 4,170 l/day (surplus 139%).

8. ECONOMY

The economical calculations are in Tables 1 and 2. There is one example of price comparison of fixed and tracking solar pumping systems with an equal pumping capacity. We can see the economic advantage of the tracking system. While the tracking stand is more expensive in comparison with the fixed one, the main value in the final price is the price of solar panels. That's the reason, that the advantage is 1,050 US$ in the case described in Table 1 and even 3,390 US$ in the case of a large system with the ridge concentrator discussed in Table 2.

Table 1. Price comparison of two solar pumping systems with equal pumping capacity (tracking stand without concentrator)

System with fixed stand	Price (US$)	System with tracking stand	Price (US$)
10x 55W PV panel	3,200	6x 55W PV panel	1,920
1x fixed stand	310	1x tracking stand	540
1x water pump	770	1x water pump	770
Total	4,280	Total	3,230

Table 2. Price comparison of two solar pumping systems with equal pumping capacity (tracking stand with soft ridge concentrator)

System with fixed stand	Price (US$)	System with tracking stand	Price (US$)
12x 120W PV panel	8,400	6x 120W PV panel	4,200
2x fixed stand	780	1x tracking stand	1,590
1x water pump	1,770	1x water pump	1,770
Total	10,950	Total	7,560

9. DURABILITY TESTING

The functionality of the new construction had been verified by real operation of many our systems, but it was necessary to perform a durability test to establish its robustness. The tracker must work for a minimum of 20 years with minimum maintenance. Our small tracking stand TRAXLE™ with two PV panels was tested. A simple device was constructed, enabling us to perform and count accelerated cycles. We let our solar tracker perform 10,000 cycles, the equivalent of 30 years of normal operation. Then we took the tracker apart to measure the wear on its parts. The brushes in the motor had, in-fact worn down by two percent. This was done after 10,000 cycles (which is about 6.10^6 rotations). This wear corresponds to the declared life time of the motor, which is 2,000 hours at a speed of 5,000 rot./min (equivalent to 6.10^8 rot.) with the brushes fully worn down. From this we can see, that the slowly cycles do not increase the speed of wear-down by a great deal as opposed to the standard demand. The motor performs approximately 600 rotations per cycle, one cycle is equal to one day. The movable nut M4 had initially a crest clearance of 0.05 mm. Final crest clearance was 0.2 mm. This value is well within the requirements of the tested device, because the device is not sensitive to the crest clearance of the movable nut. The test was performed, because we needed to know whether, after 10,000 cycles, the movable female nut would be totally worn down, or if it would be worn down to the point of decreasing the carrying-capacity of the movable female screw, which is in this case 50 N. Wear-down of other components in the kinetic mechanism are not crucial and, in any case, our tests have not revealed any flaws. Therefore the whole kinetic mechanism has passed the durability test.

Because we are dealing with a device which is mounted outdoors, it has to meet international safety requirements, including wind speed effects. The international norm CSN P ENV 1991-2-4 „The fundamentals of designing and loading of the constructions, part 2-4 Loading of the constructions by wind" speaks about resistance of constructions to wind. The whole territory of Europe is never subject to speeds exceeding $v = 160\text{km.h}^{-1}$ near the ground level. That is why our device was tested in a wind tunnel up to this particular wind speed. The forces acting upon the solar system were to this date either theoretically calculated, or were submitted to static or dynamic testing by simulated weights. We have not yet found any

Figure 15. The solar tracking system during the tests in the wind tunnel

reference in the literature about testing any stand with mounted solar panels in a wind tunnel. Our stand is protected against wind by a self-locking transmission with a maximum torque of $M = 500 \div 3{,}500\,\text{N.m}$ (it depends on the size), and is designed in a way that lets it withstand wind of more then $v = 160\text{km.h}^{-1}$. We performed a factual test of durability in a large wind tunnel with a diameter of 3 meters, which is located in the Aircraft Research and Testing Institute in Prague. The whole solar system was placed successively in four positions relative to the direction of wind. These directions were: perpendicular to the front, perpendicular to the back, perpendicular to the side, sideways to the back with an angle of 45°. In each position we slowly increased the wind speed up to $v = 160\text{km.h}^{-1}$, and then let it affect the solar tracker for the time $t = 3\text{min}$. The whole structure was very stable and had minimum vibrations. The system kept its stability, even when the wind speed was increased up to $v = 180\text{km.h}^{-1}$ for a period of time $t = 1\,\text{min}$. The tests did not damage the solar tracker in any way and the system was working properly after the test. Fig.15 shows the solar system during the tests in the wind tunnel.

According to (White, 1994), a wind of speed v in air of density ρ and acting perpendicularly upon rectangular area *S*, produces a force *F* given by Eq. (4)

$$F = c_x S \rho \frac{v^2}{2} , \qquad (4)$$

where c_x is the coefficient for the resistance of air for a rectangular planar area *S*, which is oriented perpendicularly to the flow. Values c_x are provided in Table 3, where *a/b* is the ratio of the sides of the rectangle.

Table 3. The coefficient c_x for the resistance of air for a rectangular planar area which is oriented perpendicularly to the flow in dependence on the ratio a/b of the sides of the rectangle

a/b	c_x
1	1.18
5	1.2
10	1.3
20	1.5
∞	2

10. CONCLUSIONS

The described newly designed and patented device (Poulek at all, 1998, 2000a, 2001) for automatic orientation of the solar energy collectors with the soft concentrator of radiation can double the solar energy harvest and substantially reduce price of PV energy. It has a wide range of advantages in comparison with existing devices. First of all, its simplicity has positive influence on the reliability and price. In the near future, we will use our device for the construction of the solar pumping systems which can be used above all in agriculture for irrigation, feeding and so on. We expect the pumping capacity surplus up to 150% in comparison with conventional solution which corresponds with our experiments. The new solar tracker was also shown at several important conferences and exhibitions, for example on 16th European Photovoltaic Solar Energy Conference and Exhibition in Glasgow 2000 (Great Britain) and on 17th European Photovoltaic Solar Energy Conference and Exhibition in Munich 2001 (Germany).

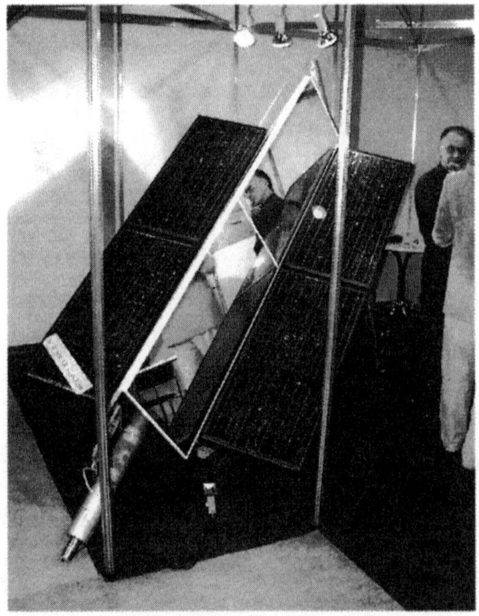

Figure 16. Photovoltaic solar system with the soft ridge concentrator installed at the exhibition stand (PV exhibition in Great Britain – Glasgow 2000)

Figure 17. Portable solar tracking system with the soft ridge concentrator (600 W)

Examples of our solar tracking systems are on the following figures. Fig.15 shows the PV solar system with the soft ridge concentrator installed at the exhibition stand (PV exhibition-Glasgow 2000). Fig.17 shows the portable solar tracking system with the soft ridge concentrator (600 W). Fig.18 shows the photovoltaic solar pumping system 3 kW installed in Great Britain and Fig.19 shows the PV solar tracking system 1 kW installed in Spain (Canary Islands).

More information: http://www.traxle.cz.

Figure 18. The solar pumping system 3 kW installed in Great Britain

Figure 19. The solar tracking system 1 kW installed in Spain (Canaria Islands)

REFERENCES

Energy Statistic Yearbook, 1997, UNO.

Poulek,V., 2000a, Apparatus for Orientation of Solar Radiation Collectors, *US Patent* No. 6,089,224.

Poulek,V., 2001, Motion Transforming Apparatus, *US Patent* No. 6,253,632 B1.

Rauth, H.U., Dress, A., Pruschek, R., Weidele, T., 1995, Annually generated electricity of one and two axes solar tracking systems, *Proc. 13th European PV Solar Energy Conference*, Nice, 23-27 October 1995: 1015-1018.

Poulek,V., Libra,M., 1998, New solar tracker, *Solar Energy Materials and Solar Cells*, **51**, 2: 113-120.

Poulek,V., Libra,M., 2000b, A New Low Cost Tracking Ridge Concentrator, *Solar Energy Materials and Solar Cells*, **61**, 2: 199-202.

Klotz, F.H. 1995, PV systems with V-trough concentration and passive tracking concept and economic potential in Europe, *Proc.13th European PV Solar Energy Conference*, Nice 23-27 October 1995: 1060-1063.

Nann, S., 1991, Potentials for tracking photovoltaic systems and V-troughs in moderate climates, *Solar Energy*, **45**: 385-393.

Nostell, P., Roos, A., Karlsson, B., 1998, Ageing of solar booster reflector materials, *Solar Energy Materials & Solar Cells*, **54**: 235-246.

Perers, B., Karlsson, B., Bergkvist, M., 1994, Intensity Distribution in the Plane From Structured Booster Reflectors With Rolling Grooves and Corrugations, *Solar Energy*, **53**: 215-226.

Schissel, P., Jorgensen, G., Kennedy, C., Goggin, R., 1994, Silvered PMMA reflectors, *Solar Energy Materials & Solar Cells*, **33**: 183-197.

Stacey, R.W. and McCormick, P.G., 1984, Effect of Concentration on the Performance of Flat Plate Photovoltaic Modules, *Solar Energy*, **33**, 6: 565-569.

White, F.M., 1994, *Fluid Mechanics*, Mc. Graw-Hill, inc.

Precision Farming - A Multidisciplinary Approach for Cereal Production

RICHARD J. GODWIN
Precision Farming Centre, Cranfield University at Silsoe, Silsoe, UK

1. INTRODUCTION

This paper provides an overview of the approach taken by Cranfield University and its partners, namely: Arable Research Centres, AGCO Ltd (Massey Ferguson), Hydro Agri and Shuttleworth Farms, in a Home-Grown Cereals Authority funded project to develop practical guidelines for implementing precision farming technology for the UK cereal industry by:

i. developing a methodology for identifying the causes of within-field variation,
ii. exploring the use of remote sensing methods to enable management decisions to be made in "real-time" during the growth of the crop,
iii. determining the potential economic benefits of precision farming,
iv. collaborating with farmers to ensure that the research findings are appropriate for adoption.

Reference will be made to parallel work undertaken by the team as appropriate. The emphasis of this work is placed on practical application that shows a benefit to the farmer by providing tools that can help him manage an ever-increasing size of enterprise when the economic margins are under great pressure. The concept of managing parcels of land with different soil and agro-climatic factors in a spatially variable manner to optimise the economic returns is not new, as our forefathers used the same approach when allocating field boundaries. It is the technology that assists in recognizing the spatial boundaries together with equipment for yield recording and the variable application of agronomic inputs that has created much excitement in recent years. The advent of affordable differential global positioning systems started the development, with a number of yield mapping systems appearing on the market from 1990. A conceptual model of the principles of precision farming is shown in Fig. 1.

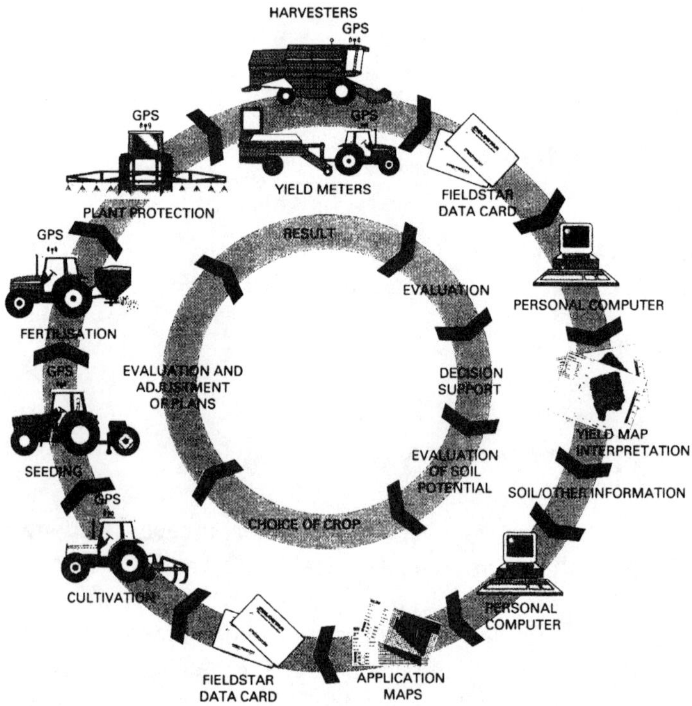

Figure 1. Overview of Precision Farming system (courtesy of AGCO Fieldstar)

Agricultural engineers have been the core to these developments, where engineers with expertise in instrumentation and control, information technology, soil management, soil and water engineering, remote sensing, geographical information systems, and field mechanization have joined forces with soil surveyors, agronomists and crop nutritionists. Whilst there have been, and still are, challenges to the hardware and software aspects of the precision farming system, the single greatest challenge is in combining the information from yield maps, crop performance records (both historic and "real time") and soil analysis into a practical strategy for the variable application of crop treatments for a particular field.

2. APPROACH

The factors in Table 1, after Earl et al (1996), show those which could influence the yield of crops in a given location. Those on the left of the table have to be considered as they can have major effects upon yield yet there is little that can be done to control them. The factors on the right, however, can be manipulated in a spatially variable manner and could lead to economic benefits from either (i) yield improvements due to increases in input or (ii) savings in input costs without an adverse effect upon crop yield.

Table 1: Factors influencing yield variation

Little control	Possible control	
Soil texture	Soil structure	pH levels
Climate	Available water	Trace element levels
Topography	Water-logging	Weed competition
Hidden features	Nutrient levels	Pests and diseases

The main HGCA funded study was planned for 5 cropping seasons with harvests in 1996-2000, where each of four fields (typical of approximately 55% of the land used for cereal production in England and Wales) were selected as they had been in continuous cereals for several years prior to the experimental work. AGCO Ltd. harvested and produced yield maps for the fields during the 1995 harvest.

At the outset of the project it was agreed that the reasons for the underlying field variation needed to be established prior to managing the crop in a spatially variable manner. Hence, uniform 'blanket' treatments were applied in the 1995-6 and 1996-7 seasons. The yield maps for these two seasons, together with that of the 1995 harvest, provided an indication of crop yield variation both in space and time. Since the 1997-8 cropping season, variable inputs of both nitrogen and seed have been studied, together with other significant corrective factors.

3. BACKGROUND VARIATIONS

3.1 Crop Yield

Typical variations in crop yield can be seen in Fig. 2, which shows that there is some similarity over the three-year period. The spatial trend map (Larschied et al., 1997) for the period shows that, on average, the yield range is in excess of ±20% of the mean, with the higher yielding zones to the west and the lower yielding zones to the east of the 100% (or mean) contour.

These maps have been corrected using the algorithms developed by Blackmore and Moore (1998) to compensate for field operational problems of combine harvester grain filling at the headlands and crop harvesting widths of less than the full width of the combine harvester cutter bar.

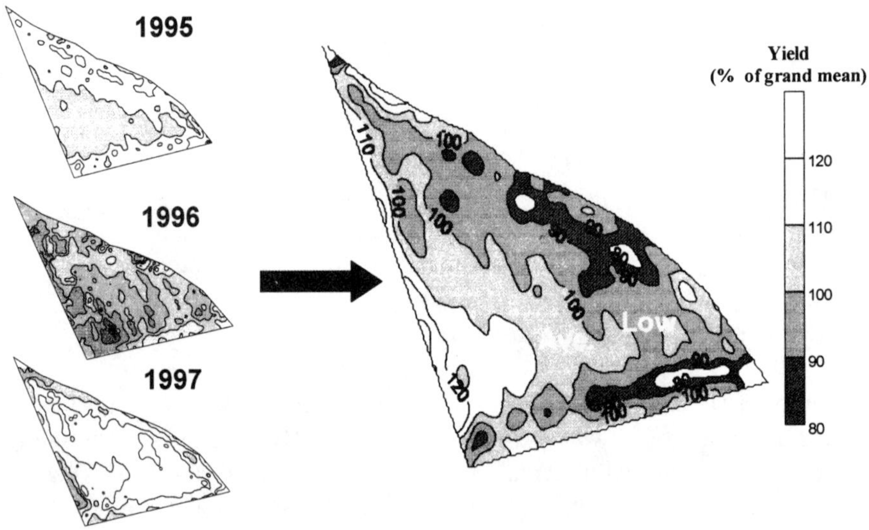

Figure 2. Spatial trend map for yield at Trent Field, Andover 1995 - 1997

3.2 Soil Types

The fields were initially surveyed at a commercial detail level of approximately 1 auger hole/ha to give an overview of soil textural and profile variation. These were followed by "targeted" profile pit observations. The positions of the profile pits, 3 m long x 1 m wide x 1.5 m deep, were selected to encompass the range of yields observed in the spatial trend map and from the results of auger sampling. These pits were to provide detailed information for soil classification and information on crop rooting depth and soil drainage status. The farmers were present in the field during this activity and in each case it was the first time they had seen the soil below a depth of 0.3 m in a number of locations in the same field. They were intrigued by the results. Photographs taken of these geo-referenced soil profiles can be passed to successive generations and have greater impact than traditional written profile descriptions.

Further studies with both soil coring apparatus (to a depth of 1 m) and electromagnetic induction (EMI) equipment (Waine, 1999) gave improved resolution to soil textural boundaries. The latter technique is particularly applicable in areas without saline conditions, which are at field capacity for a period of the year. This enables the soil texture to be estimated and can also provide an indication of the available water between field capacity and permanent wilting point. Table 2 provides the details of the soil types referred to in this paper.

Table 2. Description of Soil Series

Andover Series:	Topsoil:	Shallow calcareous silty clay loam
	Subsoil:	Brown silty clay loam over chalk
Panholes Series:	Topsoil	Brown strong silty clay loam
	Subsoil:	Brown silty clay loam over chattered chalk
Moreton Series:	Topsoil:	Calcareous silty clay loam
	Subsoil:	Brown silty clay loam over brashy oolitic limestone
Sherborne Series:	Topsoil:	Calcareous silty clay loam
	Subsoil:	Brashy oolitic limestone
Didmarton Series:	Topsoil:	Deep silty clay with small limestone
	Subsoil:	Deep structureless silty clay

3.3 Soil Fertility and Crop Nutrition

Detailed analysis of macro- and micro- nutrients in both soil water extract and plant tissue was conducted at 100 m grid spacings. These indicated that whilst there was variation in each of the fields, none were below the critical threshold for crop growth.

3.4 Crop Canopy

Variations in crop canopy occur both in space and time in the same field. To obtain consistent and reliable data to monitor crop development for 'real time' management and to explain field differences, a light aircraft was equipped with two digital cameras fitted with red (R) and infra red (IR) filters. Field images obtained from aerial digital photography (ADP of the fields from a height of 1000 m (Wood et al, 1998) give a pixel resolution of 0.5 m x 0.5 m. Normalised Difference Vegetation Index (NDVI values were estimated using Equation 1.

$$NDVI = \frac{IR - R}{IR + R} \quad (1)$$

The resulting images, such as Fig. 3 (normally in colour), show the effect of variations in crop development immediately prior to the first application of nitrogen. These images are (i) immediately valuable in recognising patterns of field variability, and (ii) provide detailed spatial data on crop tillers/shoot density. These data, when calibrated against detailed agronomic measurements at targeted locations, can be used in near "real time" to estimate crop condition and potential nutritional requirements.

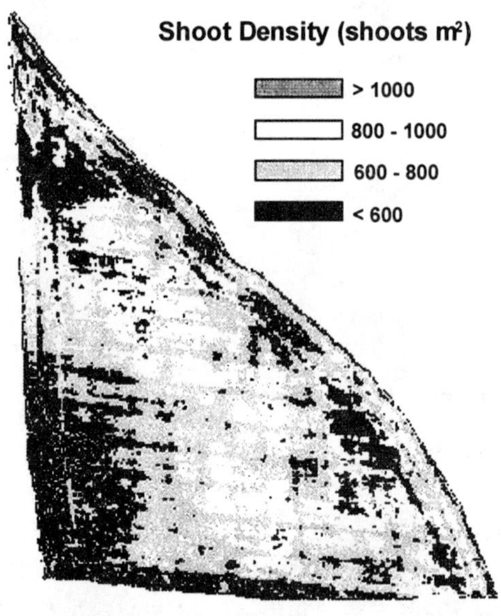

Figure 3. Normalised Digital Vegetation Index (NDVI) image of Trent Field, Andover

3.5 Interim Conclusions from the Field Variability Studies

1. Major causes of yield variation in the study fields were
   - soil and its associated water holding capacity
   - spatial variation of nitrogenous fertilizer
2. Potassium, phosphorus and the micro-nutrients were not limiting.
3. Yield map sequences can identify stable yield patterns

4. Aerial digital photography (ADP)
   - allowed variations in crop yield components to be mapped in near "real time".
   - can be used with confidence to monitor the effects of variable and uniform applications.

4. VARIABLE APPLICATION OF NITROGEN

4.1 Experimental Design

One of the aims of this project was to develop an experimental methodology that could be employed by farmers to determine an optimal application strategy for a given input in any particular field, in this case nitrogen. To achieve this, it was important to use standard farm machinery for the experiments. This resulted in a move away from the traditional small plot randomised block experimental design.

In line with the suggestions made by Thompson and Robert (1995), the proposed design comprised a series of long strips, which ran through the main areas of variation within each field, an example of which is given in Fig. 4. The width of each strip was dependent upon the existing tramline system and/or the working width of the machinery available. The treatment strips were, therefore, half the width of a tramline. The fertiliser was applied using a pneumatic fertiliser spreader that was capable of operating the left and right booms independently. The strip widths used allowed the experiments to be harvested by the combine harvester without the inclusion of the tramline wheel marks.

4.2 Treatments

Two types of treatment were conducted on each of the sites, uniform and variable.

i. Uniform treatments

These treatment strips had different rates of nitrogen applied uniformly along their complete length. The purpose of this was to provide an indication of the crop response to different levels of nitrogen in the various management zones.

ii. Variable treatments

These treatment strips were established to test the following strategies (Welsh, et al., 1999):

(i) Increasing the fertilizer application to the better, or potentially better, yielding parts of the field whilst reducing the application to the poorer yielding parts.
(ii) Reducing the fertilizer application to the better, or potentially better, yielding parts of the field whilst increasing the application to the poorer yielding parts.

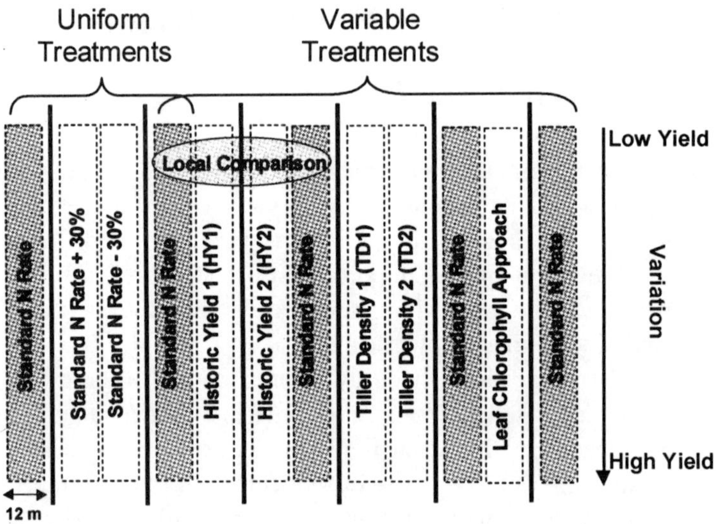

Figure 4 Plan of field experiments

However, before these strategies could be implemented, the high, average and low yielding zones had to be identified. Two methods were used:
(i) historic yield data, as shown in Fig. 2
(ii) tiller density data, estimated from NDVI data, as shown in Fig. 3.

Using this approach, experimental strips, as shown in Fig. 4, were established to give the following treatments:

Historic Yield 1 (HY1). High yield zone received 30% more nitrogen; average yield zone received the standard nitrogen rate; and the low yield zone received 30% less nitrogen.

Tiller Density 1 (TD1). High tiller density zone received 30% more nitrogen; average tiller density zone received the standard nitrogen rate; and the low tiller density zone received 30% less nitrogen.

Historic Yield 2 (HY2). High yield zone received 30% less nitrogen; average yield zone received the standard nitrogen rate; and the low yield zone received 30% more nitrogen.

Tiller Density 2 (TD2). High tiller density zone received 30% less nitrogen; average tiller density zone received the standard nitrogen rate; and the low tiller density zone received 30% more nitrogen.

Standard N rate strips were located adjacent to each of the variable treatment strips to allow treatment comparisons to be made, since classical experimental design and statistical analyses with replicated plots is not possible.

5. RESULTS

An example of the yield distribution along the historic yield strips is given in Fig. 5. This shows the benefits of adding additional fertilizer to the high yielding zone and the consequences of reducing the fertiliser application to the low yielding zone. Nitrogen response data from 2 fields are shown in Fig. 6. This indicates that the Andover series soil area of Trent field not only performs better than the Panholes series, but that it benefited from additional nitrogen. In contrast to this, the results from Twelve Acres show different maximum yields in the three soil series zones, although the optimum nitrogen application rate remains constant.

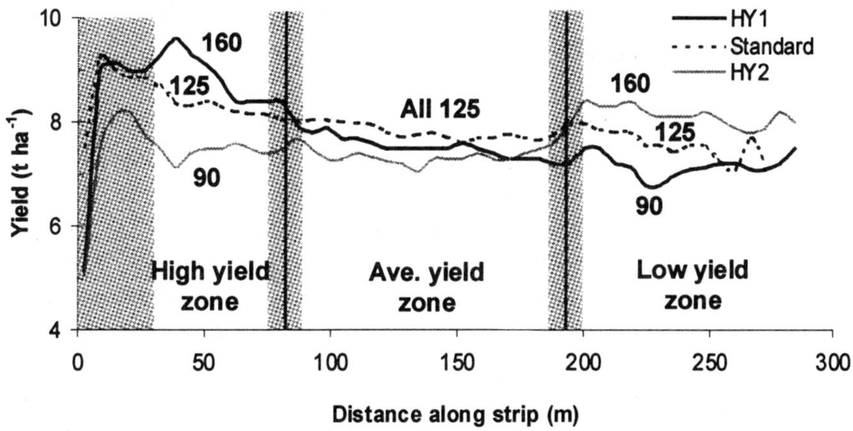

Figure 5. Combine Yield of 'Historic Yield' treatments (HY1 & HY2) compared with a standard application along the treatment strips. Numbers in bold refer to the nitrogen application rate in that section of the field. Shaded areas represent transition zones i.e. initial fill time of combine and fertilizer overlap areas between sections, which were omitted from the analysis.

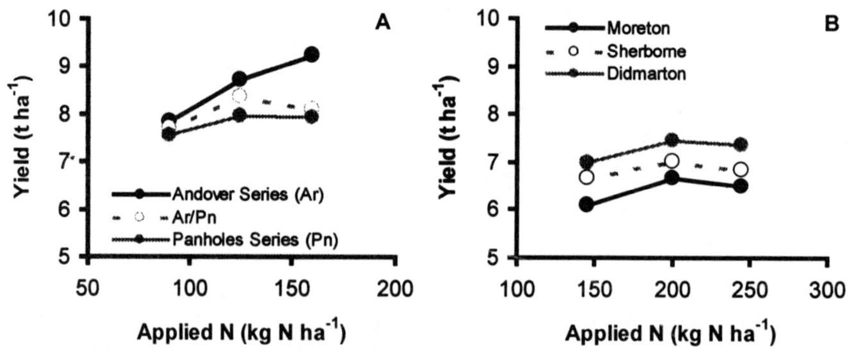

Figure 6. Localised yield responses to applied nitrogen at (A) Trent Field, Andover and (B) Twelve Acres, Cirencester.

The results of two harvests would indicate the greater reliability of the tiller density/canopy management approach to managing the nitrogen application

6. VARIABLE SEED RATE AND CANOPY MANAGEMENT STUDIES

Similar strip experiments were established to study the above by varying seed rates to provide a range of crop canopy conditions (Wood et al., 2000). The results are still under investigation, however, the immediate results support the capability of ADP as an indicator of canopy condition and show that whilst there are initial differences in crop establishment with seed rate there is significant tiller compensation. The most immediately valuable result is that farmers can reduce seed rates from the current level whilst maintaining or improving yields, thus providing economic benefits through savings in seed costs and, in some instances, improved output (Fig. 7). The results shown in Fig. 8 illustrate the very dramatic effect that localised water logging can have on the yield of a given strip as it passes from high ground on the left through the low-lying ground in the centre. The effect of crossing a single tile-drain line can be seen at (A). Localised yield penalties of up to 4t/ha resulted.

7. ECONOMIC IMPLICATIONS

The results in Fig. 6 have shown significant variation in the crop yield -v- nitrogen response curves where in certain conditions the peak yield occurs at significantly different nitrogen application rates. Hindsight analysis can then most definitely show significant economic benefits as discussed by Godwin et al., (1999). Some fields, however, show different parts having different levels of response but the "peak" yield occurs at the same nitrogen rate. Whilst the "most economic rate of nitrogen" (MERN) (Kachanoski et.al 1996) would be different for each level of response the effect of this in practice would mean only very slight modifications in N application rate of 5-10 kg/ha

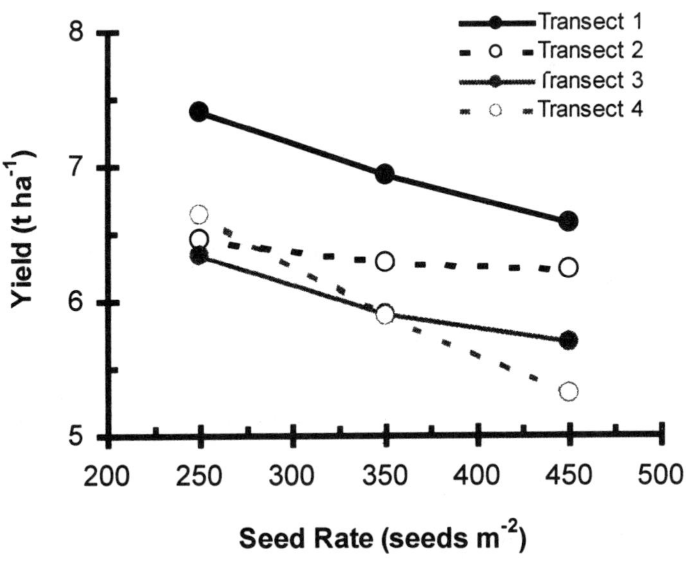

Figure 7. Yield response to varying seedrates at Onion Field, Bedfordshire in 1998

Wood et al., (2000) demonstrated that savings in seed rate and additional yield produce average benefits of £31/ha ($70/ha) while the impact of water logging reduced the return by £70/ha ($160/ha) for the field, which would be easily offset by reinstalling the network of mole drains (£15/ha/year) over the existing field drains.

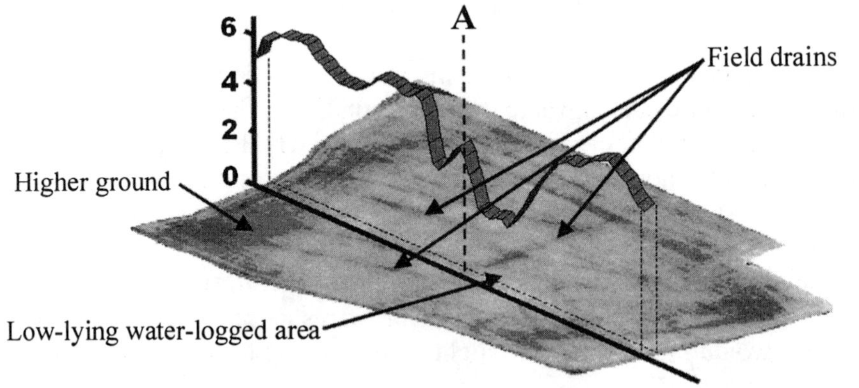

Figure 8. Yield depression resulting from severe water-logging at Onion Field, Bedfordshire in 1999

8. CONCLUSIONS

i. There is variation in yield in cereal fields in excess of ±20% about the mean.

ii. Yield response to nitrogen varies within individual fields.

iii. Yield maps are of great assistance in understanding the causes of variability by directing targeted sampling. Routine grid sampling is too expensive.

iv. Electromagnetic induction can provide positive indication of soil textural variability in areas without salinity problems.

v. Aerial digital photography is proving to be a powerful tool in spatially managing nitrogen applications in near 'real time'.

vi. Managing both nitrogen and seed rate can individually give economic returns of approximately £20/ha (≈$45/ha) depending upon soil conditions.

vii. Calculate the "take off" of P&K from yield maps and apply using a replacement strategy.

viii. Strip experiments with yield mapping combines can be implemented on farms for local evaluation of site specific problems.

ix. Get the fundamentals of crop production right - take the trouble to correct the factors that should have always been corrected!

x. Know your soils - photograph profile pits as a long term "one-off" investment.

9. FUTURE REQUIREMENTS

i. Techniques to estimate the potential benefit of precision farming for a given farm enterprise prior to investment of capital and time.
ii. Further development of aerial digital photography for crop nitrogen management in real time.
iii. Product quality mapping and crop quality management.
iv. The development of "traceability" records to satisfy consumer concerns (Miller, 1999).
v. Tilth sensing and seed placement.
vi. Application to other crop sectors (Godwin *et al.*, 1997 and 1999).
vii. New and multiple link sensors (Miller, 2000) for determining crop nutrient and disease status.

ACKNOWLEDGEMENTS

The author would like to express his thanks to:
i. HGCA, Hydro Agri, AGCO Ltd., for their support for the main project referred to and to MAFF, AGCO Ltd., and RDS Technology Ltd. for the support for associated projects.
ii. The many colleagues, in particular James Welsh, Gavin Wood, Simon Blackmore, Richard Earl and John Taylor of Cranfield University and Mike Carver and Stuart Knight of Arable Research Centres, and doctoral students whose work has been invaluable in bringing together the combined work.

REFERENCES

Blackmore, B. S. and Moore, M. R. 1998. Remedial correction of yield map data. *Precision Agriculture,* **1**, 53-66 (1999), Kluwer Academic Publishers, Netherlands.

Earl, R., Wheeler, P. N., Blackmore, B. S. and Godwin R. J. 1996. *Precision farming - the management of variability.* Institution of Agricultural Engineers, Silsoe, Bedford, UK Landwards, Winter 1996.

Godwin, R J and Wheeler P N. Yield mapping by mass accumulation rate. *ASAE Paper No. 97-1061*, ASAE Annual International Meeting, Minneapolis, Minnesota, USA, 10-14 August 1997.

Godwin, R J., James, I T., Welsh, J P and Earl, R. 1999. Managing spatially variable nitrogen : A practical approach. *ASAE Paper No. 99-1142,* ASAE/CSAE-SCGR Annual Meeting, Toronto, July 1999.

Godwin, R J., Wheeler, P N., Watt, C D. and Richards, T. 1999. Cumulative mass determination methods for yield maps of non-grain crops. *Computers and Electronics in Agriculture* : Spatial Yield Recording of Non-Grain Crops. Publ. Elsevier Sci.Pubs.

Kachanoski, R. G., Fairchild, G. L. and Beauchamp, E. G. 1996. Yield indices for corn response to applied fertilizer: application in site-specific crop management. In *Precision Agriculture: Proceedings of the 3rd International Conference (* P C Robert, R H Rust, W E Larson, Eds.), Madison, Wisconsin, ASA, CSSA, SSSA, pp 425-432.

Larscheid, G., Blackmore, B S., and Moore, M R. 1997. Management decisions based on yield maps. In *Precision Agriculture '97*, Warwick, (J.V. Stafford ed.), BIOS Scientific Publications Ltd. Oxford, UK, 895-903.

Miller, P C H. 1999. *Automatic recording by application machinery of rates and spatial distribution of field inputs.* Proceedings No. 439, The International Fertiliser Society.

Miller, P C H. 2000. Crop sensing and management. *Proceedings of HGCA 2000 Crop Management for the Millennium Conference*, Cambridge (January), Home-Grown Cereals Authority, London.

Thompson, W H. and Robert, P C. 1995. Evaluation of mapping strategies for variable rate applications. In *Site-Specific Management for Agricultural Systems* (P C Roberts, R H Rust, W E Larson, eds), Madison, Wisconsin, ASA, CSSA, SSSA, pp303-323.

Waine, T. 1999. *Non-invasive soil property measurement for Precision Farming.* Unpublished EngD Thesis, Institute of AgriTechnology, Cranfield University at Silsoe, Bedford UK.

Welsh, J. P., Wood, G. A., Godwin, R. J., Taylor, J. C., Earl, R., Blackmore, B. S., Spoor, G., Thomas, G and Carver, M. 1999. Developing strategies for spatially variable nitrogen application. In *Proceedings 2nd European Precision Farming Conference*, Denmark, July 1999.

Wood, G A., Godwin, R J., Taylor, J C., Earl R., Knight, S., Carver, M F. Seed rate and nitrogen interactions. In *Proceedings of HGCA 2000 Crop Management for the Millennium Conference*, Cambridge (January), Home-Grown Cereals Authority, London.

Monitoring Frost Injury to Winter Crops:
An Intelligent Geo-Information System Approach

VICTOR YAKUSHEV, DMITRY KURTENER, and VLADIMIR BADENKO
Agrophysical Research Institute, Russian Academy of Agricultural Sciences
St. Petersburg, Russia

1. INTRODUCTION

Intelligent technologies have become a hot topic in recent years, and they have become the core of precision agriculture. They are used frequently for agricultural development and for transforming traditional agro-technologies into modern agro-technologies. Currently, the systems used for monitoring of agricultural fields are built on intelligent technologies. For example, Wheat Cultivation Simulation, Optimisation, and Decision-Making System (WCSODS) is a software package that applies intelligent technologies to monitor wheat cultivation and management (Hongxin et al., 2001).

The monitoring of frost injury to winter crops is the most important aspect of the monitoring of agricultural fields. The technique for monitoring frost injury to winter wheat was developed using remote sensing data (Bangjie et al., 2001).

An intelligent geo-information system approach is based on a combination of Geographic Information Systems (GIS) and intelligent technologies. GIS is a comprehensive system that can manage geographic information. GIS not only shows geographic information with a digital map system, it also analyses the data visually. GIS is used more widely than a general information query system because GIS offers decision-making support, dynamic simulation, statistical analyses, and forecasts (Wang et al., 2001).

Intelligent technologies are high-tech since they are based on the use of various combinations of component models. The expert system technique, the application of fuzzy set theory and fuzzy logic, the artificial intellect technique, and the simulation of a human's thinking are some of the component models. Intelligent technologies play an important role in optimally utilizing agricultural resources, precisely organizing agricultural production, and improving the environment (Fanlun, 2001).

An intelligent geo-information system approach (IGIS) in the agriculture domain can be used for a variety of purposes. This approach was applied to assess polluted agricultural fields in order to design a strategy for territorial prophylactic actions, for land suitability evaluation in the process of agricultural experimentation, for the assessment of agricultural lands to plan site-specific residue management, and for the multi-dimensional evaluation of areas on the land market by Kurtener and Badenko (2000 a,b,c) and Yakushev *et al.* (2000).

The aim of this chapter is to describe the IGIS approach as an appropriate methodology for the monitoring of frost injury to winter crops and to present an example of its application in the analysis of a particular agricultural field situation.

2. CONCEPTION

An intelligent geo-information system approach that aims to monitor frost injury to winter crops is built using different ways of *interfacing* GIS and component models. According to Hartkamp *et al.* (1999), the term *interfacing* may be used to indicate the simultaneous use of GIS and modelling tools. Hartkamp *et al.* suggests the terms *linking*, *combining*, and *integrating* to indicate the degree of interfacing. In this context, the term *linking* is defined as merely the exchange of input and output data between GIS and a component model. The term *integrating* implies incorporating one system into the other. The term *combining* can be defined as a special case when one part of the component model is linked with GIS and a second part of the component model is integrated into GIS.

In our case, the intelligent geo-information system is designed with linking and integrating. The case-based system technique is linked with GIS. The other component models are integrated into the GIS environment. These models are: 1) a model for the description of the soil temperature of the root area; 2) a verbal model for the estimation of frost injury to winter crops; 3) a fuzzy model for the estimation of frost injury to winter crops; and 4) a fuzzy model for the estimation of economical losses caused by frost injury to winter crops. The structure of this system is shown in Fig. 1.

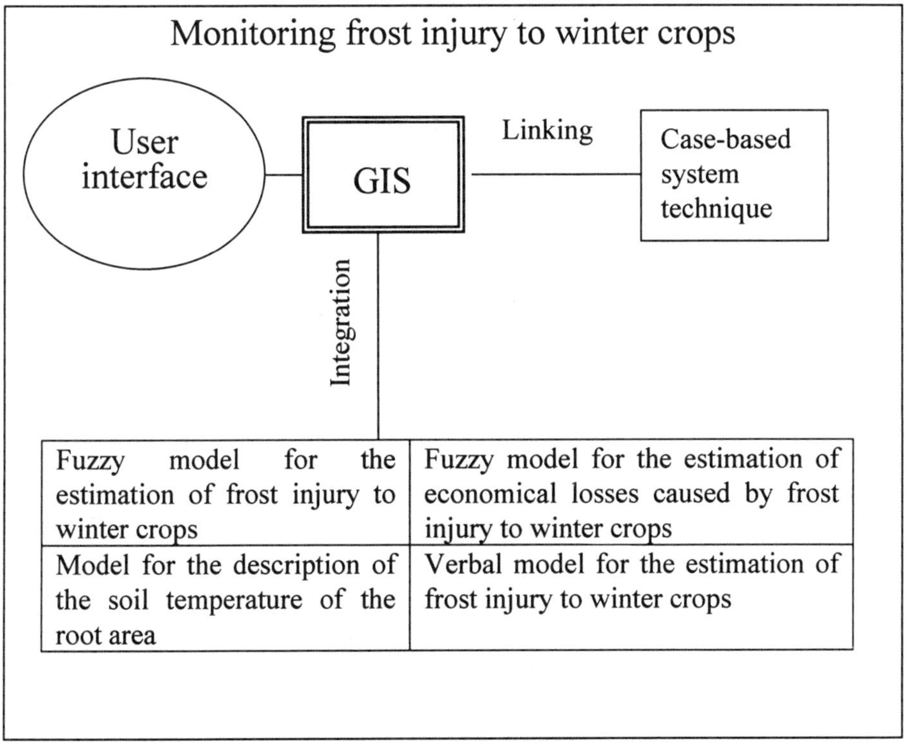

Figure1. Structure of an intelligent geo-information system for the monitoring of frost injury to winter crops.

Intelligent geo-information system software is built with a special kind of interfacing with GIS. When integrating a component model into a GIS environment, the same programming language should be used both in the GIS software and in the programming of the fuzzy model. For example, if we use MapInfo® software, then we should use MapBasic® as the language to program the component models (Wang et al., 2001).

3. A VERBAL MODEL FOR THE ESTIMATION OF FROST INJURY TO WINTER CROPS

An analysis of the empirical data (Moiseychik, 1975) shows that frost often destroys winter crops. Several factors such as the soil temperature of the root area (at 2 cm depth), the depth of the boundary between the frozen and melted soil layers, the thickness of the snow cover, the kind of crops, the crop's resistance to frost, and the density of the planting (the number of stems) determine the amount of thermal damage to winter crops.

Also, frost injury is related to the crop's growth stage. For example, if at the end of autumn the development of the winter crop is normal, it's resistance to frost will be strong. If the winter crop is underdeveloped or outgrowing, it's resistance to frost will be decreased.

Current analysis of the empirical data (Moiseychik, 1975) shows that we can identify the critical soil temperatures in the root area (T_{root}). The critical temperatures in T_{root} that will destroy winter barley crops are

$-7 < T_{root} < -12$ *for an underdeveloped crop*
$-14.8 < T_{root} < -19.2$ *for a normal crop*
$-7 < T_{root} < -13$ *for an outgrowing crop*

where T_{root} has dimensionality centigrade.

The critical soil temperatures in the root area that destroy winter wheat crops with mean frost resistance are:

$-11 < T_{root} < -17$ *for an underdeveloped crop*
$-14 < T_{root} < -19.4$ *for a normal crop*
$-10 < T_{root} < -15$ *for an outgrowing crop*

The critical soil temperatures in the root area that destroy winter rye crops are:

$-11 < T_{root} < -22$ *for an underdeveloped crop*
$-14 < T_{root} < -25$ *for a normal crop*
$-11 < T_{root} < -22$ *for an outgrowing crop*

4. A MODEL FOR THE DESCRIPTION OF THE SOIL TEMPERATURE IN THE ROOT AREA

Clearly, frost injury to winter crops is related to the soil temperature in the root area. The thickness of the snow cover and the density of the planting (the number of stems) also affect the soil temperature.

A description of the soil temperature in the root area is carried out by models of the heat flow in the soil, which are process-oriented, semi-process-oriented, or non-process-oriented (Horton, 1989). Process-oriented models are based upon physical principles and generally consist of partial differential equations that require detailed initial and boundary inputs. The semi-process-oriented models are partly based upon physical principles and generally require some regression relationships that are site specific. Non-process-oriented models are generally based upon site specific, regression-type relationships and are used to estimate the soil temperature.

Many studies have been performed using the semi-process-oriented approach. (Nerpin and Chudnovsky, 1970, Chudnovsky, 1976). Several studies have been done on the description of the soil temperature during the winter period (Kurtener and Chudnovsky, 1979, Romanov et al., 1990). Two processes characterize this period: soil freezing and soil thawing. The soil has two layers: a frozen soil layer and a melted soil layer.

Using the semi-process-oriented approach, the theoretical description of the soil temperature during the winter period is based on the analysis of the processes of energy and mass transfer in the system of the soil, the plants, and the active layer of the atmosphere. The snow and plant layers are considered to be uniform units. The basic statements of the model include:

1. Energy and mass transfer within the atmospheric, snow, and plant layers are considered to be quasi-steady-state processes.
2. Heat transfer in the frozen soil layers and the melted soil layers is defined as a dynamic process.
3. Soil water freezes in a narrow range of temperatures.
4. The thickness of the snow layer is varying.

The following model of the heat flow is an equation describing one-dimensional conductive heat transfer in the soil layer:

$$-\lambda \frac{\partial^2 T}{\partial x^2} = C(T)\frac{\partial T}{\partial t}, \qquad C(T) = c + L\rho W \frac{di}{dT} \tag{1}$$

where T is temperature, t is time, x is the perpendicular distance to the soil surface, λ is the soil thermal conductivity, c is the soil volumetric heat capacity, ρ is the soil bulk density, W is the volumetric soil water content, i is the ratio of ice to water ($0<i<1$), and L is the latent heat of phase transition. If $i=0$, then Equation (1) describes the heat transfer in a melted soil layer.

The boundary condition for Equation (1) is formulated by the balance method (Kurtener et al., 1998) as a boundary condition of the third generalized order

$$- \lambda \frac{\partial T}{\partial x} = N (T_E - T), \; x = 0 \tag{2}$$

where T_E is the equivalent temperature of the environment and N is a generalized heat exchange coefficient. Kurtener et al. (1998) worked out an algorithm for the determination of the values of T_E and N as functions of all model variables (major meteorological parameters, plant characteristics, soil parameters, and agricultural or technical parameters). This algorithm foresees the following steps:

- to analyze the initial data and to write a heat balance equation for the soil surface,
- to select correlation to define the heat balance equation components,
- to transform them into a linear form relative to the temperature of the soil surface,
- to deduce formulas for T_E and N.

The balance method was tested in four experiments (Kurtener et al., 1979), performed at the Agrophysical Research Institute and Polezskaia Experimental Station. The balance method was also tested by using experimental data from the Soil Fertility Research Center, Munchberg, Germany (Kurtener et al., 1987). The experiments gave satisfactory results. Kurtener and Hoffmann (1997) have modified the balance method.

Many studies have been performed using the non-process-oriented approach. The non-process-oriented model is generally based upon site specific, regression-type relationships and is used to estimate the soil temperature. Moiseychik (1975) uses the following formula to describe the soil temperature during the winter period

$$T_{root} = 0.372\, T_{air} - 0.057\, H_{frost} + 0.425\, H_{snow} - \\ -0.003\, H^2_{snow} - 0.0005\, N_{stem} \tag{3}$$

where T_{root} is the soil temperature of the root area (at 2 cm depth), T_{air} is the air temperature, H_{frost} is the depth of the boundary between the frozen and melted soil layers, H_{snow} is the thickness of the snow cover, and N_{stem} is the density of the planting (the number of stems).

5. A FUZZY MODEL FOR THE ESTIMATION OF THE DAMAGE TO WINTER CROPS BY FROST

A fuzzy model for the estimation of the damage to winter crops by frost is based upon the fuzzy set theory. The central concept of the fuzzy set theory is the membership function, which represents the relationship of an element to a set. The membership function of a fuzzy set is expressed on a continuous scale from one (full membership) to zero (full non-membership). One of the principal benefits of the application of the fuzzy set theory in real world situations is that most indices used in the monitoring of frost injury to winter crops can be estimated by using membership functions. In particular, by setting the membership function to one, we symbolize a maximum value of the indicator of the damage to winter crops by frost, and by setting the membership function to zero, we represent a minimum of value of the indicator.

The distinguishing feature of the fuzzy modeling approach is that indicators are defined using the membership functions. The choice of a membership function is somewhat arbitrary and should mirror the subjective expert opinion. The starting positions for the building of a membership function in our case are:

- It is assumed that the indicator of the damage to winter crops by frost can be expressed as function of the soil temperature of the root area (at 2 cm depth), the kind of crops, the crop's resistance to frost, and the density of the planting (the number of stems).
- The thermal effect can be very small and the damage to winter crops is negligible.
- The thermal effect can be greater and the damage to winter crops by the thermal effect is considered to be impermissible.
- The thermal effect is neither very small nor very large, and it is not quite clear whether the damage to winter crops is serious or not.

Mathematically, all of these situations can be described by the following membership function $\mu_d(x)$

$$\mu_d(x) = \begin{cases} 0; & x < th_{low} \\ f(x); & th_{low} \leq x \leq th_{up} \\ 1 & x > th_{up} \end{cases} \qquad (4)$$

where th_{low} and th_{up} are the lower and upper threshold values of the thermal effect, and x is the current value of the thermal effect. It is easy to see that

$\mu_d(x)$ has grades in the interval [0,1]. $f(x)$ can be described by different mathematical expressions. In this study, a linear function is selected

$$f(x) = (th_{up} - x)/(th_{up} - th_{low}) \tag{5}$$

where x is the current thermal effect (the soil temperature at 2 cm depth), and th_{low} and th_{up} are defined by Table 1

Table 1. Mean values of a and b coefficients into Equation (5)

Plants	th_{low}	th_{up}
Barley:		
– *Underdeveloped crop*	-7	-12
– *Normal crop*	-14,8	-19,2
– *Outgrowing crop*	-7	-13
Rye:		
– *Underdeveloped crop*	-11	-22
– *Normal crop*	-14	-25
– *Outgrowing crop*	-11	-22
Wheat with mean frost resistance:		
– *Underdeveloped crop*	-11	-17
– *Normal crop*	-14	-19,4
– *Outgrowing crop*	-10	-15

The membership function in graphical form is shown in Fig. 2, and one can see that if the thermal effect is below the lower threshold values, the damage to winter crops is negligible. However, if the thermal effect is above the upper threshold values, the damage is fully classified as not permissible.

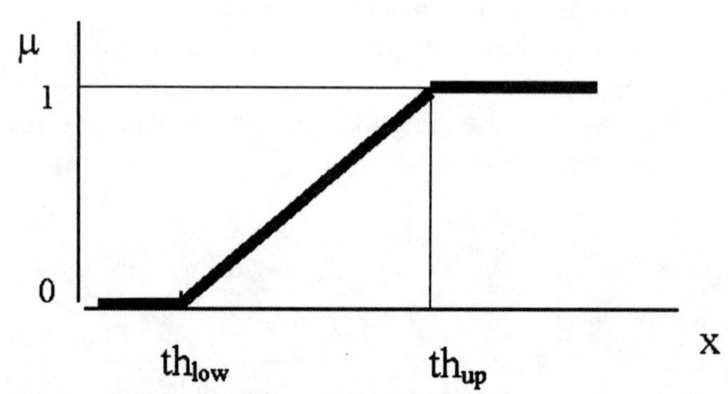

Figure 2. Fuzzy linear membership function

6. A FUZZY MODEL FOR THE ESTIMATION OF ECONOMICAL LOSSES CAUSED BY THE DAMAGE TO WINTER CROPS BY FROST

Economical losses caused by frost damage to winter crops are determined by two main factors: the fall in the yield and the expenditure to re-sow the land. It is easy to see that the cost to re-sow the land will be quite small or zero if the damage to the winter crops is negligible. Also, it is obvious that the expenditure to re-sow the land will be the greatest if the damage to the winter crops is great. Let us denote the index of the fall in yield as Y and the index of the expenditure to re-sow the land as S. The ratio of Y to S increases with an increase in frost damage to winter crops.

The starting positions for the building of a membership function follow:

- The indicator of economical losses caused by frost damage to winter crops μ_e will equal 0 when Y/S = 0.
- μ_e will equal 1 when $Y/S > 1$.
- Y/S is neither very small nor very large, and it is not quite clear whether μ_e is big or not.

Mathematically, all of these situations can be described by the following membership function $\mu_e(x)$:

$$\mu_e(x) = \begin{cases} Y/S, & 0 \le Y/S \le 1 \\ 1, & Y/S > 1 \end{cases} \tag{6}$$

It is easy to see that $\mu_e(x)$ has grades in the interval [0,1]. The fall in the yield Y is dependent upon the damage to winter crops and can be expressed in the term μ_d.

7. CASE-BASED REASONING

The main goal of case-based reasoning (CBR) is to predict future events by studying and reusing past cases. The combination of this approach with expert system techniques has many uses in the agricultural domain. By using CBR, a series of systems to forecast insect/pests have been developed (Fanlun, 2001).

In our case, CBR is used as an alternative way to monitor frost injury to winter crops. CBR is used through data preprocessing with rough sets theory and through the case matching strategy based on fuzzy theory.

8. AN EXAMPLE OF THE APPLICATION

8.1 Area of Study and the Dataset

We applied this approach to study the territory located in the suburbs of the Saint-Petersburg agricultural area. The territory contains 26 homogeneous plots. The plots differed in their soil physical properties, in their types of vegetation (Fig. 3), in their planting densities (Fig. 4), and in their orientations and inclinations of the soil surface (Figs. 5 and 6). All the data are stored in MapInfo® tables.

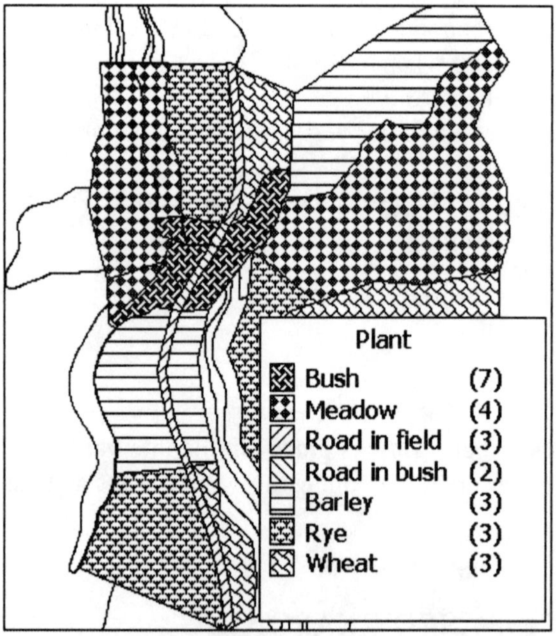

Figure 3. Distribution of the types of vegetation.

8.2 Results and Discussion

The soil temperature of the root area (at 2 cm depth) is calculated and mapped automatically and is given in Fig. 7. It is easy to see that the soil temperature changed from -2 to -12 Celsius.

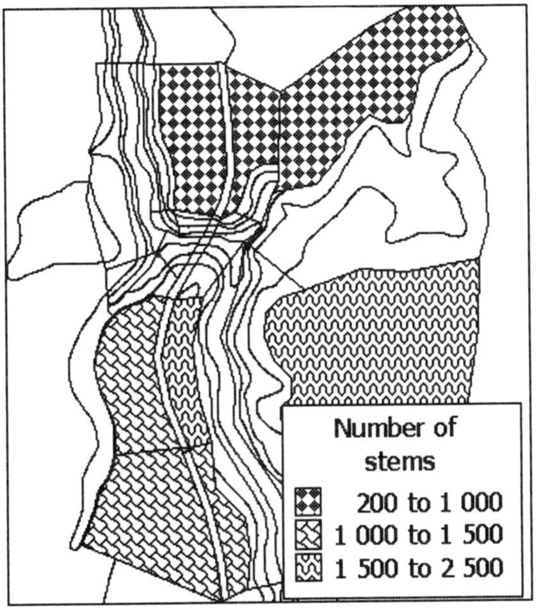

Figure 4. Density of the planting (the number of stems).

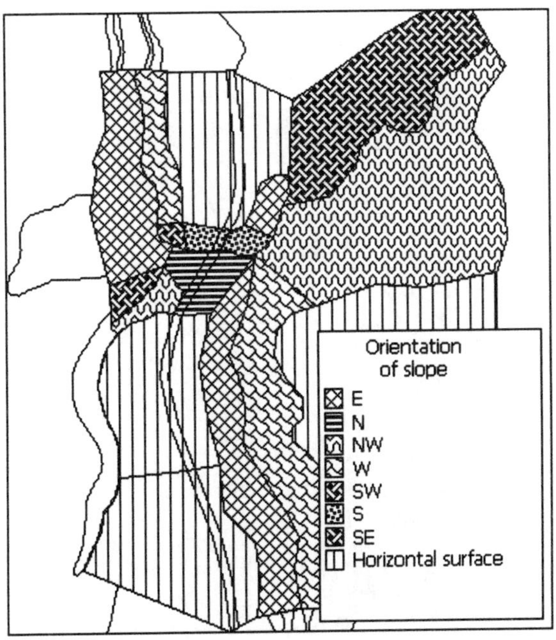

Figure 5. Orientation of the soil surface

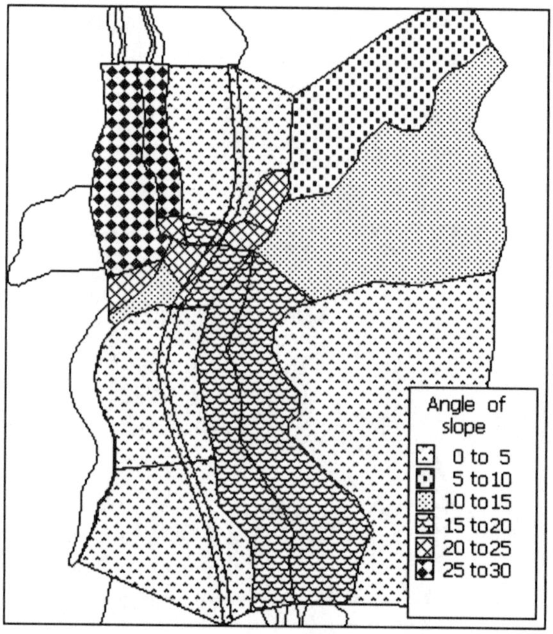

Figure 6. Inclination of the soil surface

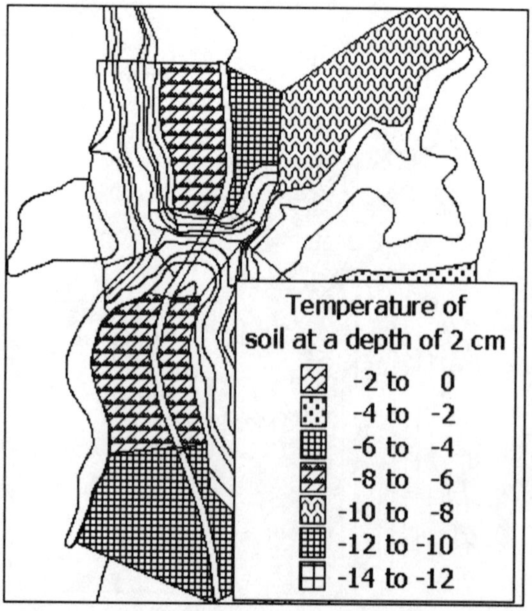

Figure 7. Soil temperature of the root area (at 2 cm depth)

The index of the frost damage to winter crops is also calculated and mapped automatically. The result obtained (Fig. 8) shows the current state of the frost damage. The application of the fuzzy set membership functions as a specific indicator permits an end user: 1) to see readily the more destroyed agricultural plots; and 2) to estimate quickly the extent of the frost damage.

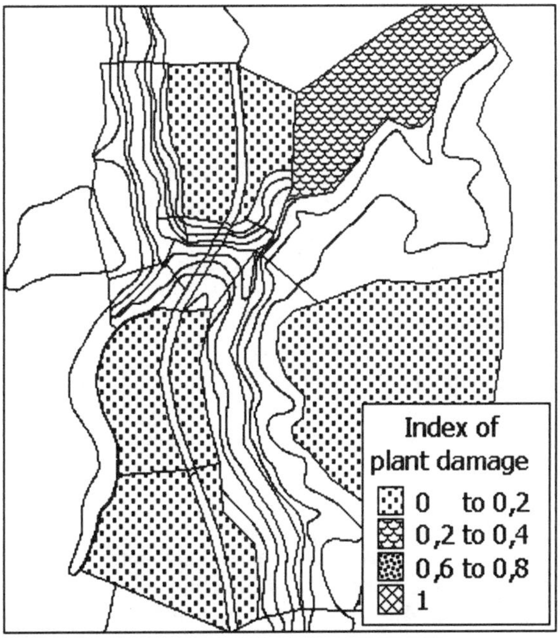

Figure 8. Indicator of the frost damage to winter crops

9. CONCLUSIONS

An intelligent geo-information system approach is an appropriate methodology to support the monitoring of frost injury to winter crops. It is based on the combination of GIS with intelligent techniques. We have presented an example of a practical implementation of a developed technique.

The material presented in this chapter demonstrates the advantages of the intelligent geo-information system approach. Due to this approach, it is possible to describe both subjective (knowledge-driven) and deterministic (data-driven) information.

An intelligent geo-information system approach can be applied in other ways. For example, it is an effective methodology to support the monitoring of agricultural fields.

REFERENCES

Bangjie, Yang, 2001, Monitoring frost injury to winter wheat using remote sensing. In *Proceedings of ICAST on Promoting global innovation of agricultural science & technology*, Novenber 7-9, Beijing China, p.245-249.

Chudnovsky, A.F. 1976, *The Heat physics of soil.* (In Russian.) Nauka Publ., Leningrad, 352p.

Fanlun, Xiong, 2001, Intelligent technology for sustainable development of agriculture. In *Proceedings of ICAST on Promoting global innovation of agricultural science & technology*, Novenber 7-9, Beijing China, p.64-72.

Hongxin, Cao, Sun Lirong, Dong Yuhong, Wang Zhiliang, Gao Liangzhi, 2001, Studies for applying intelligent information technology in wheat cultivation. In *Proceedings of ICAST on Promoting global innovation of agricultural science & technology*, November 7-9, Beijing China, p. 267-269.

Horton, R., 1989, Canopy shading effects on soil heat and water flow. *Soil Sci. Am. J.* **53**:669-679.

Hartkamp, A., White, J. and Hoogenboom, G., 1999, Simulation and modeling: interfacing Geographic Information Systems with agronomic modeling. *Agronomy Journal*, **91**:761-772.

Kurtener, D.A., A.F. Chudnovsky, 1979, *The agrometeorological foundation of thermal melioration of soil.* (In Russian) Hydrometeorolog. Publ., Leningrad, 235p.

Kurtener, D., F. Hoffmann, G., Trubacheva, M. Bersirov, K. Kunkel, I. Klank, 1987, Eine Methode zur Berechnung der Tagesmitteltemperatur des Bodens. (In German) *Z. Meteorol.* Vol. **37**. 4:199-203.

Kurtener, D., F. Hoffmann, 1997, A selection of boundary condition at the soil surface, in *6th Intern. Conference on Agrophysics*. Lublin. Poland. Vol. 1, p.93-95.

Kurtener, D., Richard G., Badenko, V., 1998, Theoretical description of heat and water flow in soil taking into account soil tillage. In *Agrophysical and Ecological Problems of Agriculture in the 21st Century*. Publ. of St.Petersburg Branch of ISSTRO. 1:29-53.

Kurtener, D., Yakushev, V., Badenko, V. and Pourabbas, E., 1999, Development of methodology of multiple assessment of landscape parcels on the base fuzzy models integrated into GIS environment. *SPBISTRO Special Publ.* No 1, St. Petersburg, Russia.

Kurtener, D. and Badenko, V., 2000a, Precision agriculture experimentation on the base of fuzzy models synthesised with GIS. *Aspects of Applied Biology*, **61**:139-143.

Kurtener, D. and Badenko, V., 2000b, Development of the methodology of assessment of site-specific residue management actions on the basis of fuzzy models integrated into a GIS environment. In *Proceedings of 15th ISTRO conference on agroecological and ecological aspects of soil tillage* (Morrison, J. ed.), ISTRO Press, Fort Worth, Texas CD-ROM.

Kurtener, D. and Badenko, V., 2000c, Applications of GIS knowledge management for decision making in the field of land market and land consolidation. In *Proceedings of UDMS'2000*, UDMS Press, Delft, The Netherlands, CD-ROM.

Moiseychik, V.A., 1975, *Agrometeorological conditions of winter crops wintering*. (In Russian.) Hydrometeorolog. Publ., Leningrad.

Nerpin, S.V., A.F. Chudnovsky, 1970, *Physics of the soil.* Israel Program for Scientific Translations. Keter Press, Jerusalem, 466p.

Romanov, P.G., D.A. Kurtener, I.B. Uskov, 1990, Optimum methods of thermal melioration for a Northern climate condition. (In Russian.) *Trans. of the Congress: Rationalization use of nature in permafrost zone.* Yakutsk.

Wang, Xiao-feng, Ru-jing Wang De-shuang Huang, 2001. Integrative technique research of GIS and agriculture intelligent decision support systém. In *Proceedings of ICAST on Promoting global innovation of agricultural science & technology,* November 7-9, Beijing China, p. 233-234.

Yakushev V.P., Kurtener D.A., Badenko V.L., Kudashev E.V., Cudlip W., 2000, Methodology of multiple assessment of landscape parcels on the base of fuzzy set theory models integrated into geographic information systems (GIS). *Russian Agricultural Science*, **4**: 42-43.

PART 2

PROPERTIES AND QUALITY

Stress Relaxation in Potato

Before and after Cooking

JIŘÍ BLAHOVEC
Czech University of Agriculture in Prague, Prague, Czech Republic

1. INTRODUCTION

Most part of vegetable flesh is formed by soft and juicy parenchyma. The big cells with thin cell walls belong to the important characteristics of its structure. The vegetable deformation has partly irreversible character (Mohsenin, 1970), even at low deformation levels. It means that the simple mechanical tests, originally developed for testing the solid products of elastic character, can be used for testing the vegetables only in some limited cases. Instead of elasticity the rheological methods, models and theories have to be used. They are based on an important role of time in deformation process. The simple Hooke's Law:

$$\sigma = E\varepsilon \tag{1}$$

where σ is stress, E Young's modulus, and ε strain, is then rewritten into more general form (Dotsenko, 1979):

$$d\sigma/dt = E\ d\varepsilon/dt - f(\sigma, t) \tag{2}$$

where f is a positive function of time and stress. In *stress relaxation tests,* the strain is kept constant in time and the first term on the right hand side of Eq. (2) equals then zero. If the function f depends on stress only and using parameter x, the solution of Eq, (2) is obtained in simple form (Dotsenko, 1979):

$$t = \int_{\sigma}^{\sigma_0} \frac{dx}{f(x)} \tag{3}$$

where σ and σ_0 is the actual stress and the initial stress, respectively. We will term the function $f(\sigma)$ as the *source f-function*. Mathematical equations used for the source *f*-functions in some simple relaxation models are given in Appendix A. In the previous paper I have published the results of relaxation tests I have obtained by compression on the vegetable flesh of potato tubers, roots of parsley and carrot and on the bulb tissues of kohlrabi (Blahovec, 1996a). The obtained data were evaluated using the simple relaxation models of power and logarithmic forms. It was shown that the models could be rewritten using following parameters: the initial slope of the relaxation curve and the time factor. Both the simple models gave the comparable results that differ systematically only due to the different structure of the models. Source *f*-function plotted against stress value seems to be a better tool for the study of the model in detail by regression analysis (Blahovec, 2001) than the classical relaxation curve. Equation (2) shows that source *f*-function can be obtained from slope analysis of the standard relaxation curve. The experimental functions $f(\sigma)$ for potato tissue (Blahovec, 2001) have in semi-logarithmic scale characteristic sigmoidal shape that cannot be approximated successfully by the classical simple models. New equation for description of the source *f*-functions, denoted as Sigmoidal Rate Controlled Model (Sigmoidal RCM) is given in Appendix A, Eq. (A8.2).

Sigmoidal RCM and its application to experimental source *f*-functions obtained on potato tubers led to the improvement one order of magnitude of the existing approximate models (Appendix A). In this paper I will study the experimental relaxation curves obtained of potato tubers before and after cooking. The aim is to find relations between model parameters given by approximate procedure for different pre-stress levels, pre-stress strain rates and pre- and post-cooked state.

2. EXPERIMENTS WITH PRE-COMPRESSION

The potatoes (varieties 'Nicola' and 'Panda') were supplied by the Potato Research Institute in Havlíčkův Brod, located in the east part of Bohemia. The tubers were harvested in October 1998 and kept in a cold store (at temperature 6 ^{0}C and 95 % relative humidity) until the end of January and than tested.

Compression tests were carried out on cylindrical specimens of 15 mm in diameter and 23 mm in length. They were cut from the central parts of tubers by a cork borer in such a manner that the specimen axis was approximately parallel to the tuber axis. Part of specimens were vacuum packed in plastic bags (Fatranex, Fatra a.s., Czech Republic) and cooked in the water bath 20 minutes at the temperature (95 ± 0.2) ^{0}C. Then the specimens were removed and cooled rapidly in another water bath at temperature (25 ± 0.2) ^{0}C. All tests were performed in an Instron universal testing machine, type 4464.

Twenty-five specimens were used for the usual stress relaxation test; 20 at pre-strain rate 0.167 mm.s^{-1} (5 specimens per every different pre-strain level: 5 %, 10 %, 15 %, and 20 %), and 5 specimens at level pre-strain rate 0.833 mm.s^{-1} (pre-strain level 15 %). The real relaxation time was 10 minutes. Deformation force was registered ten times per second and the obtained data, 6000 per every specimen, were used for numerical calculation. This number was reduced in some numerically difficult calculations up to 70 points. In this case, the force interval between the initial and final points was covered equidistantly by the selected data. All the data obtained for relaxation times lower than 0.3 second were excluded from the evaluation procedures.

Pre-strain parameters, e.g. pre-strain stress and strain, and also all the computed quantities, are used here in so-called "true" or Hencky's representation (Mohsenin, 1970). For incompressible media the true stress is approximated by $\sigma_{0t} = \sigma_0(1 - \varepsilon_0)$ with σ_0 and ε_0 are the engineering stress and strain, respectively. A similar simple equation holds for the true strain: $\varepsilon_{0t} = -\ln(1 - \varepsilon_0)$.

The first time derivative of the experimental stress relaxation curve was numerically determined as the first derivative of the Lagrangian interpolation polynomial of degree 2 in 3 successive points. These values were related to the middle of the three neighbour points so that the derivative was determined in every point (the first and last points were excluded) of each stress relaxation curve. The model parameters in Eq. (A8.2) were determined by minimisation of Chi-square function SFC (standard fitting coefficient) defined by the following equation:

$$SFC = \frac{100}{\sqrt{n}} \sqrt{\sum_{k=1}^{n} \left(\frac{x_k}{x_{km}} - 1 \right)^2} \qquad (4)$$

where n is number of the evaluated points, x_k is an experimental value and x_{km} is the corresponding model value of source f-function at the same point (for the same stress).

Minimisation was performed using the FORTRAN programme MINUIT that combines more different numerical methods for this purpose (James and Roos, 1975). In our case only simplex (SIMPLEX), generalised gradient method (MIGRAD) and some procedures allowing to overcome the local minima were used. The minimisation process led to the *model parameters that give the best approximation* of the experimental curves.

3. SOURCE *F*-FUNCTION MODELLING

Typical source *f*-functions of the measured stress relaxation curves are in Fig. 1. They were approximated by sigmoidal RCM – see Eq. (A8.2) - and the obtained data are in Tables 1 and 2.

Previous attempts to use some other models summarised in Appendix A led to much higher SFC values (Blahovec, 2001) than those observed for sigmoidal RCM (3-30 % -Tables 1 and 2). The main advantage of the sigmoidal RMC in comparison to the other models reviewed in Appendix A consists in the fact that the requests on source *f*-function given in Appendix B can be really fulfilled. We did not obtain any stress relaxation curve with n less than 2, so that all experimental source *f*-functions are included into the class of sigmoidal source *f*-functions. This is in agreement with our analysis in Appendix B (see Eq. (B3)).

The highest SFCs were calculated for the lowest pre-stresses applied to the raw samples. In the raw state, i.e. before cooking, the initial pre-stress is higher in Nicola than in Panda. This observation is in agreement with higher Young's modulus of Nicola in comparison to Panda (Blahovec and Esmir, 2001a). The stress in raw Nicola decreased more quickly than in Panda so that the source f-functions of both the varieties in Fig. 1 approach one to another in high-relaxed states. However, the initial stress difference between Nicola and Panda was not reduced during the whole-observed parts of relaxation. The opposite relation was observed after cooking: the source *f*-function of Panda was always lower than the source *f*-function of Nicola. There is an indication of higher "viscosity" of cooked Panda in comparison to cooked Nicola (Blahovec and Esmir 2001b).

The experimental data well determined the part of the source *f*-function at stresses higher than ≈ 0.5 σ_{0t}. This part of the source *f*-function was useful for good estimation of the model parameters *K*, *C*, and *n*. This does not apply to the parameter C_1, which strongly depends on the low-level part of the source *f*-function, where no data were obtained. The minimisation procedures with C_l as a free parameter led to results with very high variation not only of C_l values, but also of all the other model parameters. This is why the parameter C_l was fixed at the mean values of the preliminary results. For raw tubers there were $3.2/\sigma_{0t}$ and $6.4/\sigma_{0t}$ for pre-deformation strain rate 0.167 $mm.s^{-1}$ and 0.833 $mm.s^{-1}$, respectively. Only one common value $2.5/\sigma_{0t}$ was used for cooked tubers.

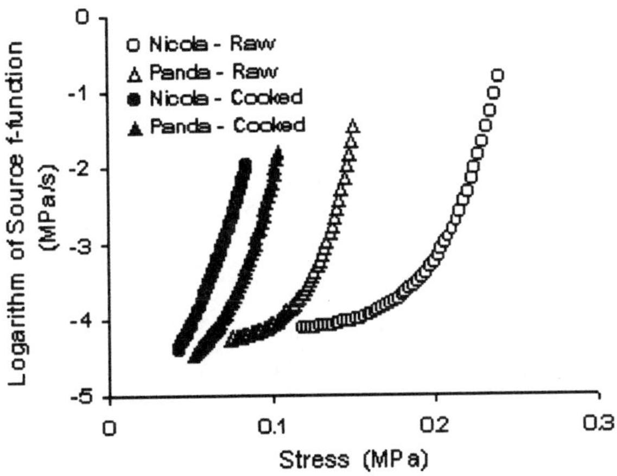

Figure 1. Typical source *f*-functions obtained at 15 % pre-strain. The observed stress relaxation began at highest stresses in pre-stress state and continued with decreasing stress. The tests were finished at the left sides of the source *f*-functions at the lowest stress.

Table 1. Mean values of model parameters (sigmoidal model) for raw tubers, LSD - least significant difference, SL – strain level, SR – strain rate

Variety	SL (%)	SR ($mm.s^{-1}$)	σ_{0t} (MPa)	SFC (%)	K ($Pa.s^{-1}$)	$C_1\sigma_{0t}$ (-)	$C\sigma_{0t}^n$ (-)	n (-)
Nicola	5	0.167	0.069	21.7	36.5	3.2	6.09	5.82
Nicola	10	0.167	0.240	8.0	88.3	3.2	7.49	7.38
Nicola	15	0.167	0.362	5.8	119.4	3.2	7.61	6.02
Nicola	15	0.833	0.387	7.9	95.2	6.4	10.6	5.54
Panda	5	0.167	0.027	30.3	21.2	3.2	7.32	3.92
Panda	10	0.167	0.150	8.4	66.2	3.2	6.32	6.84
Panda	15	0.167	0.302	5.9	104.2	3.2	7.50	6.58
Panda	15	0.833	0.316	6.3	90.3	6.4	9.70	5.75
Panda	20	0.167	0.435	5.7	126.2	3.2	7.72	5.48
		Variety	*					
Anova	+ LSD	SL	*	*	*		*	*
p = 0.05		SR			*	*	*	
		LSD	0.022	1.96	1.38	-	0.22	0.58

Table 2. Mean values of model parameters for cooked specimens, LSD - least significant difference, SL – strain level, SR – strain rate

Variety	SL (%)	SR ($mm.s^{-1}$)	σ_{0t} (MPa)	SFC (%)	K ($Pa.s^{-1}$)	$C_1\sigma_{0t}$ (-)	$C\sigma_{0t}^n$ (-)	n (-)
Nicola	5	0.167	0.023	7.52	2.54	2.5	8.27	2.38
Nicola	10	0.167	0.086	7.13	16.1	2.5	6.62	2.41
Nicola	15	0.167	0.130	2.58	42.7	2.5	6.15	2.66
Nicola	15	0.833	0.149	12.5	28.8	2.5	9.05	2.01
Panda	5	0.167	0.024	6.22	2.14	2.5	7.53	3.42
Panda	10	0.167	0.104	5.01	31.5	2.5	6.35	3.79
Panda	15	0.167	0.166	6.34	36.9	2.5	7.29	2.87
Panda	15	0.833	0.188	6.09	77.5	2.5	7.58	3.39
		Variety	*					*
Anova	LSD	SL	*	*	*		*	
p = 0.05		SR	*				*	
		LSD	0.017	2.01	9.43	-	0.475	0.35

Mean values of the model parameters obtained for raw tubers are in Table 1. It contains parameters C_1 and C in the dimensionless normalised form ($C_1' = C_1\sigma_{0t}$ and $C' = C\sigma_{0t}^n$), that did not vary so much with change of the test conditions. The parameter C_1' depended only on the pre-strain rate, the parameter C' also increased with increasing pre-strain value.

Similar results for cooked tubers are in Table 2. The results in Table 2 are nearly identical to the results for raw tubers in Table 1. The ANOVA showed that the pre-strain level influenced dominantly the parameters K and C and predetermined also the SFC. Parameter C depended also on strain rate and exponent n depended on the variety. Parameter n obtained for cooked Panda was significantly higher than the parameter obtained for cooked Nicola.

3.1 Role of Pre-stress

The obtained model parameters had some common properties. Values of the source *f*-function at the initial stress σ_{0t}, (Fig.2) it were calculated according to the equation:

$$ISF = K(1 - \exp(-C_1'))\exp(C') \qquad (5)$$

ISF increased with increasing pre-stress value rather non-linearly under the same curve, independently on potato variety and cooking (Fig. 2). The important change of the ISF-PS curve was observed at pre-stress about 120 kPa, where its slope raised approximately ten times.

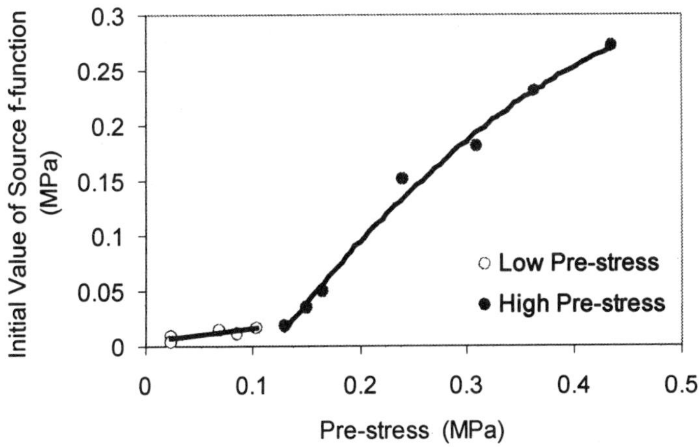

Figure 2. Initial value of the source *f*-function calculated according Eq. (5), using the data from Tables 1 and 2. The plot against pre-stress is divided into two parts: 1. for pre-stress lower than appr. 0.12 MPa, approximated by $ISF = 0.1164\,PS + 0.0041$ ($R^2 = 0.668$), 2. For higher pre-stress values, approximated by $ISF = -1.289\,PS^2 + 1.569\,PS - 0.168$ ($R^2 = 0.991$).

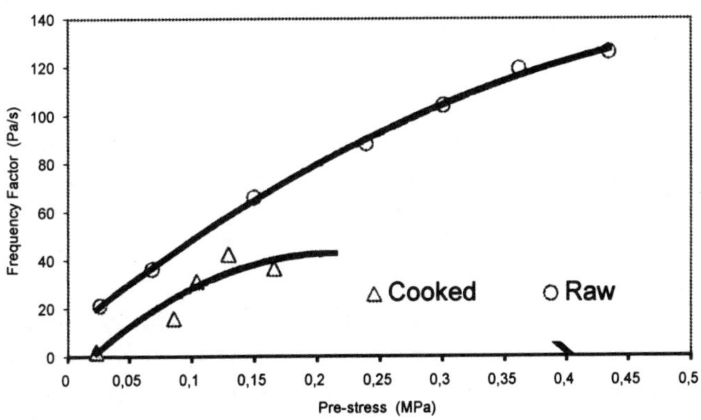

Figure 3. Frequency factor *K* - see Eq. (A8.2) - plotted against pre-stress. The values obtained at higher strain rate (0.833 $mm.s^{-1}$) are not included. Equations obtained for approximation: $K = -331.8\,PS^2 + 414\,PS + 10.17$ (raw , $R^2 = 0.998$), $K = -1161\,PS^2 + 492\,PS - 9.38$ (cooked, $R^2 = 0.906$)

The frequency parameter K increased with increasing pre-stress (Fig. 3) for both raw and cooked samples, but under the separate curves. No variety difference was observed in this case. The different values of K were observed also for different strain rates; unfortunately, trends of the changes were not clear (Tables 2 and 3).

The parameter n is in Fig. 4. No important difference between the tested varieties was observed for raw specimens. With increasing pre-stress the parameter n increased up to ≈ 7 for pre-stress ≈ 0.25 MPa and than decreased up to ≈ 5, the value observed at the lowest pre-stresses. After cooking, the observed n were significantly lower than before and differed for both the tested varieties (2.34 ± 0.42 for Nicola and 3.36 ± 0.57 for Panda). The role of pre-stress, if exists, is not significant in this case.

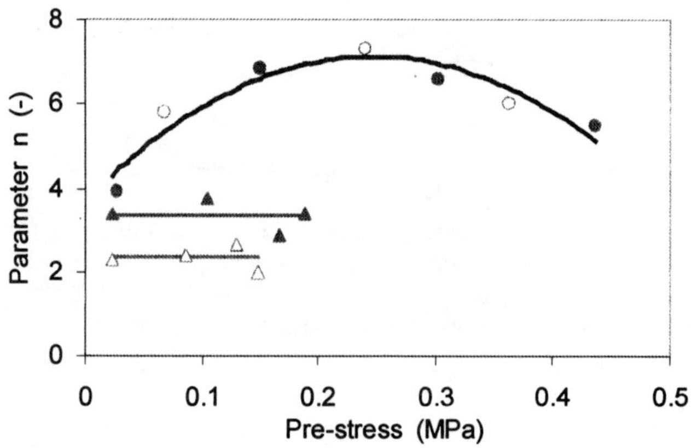

Figure 4. Parameter n plotted against pre-stress. The circles and triangles denote raw and cooked specimens, respectively. Open marks - Nicola, full marks – Panda. Mean values for cooked Nicola and Panda were 2.34 ± 0.42 and 3.36 ± 0.57, respectively. Common equation for raw specimens: $n = -56.4\, PS^2 + 27.9\, PS + 3.69$ ($R^2 = 0.871$).

With increasing pre-stress the parameter C decreased sharply (Fig. 5). The data obtained for raw tubers were well approximated by one exponential equation in contrary to data obtained for cooked tubers where individual curves exist for both the tested varieties. Part of the exponential dependence can be removed ($C' = C\sigma_{0t}^{n}$ equals 6 – 11 in Tables 1 and 2). Figure 6 shows that C' also varies with increasing pre-stress, but in different manner at small (less than appr. 0.15 MPa) and at higher pre-stresses. This observation is in agreement with calculated initial values of the source *f*-function in Fig. 2.

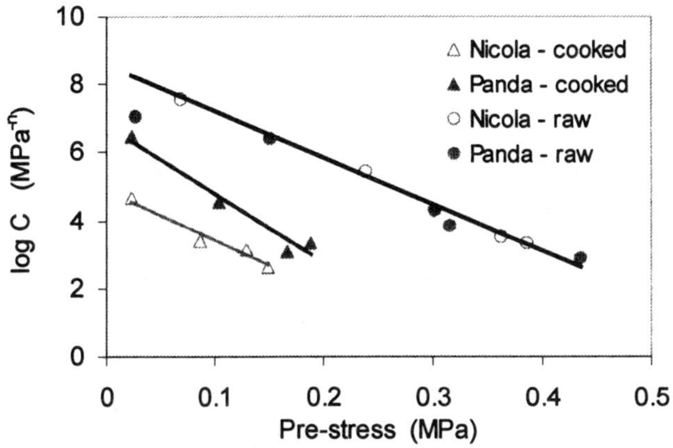

Figure 5. Parameter C plotted against pre-stress. The circles and triangles denote raw and cooked specimens, respectively. Open marks - Nicola, full marks – Panda. Regression equations for raw specimens: $\log(C) = -13.58\ PS + 8.55$ ($R^2 = 0.997$) and for cooked specimens – Nicola: $\log(C) = -15.08\ PS + 4.91$ ($R^2 = 0.955$), Panda: $\log(C) = -20.36\ PS + 6.81$ ($R^2 = 0.961$).

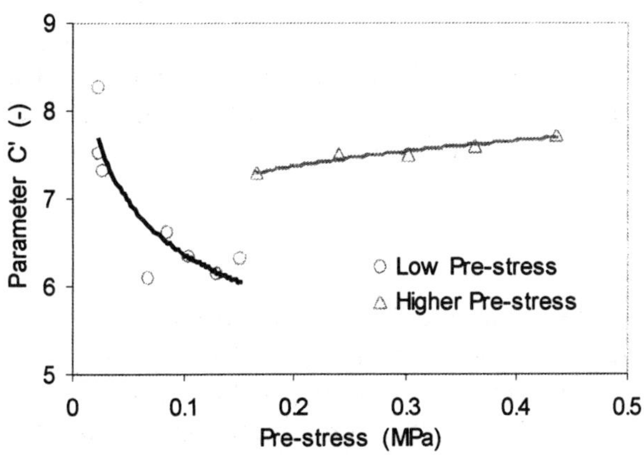

Figure 6. Parameter C' plotted against pre-stress. Regression equations: for low pre-stress – $C' = 4.75\ PS^{-0.1283}$ ($R^2 = 0{,}801$), for higher pre-stresses – $C' = 8.07\ PS^{0.056}$ ($R^2 = 0.955$).

3.2 Activation Volume

Apparent activation volume for sigmoidal RMC is given by Eq. (A8.3) that can be rewritten in the following form:

$$V(\sigma_{0t},\delta)=\frac{kT}{\sigma_{0t}}\left(\frac{C_1'e^{-C_1'\delta}}{1-e^{-C_1'\delta}}+C'n\delta^{n-1}\right) \tag{6}$$

where $\delta = \sigma_t/\sigma_{0t}$ is stress relaxation ratio and σ_{0t} pre-stress level. The parameter K from Eq. (A8.2) does not appear in Eq. (6). The pre-stress level and stress depression during a test were separated one from another in Eq. (6). The expression in the parenthesis of Eq. (6) is a U-shape function of δ. It approaches to infinity when δ approaches to zero. With increasing δ the expression decreases to the minimum at point of inflexion in the plot ln(f) - δ and then increases up to the initial value for $\delta = 1$. This indicates that at least one additional activation process takes part at low δ. The stress relaxation ratio at the point of inflexion (δ_{it}) is given by the following equation:

$$\delta_{it}^{\,2}\left(1-e^{-C_1'\delta_{it}}\right)^2=\frac{C_1'^2}{n(n-1)C'} \tag{7}$$

The calculated values δ_{it} were in the range 0.24-0.42 and 0.11-0.23 for raw and cooked specimens, respectively. Our experiments were performed for stress relaxation ratio higher than δ_{it} (higher than 0.5). In this case the apparent activation volume in Eq.(6) increased with increasing δ because the other parameters in Eq. (6) are constant (in one test). The observed trend can be approximated by a power equation:

$$V=V_1\delta^x \tag{8}$$

where V_1 is the initial apparent activation volume ($\delta = 1$) and x is the parameter. High correlation coefficients (R^2 higher than 0.988) were obtained for approximation of activation volume calculated from Eq.(6) by Eq. (8) in the individual tests. The parameters x are displayed in Table 3.

Table 3. Parameter x from Eq. (8) for the tested groups

Group	Mean Value	Standard Deviation	CV (%)
Raw Samples			
Nicola	4.582	0.528	11.5
Panda	4.197	0.925	22.0
Nicola + Panda	4.368	0.757	17.3
Cooked Samples			
Nicola	1.162	0.229	19.7
Panda	2.108	0.330	15.7

Raw samples, cooked Nicola, and cooked Panda are the three groups of the tested samples for which the parameter x in Eq. (8) can be approximated successfully by one common value. It is important that no significant influence of pre-stress was observed in any of these groups. It means that the observed decrease of apparent activation volume during the stress relaxation tests is controlled rather by stress relaxation ratio than by the pre-stress level.

The role of pre-stress level expressed by σ_{0t}, C', and n in Eq. (6) is hidden in the parameter V_1 of Eq. (8). For constant parameters C' and n, the parameter V_1 should be proportional to $1/\sigma_{0t}$ as given by Eq. (6). We tried to approximate it by power function, too:

$$V_1 = A(\sigma_{0t} / \sigma_{00})^{-B} \tag{9}$$

where A and B are parameters and σ_{00} dimensional constant (1 MPa). The experimental values are in Table 4. The sigmoidal RCM gave for raw potato results (see A and B) that differed from those given by logarithmic model in our previous experiments (Blahovec, 1996a) and in experiments of Giessmann and Szot (1982). Other possible sources of differences could be found in different stress definitions. In this paper as well as in our previous paper (Blahovec, 1996a), true stress was used. Giessmann and Szot's results could be recalculated into the simple engineering form only, because the authors did not give any information on the corresponding pre-strains. The apparent activation volumes obtained for raw potatoes of both varieties (this paper) are so closed one to another that they could be included into one group. Different results were obtained for cooked tubers, and differences between both cooked varieties were also significant. B was very close to 1 for cooked samples, so that any influence of pre-stress on C' and n, if existed (Figs. 4 and 6) had to be weak - see Eq. (6).

The strain rate sensitivity analysis (Dotsenko, 1979) of the compression tests with strain rate changes (Blahovec, 1996a) gave the apparent activation volumes (Table 4) that were close to those obtained from relaxation test by logarithmic model with constant apparent activation volume. It seems that Eqs. (8) and (9) express two different aspects of the apparent activation volume given by Eq. (6). The initial apparent activation volume V_1 in Eq. (9) seems to be activation volume in a real mechanism that controlled the initial part of stress relaxation in the stressed potato. It decreased with increasing stress level similarly as it is proposed by most of the stress relaxation models (see Appendix A). On the other hand the Eq. (8) describes loss of stress sensitivity of the controlling mechanism during the relaxation test. This behaviour could rather reflect *"exhausting effects"* replacing one mechanism by the other one. Such an exhausting effect accompanies every process that is based on some form of the real transport of some mass component.

Table 4. Parameters *A* and *B* from Eq. (9)

Group	*A* (nm^3)	*B* (-)	R^2	Notes
Raw Tubers				
Nicola + Panda	309.03	0.741	0.994	This Paper
Sokol	84.20	0.921	0.996	Log. Model – Giessmann and Szot (1982)
Pola	78.30	0.931	0.989	Log. Model – ibid.
Narew	80.00	0.890	0.997	Log. Model – ibid.
Andretta	64.80	1.074	0.999	Log. Model – ibid.
Boubin	59.00	1.250	0.991	Log. Model – Blahovec (1996a)
Karin	53.20	1.330	0.963	Log. Model – ibid.
Resy	42.30	1.580	0.963	Log. Model – ibid.
Boubin	110.40	1.130	0.976	Change of Strain Rate – ibid.
Karin	78.60	1.630	0.970	Change of Strain Rate – ibid.
Resy	93.70	1.460	0.962	Change of Strain Rate – ibid.
Cooked Tubers				
Nicola	63.1	1.051	0.994	This Paper
Panda	91.2	1.048	0.992	This Paper

3.3 Simplified Source *f*-function

The regression equations (8) and (9) can be unified into one equation, the simplified equation for the apparent activation volume is:

$$V = A\left(\frac{\sigma_{0t}}{\sigma_{00}}\right)^{-B}\delta^{x} \tag{10}$$

This equation was obtained for stresses well above δ_{it}. Equation (A0) after inserting *V* from Eq. (10) can be rewritten into the following form:

$$\ln f = \frac{A}{kT}\int\left(\frac{\sigma_{0t}}{\sigma_{00}}\right)^{-B}\delta^{x}d\sigma \tag{11}$$

that after rearrangements leads to the simplified source *f*-function f_s:

$$f_s = K\exp\left[D\left(\frac{\sigma_{0t}}{\sigma_{00}}\right)^{(1-B)}\delta^{(x+1)}\right] \tag{12}$$

where K is an integration constant and D constant parameter:

$$D = \frac{A\sigma_{00}}{kT(1+x)} \tag{12a}$$

The parameters D, $(1 - B)$ and $(x + 1)$ were given by the regression equations (9) and (10) – see Table 5. They are stress and pre-stress independent so that the role of pre-stress in Eq. (12) is given directly by σ_{ot} and indirectly by the constant K that could be pre-stress dependent. The change of stress during a test is expressed in Eq. (12) by the relative stress δ only. The simplified source f-function (12) was obtained from the sigmoidal RCM – Eq. (A8.2) if $C_1' \to \infty$, $n = 1 + x$, and $C' = D(\sigma_{0t}/\sigma_{00})^{(1-B)}$. The parameters K in both equations should be the same. Numerical calculations showed that the expected agreement existed at the highest stress levels only. For fully relaxed state ($\delta = 0$) Eq. (12) gives $f_s = K$ so that the simplified source f-function cannot be interpreted as the sigmoidal model and the obtained nonzero asymptotic value is wrong.

Table 5. The parameters of Eq. (12) obtained by regression analysis of apparent activation volume

Group of samples	D (-)	$1 - B$	$1 + x$
Raw samples	13.70	0.259	5.58
Cooked Nicola	7.21	- 0.051	2.16
Cooked Panda	7.25	- 0.048	3.11

4. DIFFERENCES BETWEEN VARIETIES

The model parameters in raw samples did not depend significantly on variety (Figs. 3-5, Tables 3-5). The main source of the differences between source f-functions for Nicola and Panda (Fig.1) was in different pre-stresses at the same pre-strain value. Those were obtained as a consequence of different elastic moduli in both varieties.

The cooking changed the relaxation behaviour significantly; practically decreasing all the model parameters. The initial values of the source f-function after cooking decreased also but the new values were located on the same pre-stress plot as before cooking (Fig. 2). The separate plots versus pre-stress for raw and cooked potato were obtained for K, C, and n. The similar differences were obtained also for the simplified source f-function in Table 5. The parameters D, B, and x of the cooked samples were always lower than for the raw ones. The positive B-value of raw potato decreased during cooking to negative values after cooking.

Differences between the tested varieties were observed in cooked state. For sigmoidal source *f*-function the significant differences occur in parameters C and n (Figs. 4 and 5). The similar results were also for the simplified source *f*-function, where significant differences were obtained for parameter $(1 + x)$. The ratio of the quantities – $(1 + x)_{\mathrm{Panda}}/(1 + x)_{\mathrm{Nicola}}$ – was about 1.44, very close to 1.43, the ratio for the parameter n. The inter-variety differences in D and $(1 - B)$ were small, but their combination could be significant, similarly as the inter-variety differences in C (Fig. 5).

The higher values of n and/or $(1 + x)$ for Panda in comparison to Nicola indicated that the stress in cooked Panda relaxed more slowly then in cooked Nicola. It could be interpreted as higher viscosity and lower fluidity of the cooked Panda in comparison to Nicola. This difference could correlated with their textural properties, especially with mealiness (Blahovec *et al.*, 2000, Blahovec and Esmir, 2001b).

5. CONCLUSIONS

Stress relaxation in raw and cooked potatoes can be well described by the sigmoidal RCM with sigmoidal source *f*-function. For shorter tests that end at higher relative stress, the simplified source *f*-function is used, too.

The apparent activation volume cannot be interpreted as a result of one real thermal activated process. The process responsible for the initial part of the relaxation curve has an exhausting character and the apparent activation volume in this part of the relaxation curve is approximated by a power function of the relative stress (σ_t/σ_{0t}) with exponent higher that 1. The initial apparent activation volume decreases with increasing pre-stress. This dependence is described also by a power function with negative exponent.

The differences between model parameters are very low for different varieties of raw potato. The main source of the difference between the potato varieties in the raw state consists in different elastic moduli of different varieties having result in different pre-stress values. Significant differences between the model parameters of different potato varieties are observed after cooking, especially in the model exponents n and x, which express some aspects of the potato cooking quality. The model parameters of cooked potato are lower than before cooking. The only one exception is the initial value of the source *f*-function, which seems to be pre-determined by pre-stress only.

REFERENCES

Blahovec, J., 1996a, Stress relaxation phenomena in vegetable tissue. Experimental results. *Journal of Materials Science* **31**: 1729-1734.

Blahovec, J., 1996b, Thermal activated theories of stress relaxation in vegetable flesh. In *Stress relaxation in solids of biological origin* (J. Blahovec, edt.). Czech University of Agriculture Prague, pp. 30-39.

Blahovec, J., 2001, Improved rate controlled model for stress relaxation in vegetable tissue. *International Agrophysics* **15**: 73-78.

Blahovec, J., and Esmir, A.A.S., 2001a, Precise study of cooked potato texture. *J. Texture Studies* **32**: 165-184.

Blahovec, J., and Esmir, A.A.S., 2001b, Stress relaxation in cooked potato tubers expressed by improved rate controlled model. *International J. Food Properties* **4**: 485-500.

Blahovec, J., Esmir, A.A.S., and Valentova, H., 2001, Simplified texture profile analysis of cooked potatoes. *International J. Food Properties* **3**: 193-206.

Dotsenko, V. I., 1979, Stress relaxation in crystals. *physica status solidi* **93**: 11-43.

Fujihara,, S., Yamamoto, R., and Masuda, Y., 1978, Viscoelastic properties of plant cell walls I. Mathematical formulation for stress relaxation with consideration for pre-extension rate. *Biorheology* **15**: 63-75.

Giessmann, E. J., Grau, P., 1980, On representation of stress relaxation behaviour of agricultural materials. In *Proceedings 2nd Int. Conference „Physical Properties of Agricultural Materials", 26-28 August 1980*, Gödöllö, Vol. 1, pp. 81-88.

Giessmann, E. J., and Szot B., 1982, Einfluss der Verformungs-geschwindigkeit auf den Wert de Relaxationskonstanten. *Tag.-Ber. Akad. Lanndwirtsch. Wiss. DDR* **208**: 63-67.

James, F, and Roos, M., 1975, Minuit - A system for function minimization and analysis of the parameter errors and correlations. *Comp. Physics Comm.* **10**: 343-367.

Mohsenin, N. N., 1970, *Physical Properties of Plant and Animal Materials.* Vol. 1. New York: Gordon and Breach.

Peleg, M., 1979, Characterization of the stress relaxation curves of solid foods. *J. Food. Sci.* **44**: 277-281.

APPENDIX A. BASIC RELAXATION MODELS

Every model in this Appendix is characterised by stress-time dependence $\sigma(t)$, source f-function $f(\sigma)$ and apparent activation volume $V(\sigma)$ given in each model by the following equation (Dotsenko, 1979):

$$V(\sigma) = kT\left[\frac{\partial \ln\left(-\frac{d\sigma}{dt}\right)}{\partial \sigma}\right]_T = kT\left[\frac{\partial \ln(f)}{\partial \sigma}\right]_T \tag{A0}$$

where k is Boltzmann constant, T absolute temperature, σ actual stress and f is the source f-function. Symbol T at bracket means that the partial derivative is calculated at constant temperature. The activation volume is an important parameter of the rate controlled models. Other symbols in the following equations are the model parameters, σ_0 denotes the initial stress in relaxation.

Maxwell's Exponential Model (Dotsenko, 1979)

$$\sigma(t) = \sigma_0 e^{-Kt} \tag{A1.1}$$

$$f(\sigma) = K\sigma \tag{A1.2}$$

$$V(\sigma) = \frac{kT}{\sigma} \tag{A1.3}$$

Maxwell's Generalised Model (Mohsenin, 1970)

In this case the source f-function cannot be expressed analytically as a function of stress so that both source f-function f and activation volume are expressed as a function of time.

$$\sigma(t) = \sigma_1 e^{-k_1 t} + e^{-k_2 t} + .. \tag{A2.1}$$

$$f(t) = k_1 \sigma_1 e^{-k_1 t} + k_2 e^{-k_2 t} + .. \tag{A2.2}$$

$$V(t) = kT \frac{k_1{}^2 \sigma_1 e^{-k_1 t} + k_2{}^2 e^{-k_2 t} + ..}{(k_1 \sigma_1 e^{-k_1 t} + k_2 e^{-k_2 t} + ..)^2} \tag{A2.3}$$

Power Model (Dotsenko, 1979)

$$\sigma(t) = \sigma_0 (1 + t / A)^{-n} \tag{A3.1}$$

$$f(\sigma) = \frac{n\sigma_0}{A} \left(\frac{\sigma}{\sigma_0} \right)^{\frac{n+1}{n}} \tag{A3.2}$$

$$V(\sigma) = \frac{kT}{\sigma} \frac{(n+1)}{n} \tag{A3.3}$$

Logarithmic Model (Dotsenko, 1979, Giessmann and Grau, 1980)

$$\sigma(t) = \sigma_0\left[1 - \frac{1}{C}\ln(bt+1)\right] \tag{A4.1}$$

$$f(\sigma) = \frac{\sigma_0 b}{C} e^{-C} e^{\frac{C\sigma}{\sigma_0}} \tag{A4.2}$$

$$V(\sigma) = \frac{kTC}{\sigma_0} \tag{A4.3}$$

Logarithmic Modified Model (Blahovec, 1996b)

$$\sigma(t) = \frac{\sigma_0}{C}\ln\frac{1+\tanh\frac{C}{2}e^{-\frac{KCt}{\sigma_0}}}{1-\tanh\frac{C}{2}e^{-\frac{KCt}{\sigma_0}}} \tag{A5.1}$$

$$f(\sigma) = K.\sinh\left(\frac{C\sigma}{\sigma_0}\right) \tag{A5.2}$$

$$V(\sigma) = \frac{KTC}{\sigma_0}\coth\frac{C\sigma}{\sigma_0} \tag{A5.3}$$

Peleg's Equation (Peleg, 1979)

$$\sigma(t) = \sigma_0\left(1 - \frac{abt}{1+bt}\right) \tag{A6.1}$$

$$f(\sigma) = ab\sigma_0\left(1 - \frac{1}{a} + \frac{\sigma}{a\sigma_0}\right)^2 \tag{A6.2}$$

$$V(\sigma) = \frac{2kT}{\sigma_0(a-1)+\sigma} \tag{A6.3}$$

Yamamoto's Model for c ≥ 0 (Fujihara *et al.*, 1978)

$$\sigma(t) = b.\ln\frac{t+T_m}{t+T_0} + c \qquad \text{(A7.1)}$$

$$f(\sigma) = \frac{b}{T_m - T_0}\frac{\left(1-e^{\frac{\sigma-c}{b}}\right)^2}{e^{\frac{\sigma-c}{b}}} \qquad \text{(A7.2)}$$

$$V(\sigma) = \frac{kT}{b}\frac{e^{\frac{\sigma-c}{b}}+1}{e^{\frac{\sigma-c}{b}}-1} \qquad \text{(A7.3)}$$

Sigmoidal RCM (Blahovec, 2001)
σ(t) cannot be expressed in analytical form, Eq. (3) has to be solved numerically

$$f(\sigma) = K(1-e^{-C_1\sigma})e^{C\sigma^n} \qquad \text{(A8.2)}$$

$$V(\sigma) = kT\left(\frac{C_1 e^{-C_1\sigma}}{1-e^{-C_1\sigma}} + Cn\sigma^{n-1}\right) \qquad \text{(A8.3)}$$

APPENDIX B PROPERTIES OF SOURCE *F*-FUNCTION FOR STRESS RELAXATION IN POTATO

The logarithm of experimental source *f*-function for potato expressed as a function of stress is a convex function (Blahovec, 2001), i.e. its derivative has to be *increasing positive function.* Equation (A0) shows that the activation volume is proportional to this derivative. For almost all of the models reviewed in Appendix A (the logarithmic model and the sigmoidal RCM are exceptions) the activation volume decreases with increasing stress value. Activation volume in the logarithmic model is constant and this is why that this RCM model was used as a basis for construction of the Sigmoidal RCM.

Equation (A4.2) for source *f*-function of the logarithmic model was modified under the following scheme:

$$f(\sigma) = K.e^{C\sigma} = g_1.e^{g_2\sigma} \tag{B1}$$

where g_1 g_2 are the stress functions replacing the constants K and C, respectively. The function g_1 modifies frequency factor K so that $f(0) = 0$. This function is concave: $g_1(0) = 0$ and being nearly constant for higher stresses. For sigmoidal RCM's $g_1 = K(1 - \exp(-C_1\sigma))$. This function is approximately equal K for high stresses.

The concave function g_1 results in concave form of $\ln(f(\sigma))$ for lower σ. The convex character of $\ln(f(\sigma))$ at higher stresses is saved by a suitable form of g_2 and the inflexion point separats the convex and the concave parts of the source f-functions. In convex part of the curve both the first and second derivatives are positive:

$$\frac{d\ln f}{d\sigma} = \frac{1}{g_1(\sigma)}\frac{dg_1(\sigma)}{d\sigma} + \frac{dg_2(\sigma)}{d(\sigma)} \tag{B2}$$

$$\frac{d^2\ln(f)}{d\sigma^2} = -\left(\frac{1}{g_1(\sigma)}\frac{dg_1(\sigma)}{d(\sigma)}\right)^2 + \frac{1}{g_1(\sigma)}\frac{d^2g_1(\sigma)}{d\sigma^2} + \frac{d^2g_2(\sigma)}{d\sigma^2} \tag{B3}$$

The first two members on the right side of the Eq. (B3) are negative saving the negative values of Eq. (B3) for stresses below the inflexion point. The last member in the right side of Eq. (B3) is positive at least for stresses higher than the inflexion point. The function $g_2(\sigma)$ is the positive increasing function with positive and increasing second derivative in this area. For special function $g_2(\sigma) = C\sigma^n$ the second derivative is $n(n-1)\sigma^{n-2}$ and such an expression is positive increasing function only if n is greater than 2. This form of $g_2(\sigma)$ was used in the sigmoidal RCM.

Determination of Hen's Eggshell Elastic Properties under Quasistatic Compression

JAROSLAV BUCHAR, LIBOR SEVERA and JANA SIMEONOVOVÁ
Mendel University of Agriculture and Forestry, Brno, Czech Republic

1. INTRODUCTION

The elastic properties belong to key engineering properties of eggshell material. Their knowledge enables e.g. the evaluation of the eggshell strength which is the important parameter of the egg quality (Lin J. et al 1993, Carnarius et al., 1996, Boushy and Ratering, 1993, Rezac et al, 2000, Nirasawa et al. 1998, Picman and Pripil 1997). There are different experimental technique of the eggshell strength measurement see e.g. Bain (1997) for a review. One of them is the quasi – static compression test that corresponds to the loading of eggs during their packing. During this test, eggs are compressed between two parallel plates by a steadily increasing load until failure results. This procedure enables to determine the force and the shortening of the egg at the moment of failure. This force is affected by many factors as e.g. by the eggshell geometry (shape and thickness), by the velocity of loading and by many other factors which more or less cover the influence of the intrinsic material properties. The exact evaluation should lead to the values of stress and strain at the failure. These parameters may be easily obtained from test mentioned above if we use flat testing specimens. The eggs exhibit a more complicated shape but they may be considered as a shell of revolution. The evaluation of the eggshell strength thus must involve the following steps:

- experiment described above
- selection of the eggshell material model

evaluation of the stress and strain in the eggshell at the moment of the eggshell failure

The eggshell is mostly considered as isotropic elastic material. The elastic properties of this material are described e.g. by the Young modulus E and by the Poisson`s ratio ν and/or by the Young modulus and by the shear modulus G, respectively. Looking on the real structure of the eggshell this assumption is oversimplified. The description of the eggshell should be based on the model of anisotropic layered shell. The use of this model is connected with many difficulties in comparison with the use of isotropic one. The stress and strain mentioned above can be determined only by the indirect way. Some papers dealing with this problem are summarised in review (Bain, 1997).

In order to avoid the complications connected with the egg`s shape some authors used a ring cut from the eggshell near of the egg equator (Rehkugler 1963, Lin et al 1995). Some authors considered the egg as a sphere (Tung et al 1969). These simplifications enable the use of some analytical methods. The real shape of the eggshell was considered e.g. in (Kitching 1997). This author used membrane solution for the shell when it is subjected to internal pressure. Results showed a significant influence of the eggshell shape on the membrane stresses. At the same time the validity of these results was confirmed by the numerical simulation based on the finite element method. The finite element method is the most used numerical method, which is implemented in many computer software systems like ANSYS, MARC etc. The finite element method was used by Manceau and Henderson 1970, Bain 1990, Eantwistle. et al. 1995, Eantwistle and Reddy, 1996 and others.

In this paper the eggshell strength is calculated using the numerical procedure mentioned above.

2. EXPERIMENTAL PROCEDURE AND RESULTS

Hen's eggs were used for the experiments. The shape of every egg was determined from a digital photo of the egg by the procedure described by Barton and Krivanek (2001). The shape of the egg was described by the parametric equations:

$$x = a\cos(\varphi), y = a\sin(\varphi)$$

where

$$a = k_1 \frac{1 + k_2\cos(\varphi)}{\sqrt{k_3\cos^2(\varphi) + \sin^2(\varphi)}}$$

φ polar angle and k_1, k_2, and k_3 model parameters with the following mean values: k_1 = 21.079 mm, k_2 = 0.172, k_3 = 0.505 (based on 38 measured eggs).

The eggs were compressed using of the testing machine TIRAtest 27025 at deformation rate 20 mm per minute. On the surface of the eggs a strain gauge was glued - see Fig.1. The length of the individual strain gauge (Hottinger Baldwin Messtechnik type 2/120LY 11, gauge factor 2.03±1 %, resistance 120±0.20 Ω) was 2 mm.

Individual eggs were loaded by the following way:

- Loading up to 10 N and subsequent unloading
- The previous step was repeated for the forces 15, 20 , 25, 30 N
- At every value of the force a relaxation test was performed, the deformation corresponding to this force was kept constant for 240 seconds and the force versus time was recorded.
- The egg was loaded up to failure.

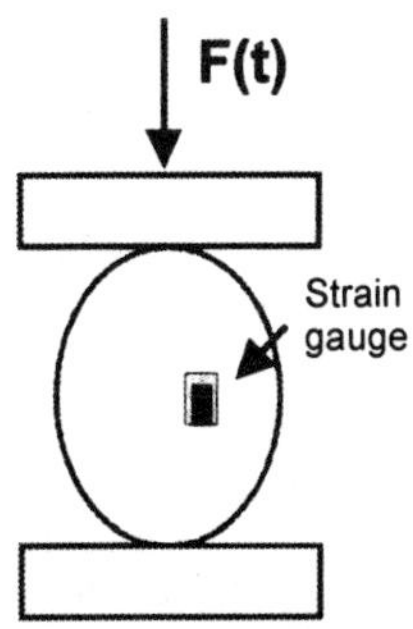

Figure 1. Schematic of the eggshell loading in compression using TIRAtest testing machine. The strain was measured in the meridian direction and then recorded via the "Spider" equipment (produced by Hottinger Baldwin Messtechnik Ltd.)

From the relaxation experiments the assumption on the elastic behaviour of eggs was approved.

In Fig.2 the record of time dependence of the force at the top of egg is displayed. Figure 3 shows the record of the force in the case of the egg failure. The force at the failure is about 49 N. The failure had a form of a thin crack propagated nearly in the meridian direction. An example of strain – time record is given in Fig.4. The record in Fig.2 corresponds to the loading to about 12 N. In this case, the eggshell was deformed only in elastic region. The strain - time record for egg failure is shown in Fig. 5. The release of the strain energy during failure can be observed in the figure. In Table 1 maximal strain and shortening observed at the maximum load force F are displayed. The strain increased linearly with maximal loading force (Fig.6).

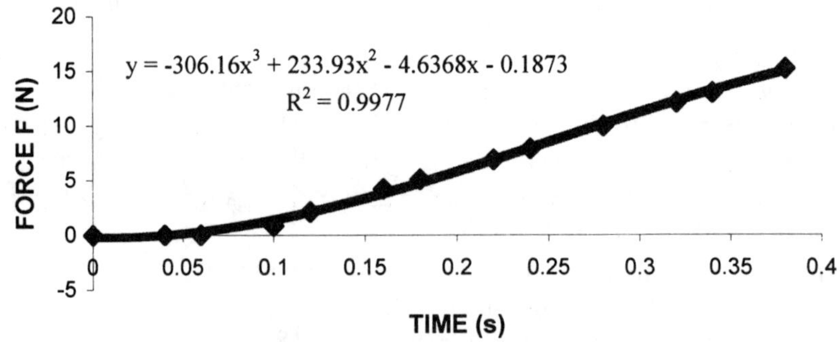

Figure 2. Loading force versus time in loading test. No damage of the eggshell was occurred.

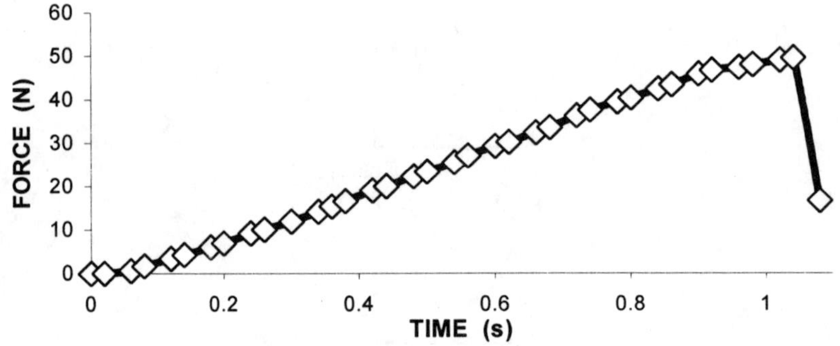

Figure 3. The experimental record loading force on the time. At the maximum of the force failure of the egg starts. The failure is in the form of the single crack. The crack propagates nearly along the meridian.

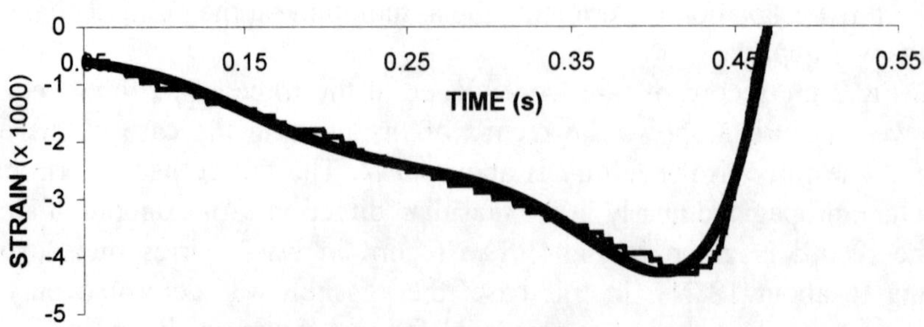

Figure 4. An example of the strain record (The smooth curve corresponds to polynomial fit of the experimental data)

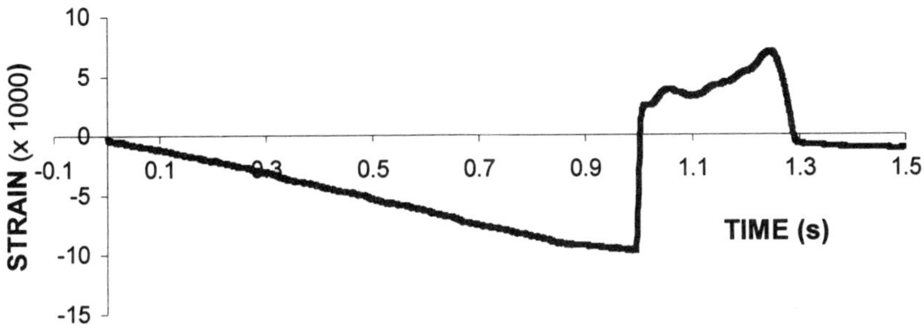

Figure 5. An example of the strain record in loading test, in which the eggshell was failed.

Table 1. Experimental values of the loading force, strain and egg shortening

FORCE F (N)	12.03	15.26	20.61	25.83	30.91	49.54
STRAIN (x 0.001)	4.32	4.68	5.76	6.96	7.68	9.60
SHORTENING (mm)	0.100	0.127	0.133	0.187	0.222	0.347

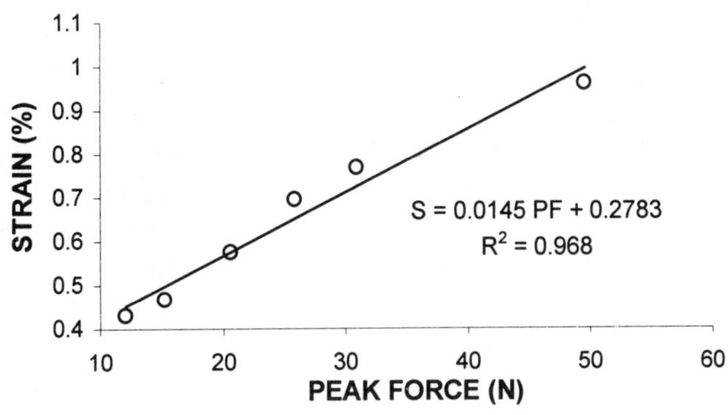

Figure 6. Strain versus maximal loading force.

The shortening of the egg depended on the loading force linearly as it is documented in Fig.7. The linearity of this dependence should be verified for another points on the surface of the egg.

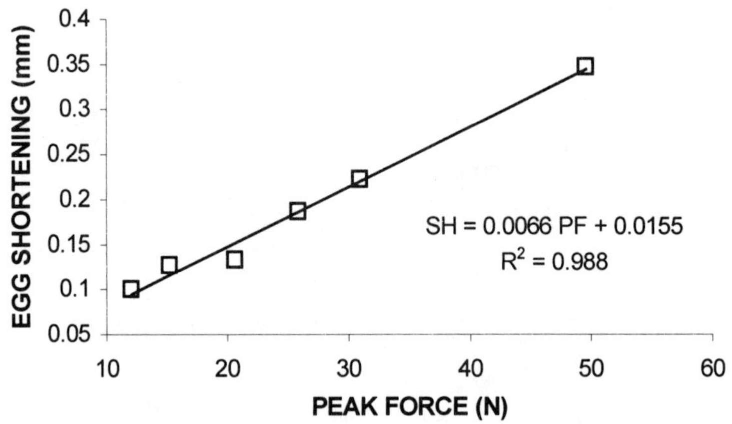

Figure 7. Egg shortening versus peak loading force

3. NUMERICAL SIMULATION

In the next step we used the model of the shell loading described in the previous chapter.

The eggshell was taken as homogeneous and isotropic body. The eggshell thickness was approximated by 0.35 mm. The elastic properties of the shell were described by the Young's modulus E and by the Poisson's ratio ν. Their values are given e.g. in review paper (Bain, 1997). According to different authors the values of E and ν varies in a broad interval. For the sake of simplicity the yolk and the white of egg were considered as a mixture with density given by mean value of both the components. The viscosity of this mixture was determined e.g. by Dawson, 1996. It was found that it does not behave as Newtonian liquid with apparent viscosity coefficient μ linearly decreases with the strain rate γ. At the temperature T = 20°C this dependence has a form:

$$\mu = 10^{-3}(9.6 - 0.0024\gamma) \qquad in \quad Pas$$

It is obvious that this approach represents a rather simplified solution, which will be improved in our forthcoming papers. For compressibility of yolk-white mixtures the values known for water were used. The air located inside the eggshell was modelled by state equation of ideal gas.

The ANSYS finite element code was used. The eggshell has been modelled using of the shell elements: 1619 elements were used. The liquid and air were modelled using of the brick elements (1814 elements). Set of computations gave the strain values for the given values of the loading force and for different values of E and ν. The obtained results are given in Table 2. The obtained results only slightly depended on egg`s liquid viscosity (Buchar and Simeonovova, 2001).

Table 2. Computed values of strains

E (GPa)	ν (1)		Strain (%)		
		F (N)	12.03	15.26	20.61
	0.3		0.428	0.435	0.482
15	0.35		0.462	0.475	0.511
	0.4		0.482	0.497	0.543
	0.3		0.396	0.418	0.461
25	0.35		0.415	0.442	0.493
	0.4		0.426	0.475	0.520
	0.3		0.327	0.403	0.443
35	0.35		0.356	0.427	0.478
	0.4		0.398	0.451	0.507

The best agreement between experimental data and theory was achieved for the following values of the elastic constants:

$$E = 23 \text{ GPa}, \nu = 0.34.$$

If we used the value of the force at which the eggshell failure occurred the numerical computation gave the value of the tensile stress near of the point of contact 19.3 MPa. This stress can be considered as the eggshell strength. The value of E was slightly higher than those reported by (Lin et al, 1995) (about 19 GPa). Real values of E seems to lie between 10.5 and 46.9 GPa (Baine, 1990). The data on the eggshell strength were given e.g. by Entswistle et al., 1995. The strength were obtained using of half eggs with blunt poles obtained by cutting the egg around the equator, which were subjected to concentric loading. The strength was about 27.3 MPa at the outer surface of the shell and 30.4 MPa at the inner surface. The experiments in which the eggs were loaded by the internal pressure (Entswistle and Reddy, 1996) led to the value of the eggshell strength 15.2 MPa. It is a problem of the used strength model. This problem should be solved in the future.

4. CONCLUSION

Egg liquids exhibited only minor influence on the egg strain and fracture behaviour. The obtained eggshell strength is in agreement with results of other authors. The main problems are:

- The choose of the proper strength criteria
- The extension of the given computation towards to the anisotropic model of the eggshell. The hopeful method may be represented by the shell theory developed by Shim et al., 1996.

The solution of these problems will be subject of forthcoming papers.

ACKNOWLEDGEMENT

The authors would like to express hearty thanks for the financial support under contract MSM: 432100001.

REFERENCES

Bain, M., 1997, A reinterpretation of eggshell strength. In *Eggs quality* (S.E.Solomon,Ed.), Manson Publishing/ The Veterinary Press / Iowa State University Press , pp. 131 – 142.

Bartoň, S. and Křivánek, I., 2001,Determination of parameters of egg`s perimeter by analysis of digital photography in Maple. In *Proceedings of the Symposium on the Quality food evaluation*, March 8, Mendel Brno.

Buchar, J and Simenovová, J.,2001, On the identification of the eggshell elastic properties under quasistatic compression. In *19th CAD – FEM USERS MEETING 2001-International Congress on FEM Technology*.17-19.October 2001, Potsdam, Berlin Germany, Vol.2, paper No 2.2.1 - 8pages. Published by CAD – Fem GmbH, Munchen.

Carnarius, K.M., Conrad, K.M., Mast, M.G. and Mac Neill, J.H.,1996, Relationship of eggshell ultrastructure and Shell Strength to the Soudness of Shell Eggs. *Poultry Science* **75** : 656- 663.

Dawson, P.L. 1996, Effects of freezing , frozen storage and thawings on eggs and egg products. In *Freezing Effects on Food Quality* (L.E.Jeremiah, Ed.) Marcel Dekker, Inc., New York .pp 337 – 366.

Eantwistle, K.M. and Reddy, T.Y., 1996, The fracture strength under internal pressure of the eggshell of the domestic fowl. *Proceedings Royal Society of London B* **263**: 433 – 438.

Eantwistle, K.M., Silyn, R.H. and Abuodha, S.O. , 1995, The relative fracture strengths of the inner and outer surfaces of the domestic fowl. *Proceedings of the Royal Society of London, Series B – Biological Sciences* **262** : 169 – 174.

El Boushby, A.R. and Raternik, 1993, Poultry Adviser **26** No 11.

Kitching, R.,1997, Pressurised eggshell analysis. *Int.J.Mech.Sci.* **39**: 391-396.

Lin, J.,Puri, V.M. and Anantheswaran, R.C.,1995, Measurement of eggshell thermal – mechanical properties. *Transaction of the ASAE* **38**: 1769 –1776.

Manceau, J.R. and Henderson, J.M., 1970, Physical properties of eggshells. *Transactions of the American Society Agriculture Engineering* **13**: 436

Nirasawa, K.,Takahashi, H., Takeda, H., Furukawa, T., Picman, J. and Pripil, S., 1997, Is greater eggshell density an alternative mechanism by which parasitic cuckoos increase the strength of their eggs ? *Journal for ornitologie* **138**: 531 – 541.

Rehkugler, G.E., 1963 Modulus of elasticity and ultimate strength of the hen's eggshell. *Journal of Agriculture Engineering Research* **8**: 352 – 354.

Rezac, P., Poschel, M. and Havlicek, Z.,2000, Relationship of plasma oestradiol – 17 beta and cholesterol levels to eggshell strength in laying hens. *Czech Journal of Animal Science* **45**: 313 – 318.

Simeonovova, J. and Buchar, J., 2001, Egg behaviour under low velocity impact. *Acta Universitatis Agriculturae et Silverculturae Mendelinae Brunensis*, **IL**: 89 – 100.

Shim, V.P.W. , Toh, S.L. and Gong, S.W.,1996, The elastic impact response of glass/epoxy laminated ogival shells. *Int.J.Impact Engng* **18**: 633 – 656.

Takeda, T. and Nagamine, Y., 1998, Restricted maximum likehood estimates of genetic parameters and genetics trends of chickens divergently selected for eggshell strength. Journal of Animal breeding and genetics **115** : 375 – 382.

Tumg, M.A., Staley, L.M. and Hunt, J.R., 1969, Estimation of the Young`s modulus and faailure stresses in the hen`s eggshells. Canadian Agricultural Engineering **11**: 3 – 5.

The Cracking Mechanisms of Grain Legume

BOHDAN DOBRZAŃSKI, JR.
B. Dobrzański Institute of Agrophysics, PAS, Lublin, Poland

1. INTRODUCTION

Legume seed quality is greatly affected by methods of harvesting and handling. In these operations, seeds are subjected repeatedly to many impacts on metal surfaces and against other seeds. Impacts may be result of threshing cylinder or rotor motion, the movement of buckets in vertical elevators, centrifugal discharging, the filling and discharging of screw conveyors, spouting and free-fall dropping (Palusen *et al.*, 1981a,b), (Pickett, 1973). In nearly all handling operations, seeds are accelerated to some velocity and then discharged onto stationary objects or other grains.

During the harvest and further handling, the seeds are subjected to impact and other type of loads, which involve the stresses in the seed coat or in the cotyledon. Mechanical and rheological properties of seeds have been tested by several researches for different purposes such as the design of processing machinery, texture and quality evaluation, and prediction and control of impact damage (Mensah et al., 1984), (Anazodo and Chikwendu, 1983; 1984), (Bilanski, 1988), (Dorell and Adams, 1969), (Fornal et al., 1990), (Hoki and Pickett, 1973), (Perry and Hall, 1966), (Rataj, 1984). The compression tests of intact biological materials provide an objective method for determining the mechanical properties significant in quality evaluation and control, the maximum allowable load for minimising mechanical damage (Arnold and Mohsenin, 1971), (Dobrzański and Szot, 1989;1992), (Paulsen, 1978). To find those values, it is necessary to carefully recognise the mechanical properties of seeds. Many researchers have already published the results of mechanical properties including a modulus of elasticity for various seeds at a given moisture content and deformation rate. The modulus of elasticity of seed is a physical property which has been suggested as a measure of the textural attribute designed as firmness or hardness (Misra and Young, 1981). It is also an important property frequently determined in the study of stress development at seed cracking. Physical properties of agricultural products are needed for adequate design of processing machines but its specific applications should be clearly understood before determining it experimentally (Misra and Young, 1981), (Mac Donald, 1985).

Large diversity of moisture is the most important factor affecting the type of deformation and origin of damage. The moisture content of the seed coat and cotyledon influences the mechanism of cracking (Dobrzański, 1998), (Misra and Young, 1981). The seed of high moisture level becomes very plastic. Frequently, large shape deformation of legume grain at compression caused the tension and breakage of the seed coat. Most studies on the mechanical properties of legume grains have been carried out for whole seeds. Only a few studies have been done on the component parts of seeds (Davison et al., 1975), (Dobrzański, 1997; 1998). However, to clarify the mechanisms of seed cracking, it is necessary to carefully recognise the mechanical properties of the seed coat and cotyledon separately.

Tension of the seed coat caused by axial compression and the shape deformation of wet seed leads the breakage, however, in two ways (Dobrzański, 1998; 2000). It's depend on the seed layout and direction of forces to the contact plane of the cotyledons.

- Firstly, a large axial displacement of non-compressibility material of wet cotyledon, caused a plastic deformation of the seed. If direction of force is parallel or perpendicular to the plane of contact of the cotyledons only, the axial compression involves the change of shape.
- Secondly, the cotyledon sliding move along the plane of contact caused a rapid displacement. This movement is involved by loading force of every differ direction to listed above.

In both cases, if a new body consists of sliding cotyledons or is created after seed deformation, the lateral area surrounding the body increases and leads to tension of the seed coat. In consequence it causes its breakage.

On the other hand, at low rate and moisture, a slightly deformation of dry seed at compression causes two types of damages as cotyledon cracks or as the seed coat splits.

- Firstly, axial compression of spherical cap of the cotyledon involved upsetting of a central part of the cotyledon. In this case, a damage appears around the flattening as crack opening displacement.
- Secondly, every impact of dry seed or a component of force parallel to the contact plane of the cotyledons, causes their relative movement and shearing of the seed coat.

Further, a major problem in drying of legume seeds with heated air is a splitting of the seed coat, affected by the temperature and moisture (Dobrzański, 2000), (Fornal et al., 1993). The drying conditions for legume seeds are dependent on the end use of the product and these vary from low to very high temperature and humidity (Misra and Young, 1981). Temperature or moisture gradients can serve stresses during drying which could crack the legume seed. Overhults et al., (1973) observed the physical damage in soybeans during drying, in the form of indentations and cracks.

The influence of rapid change of moisture during drying leads the shrinkage of the seed coat. As the most important mechanism of legume seed cracking, this mechanism will be discussed at the last part of this chapter.

2. THE SEED COAT TENSION AT COTYLEDON SLIDING MOVE OR AT PLASTIC SHAPE DEFORMATION

With the increase of moisture content the legume seed becomes as a plastic body. Even a slight force, which does not produces a negative biological result (Dobrzański, 1993b), may cause considerable deformations of the shape. Davison et al. (1979) presented a theoretical stress model of rapeseed being compressed between parallel plates. Basing on some simplifications and assumptions concerning on above model and previously reported experimental work (Davison et al. 1975), a value for the isothermal compressibility coefficient of the cotyledons was obtained. This value compared with that for water and vegetable oil tends to support the validity of the model and the assumption that the cotyledons mainly as a liquid within the seed shell. These results suggest that the compressibility of cotyledon in relation to the large shape deformation is low and in geometrical calculations could be ignore.

The model of shape deformation of legume seeds requires some simplifications and application of several assumptions, similar as were used in the rapeseed model previously reported by Davison et al., (1979).

For this model the following assumptions are use:

- The seed is a sphere surrounded by elastic shell.
- The compressibility of the legume seed's cotyledon is small and it can be assume that the volume is constant.
- The seed coat thickness is small in relation to a diameter of seed, that in geometrical dependencies of a shape deformation could be ignored.
- Compression of most legume seeds between parallel plates caused the change of shape of spherical seed into a barrel body. However, a lateral

area of the body created after small deformation is circular, while for large displacement the barrel obtains a parabolic shape.

If deformation of the seed is slight, a sphere surrounding the cotyledons do not change a spherical shape. Those, a lateral surface of the barrel created by compression of both opposite cups of the sphere should be spherical. Using the assumption of a constant volume, some geometrical calculations are need for determination of the lateral area of spherical body being compressed between parallel plates.

In initial phase of compression a shape deformation of plastic cotyledon of moist seed is observe. Larger axial compression of the seed leads a further deformation of the cotyledons, which causes the seed coat filling and its elongation. Because of the way of compression, the highest stress of the seed coat during elongation appears in the middle of distance between the plates. Hence the tension strain or elongation, is calculate at the longest barrel circumference as a ratio of the increase of barrel diameter D_z.

In this way, the problem consists in determining of a diameter of barrel formed after displacement of the value *d*. The lateral area of the barrel is describes by a circle equation of the radius *R* and centre (0,s′) co-ordinates; presented on figure 1.

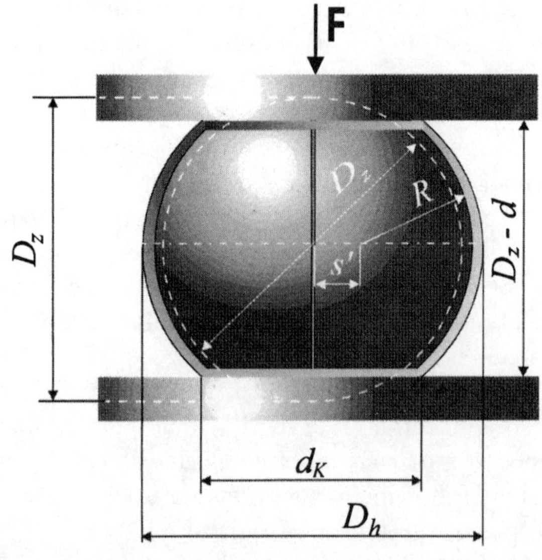

Figure 1. The shape deformation of plastic seed at axial compression. D_z – external diameter of the seed before deformation, D_h – the largest horizontal diameter of compressed seed, d_k –diameter of contact area, *R* – radius of lateral area of spherical barrel, *s'* – circle centre of a cross-section of lateral area of spherical barrel, *d* – axial displacement

Wet seeds consist of low compressible cotyledons. According to the assumption of a constant volume, a diameter of spherical barrel D_K created from the seed after shape deformation, can be determine from an equality of equivalent volume using a simple geometric dependence:

$$D_K = \sqrt{\frac{D_z^3}{D_z - d} - D_z d - \frac{1}{6}d^2} \tag{1}$$

For large axial displacement, a spherical body created from the cotyledons surrounded by the seed coat changes the shape into a barrel with parabolic lateral area. According to the model previously reported by Dobrzański, (1993a) the largest diameter D_P of parabolic barrel can be determines using following formula:

$$D_P = \frac{1}{4}\left(\sqrt{2D_z d + \frac{1}{3}d^2 + \frac{20D_z^3 + 6(D_z - d)\left(2D_z d + \frac{1}{3}d^2\right)}{D_z - d}} + \sqrt{2D_z d + \frac{1}{3}d^2}\right) \tag{2}$$

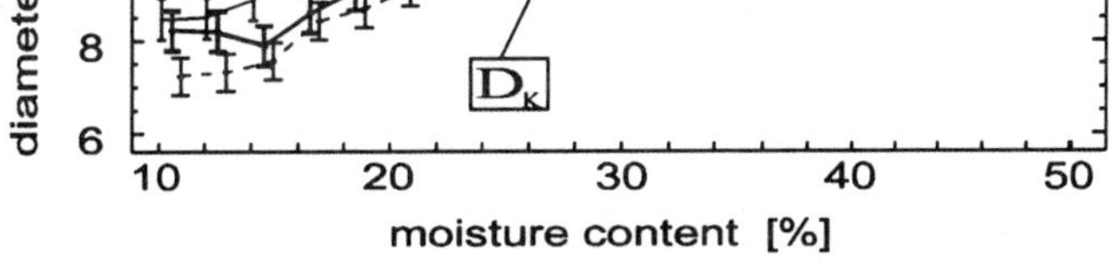

Figure 2. The largest horizontal diameter of the seed coat at cracking. D_h – seed diameter determined with the extensometer strain gauge, D_K – diameter of the seed after deformation into a circular barrel, D_P – diameter of the seed after deformation into a parabolic barrel.

In all cases, while large deformation of the seed coat caused the seed coat breakage, axial displacement d and diameter of the seed before compression D_z is sufficient in determination of barrel diameters D_K and D_P. Both diameters D_K and D_P determine in above methods and diameter D_h obtain as a result of the measurement with 2630-107 Instron strain gauge, was previously presented by Dobrzański, (2000) for various legume seeds. However, the figure 2 presents the diameters: D_K, D_P, and D_h, determined during compression of "Diament" pea seed presents as representative results for spherical seeds.

It is easy to conclude, that experimental results are well fits to the circular model at the range from 14 % to 24 % of moisture content. At the moisture in the range over those values the parabolic model is more closely to extensometric measurement. On the other hand, the barrel diameter of the seed determined at compression and plastic deformation, allows to calculate the strain of the seed coat ε_o.

The stress of the seed coat can be calculated from the Lamé equation concerning the radial strain and stress of a pipe subjected to the inner and outer pressure. Insignificant thickness *g* of the seed coat and the absence of the outer pressure allow to reduce the equation. All calculating procedures in detail was described in the previous reports (Dobrzański, 1991; 1998).

The stress at initial phase of compression of a plastic cotyledon compensates the shape deformation. Further cotyledon's deformation causes the filling of the seed coat. When this process is complete an increase of the inner pressure is observe. Assuming, that approximately pressure necessary for the shape deformation during compression is constant, the increase of the inner pressure Δp_w resulting the seed coat tensions up to the seed coat breakage, can be determine from the slope of a linear part of the force-deformation curve.

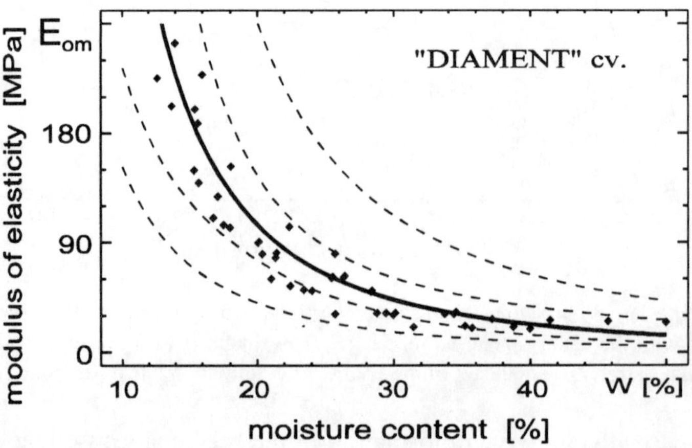

Figure 3. Modulus of elasticity of the seed coat determined at axial compression of the seed. The multiplicative model $E_{om} = 10^4\, 7.3W^{-2.18}$ (R=-0.96) presents the negative influence of moisture content on the seed coat elasticity.

Substituting characteristic values resulting form the force-deformation curve, and the diameter of barrel, calculated according to the formula (1) or (2), the modulus of elasticity of the seed coat E_{om} is determine from the following equation:

$$E_{om} = \frac{D_h \Delta p_w}{\Delta D_h}\left(\frac{(0.5D_h + g)^2 + (0.5D_h)^2}{(0.5D_h + g)^2 - (0.5D_h)^2} + \nu\right) \tag{3}$$

Figure 3 presents the multiplicative regression for "Diament" cultivar. A negative influence of the seed coat moisture on modulus of elasticity, similar for all legume seeds, is observe.

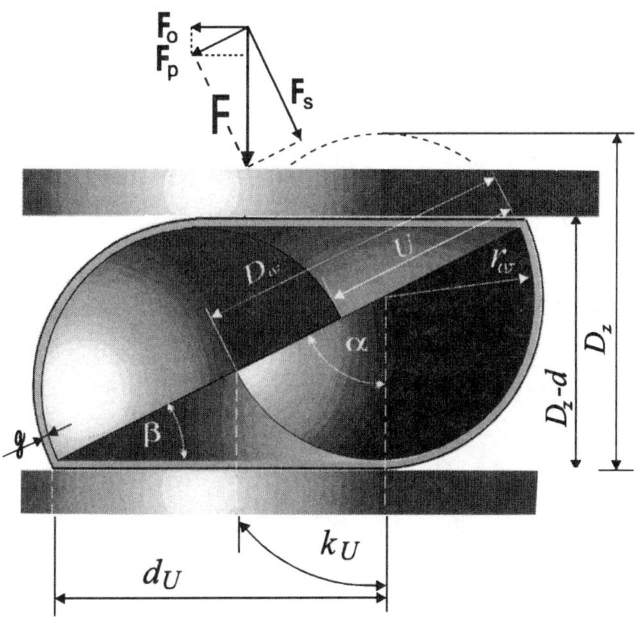

Figure 4. Geometrical dependencies of the cotyledons at sliding displacement and tension of the seed coat. α – sliding displacement angle, β – sliding plane inclination angle, D_z – external diameter of the seed before cotyledon's slide, D_w – internal diameter of the seed, r_w – internal radius of the seed coat, k_U – non-contact sector of a circle, d_U – diameter of contact area, d – axial displacement, g – thickness of the seed coat, U – segment of sliding displacement, F – axial force, F_p – the component of sliding force, F_s, F_o – other components of sliding force

Large deformation of a non-compressibility wet cotyledons involve the change of shape, however, at axial strain up to 0.6, the rupture of seed coat was not observed (Dobrzański, 1993a). Sometimes, sliding movement along the plane contact of the cotyledons caused its rapid displacement. The body created at this process covers two cotyledons and two empty spaces (Fig. 4).

The cotyledons sliding each other along the plane of contact, created the body of a new shape. The lateral area surrounded by the seed coat, is larger than lateral area of the seed before deformation. It caused elongation and in consequence tension of the seed coat. Tension of the seed coat involved the stress, which is probably the main reason of the seed coat breakage at compression of whole seed. Elongation of the seed coat and geometrical dependencies presented above in figure 4 allows to calculate a value of the strain ε_o, from the following equation:

$$\varepsilon_o = \frac{\pi r_w - k_U + d_U}{\pi r_w} \tag{4}$$

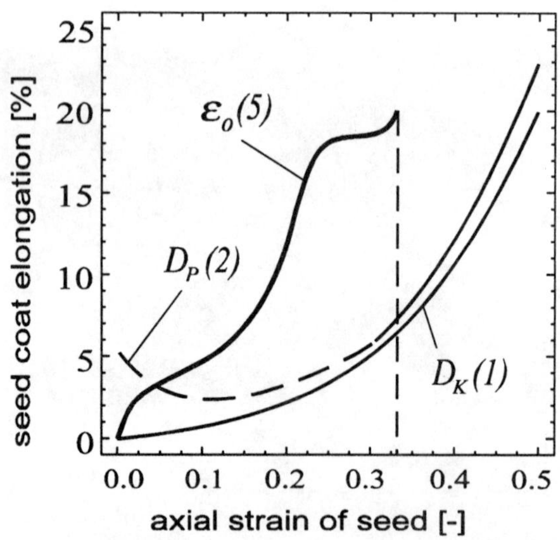

Figure 5. The seed coat elongation at axial compression of grain legume. ε_o – elongation at cotyledon's sliding displacement, D_K – diameter increase at shape deformation of the seed into a circular barrel, D_P – diameter increase at shape deformation into a parabolic barrel

Some geometrical dependencies of a cross-section of the body created by cotyledons sliding are substitute into formula (4), for determination of the elongation of the seed coat, as its strain at largest diameter:

$$\varepsilon_o = 1 - \frac{1}{180}\arccos\frac{D_z - 2d}{D_z} + 2\frac{\sqrt{\left(\frac{D_z}{2} + \frac{D_z d}{D_z}\right)^2 - \left(\frac{D_z}{2}\right)^2}}{\pi D_z} \tag{5}$$

The compression of seed leads to the shape deformation. However, a slight increase of lateral surface, expressed by elongation of diameter (Fig. 5), at low axial strain not more than $\varepsilon<0.3$ proves, that at this range the seed coat is not tensioned. In comparison, a sliding move of the cotyledons along the contact plane creates the body, which cause a tension and larger elongation over 10 %, just at the value 0.2 of the axial strain.

The above models enable the interpretation of the force-deformation curve for large displacement where the mechanical strength of non-elastic wet seeds is mainly connect with the resistance of the seed coat to tension.

3. THE SEED COAT SHEARING AT CRACKING OF DRY SEED IN TWO SEPARATE COTYLEDONS

The shearing of dry seed coat is the most frequently cracking mechanism observed in practice during processing of dry bean. It may be result of threshing cylinder or rotor motion, centrifugal discharging, the filling and discharging of screw conveyors, spouting and free-fall dropping (Palusen *et al.*, 1981a). Other grain legume such as: pea seed, french bean, soybean and fababean, as the bean consists of the seed coat and two separate cotyledons, slightly connected with a sprout only. The shearing of the seed coat being most influential component of a cracking force depends on the strength of seed coat to splitting.

Legume seed cracking model at shearing of the seed coat has been previously reported by Dobrzański, (1998). For this model the following assumptions are use:

- The seed consists of two separate elastic cotyledons surrounded by elastic sphere of the seed coat.
- A hard seed at low moisture is an elastic body. In this case, a slight deformation of the cotyledon at the seed coat shearing is insignificant and can be ignore.

After cracking, the shearing force parallel to the plane of contact of the cotyledons divides the seed into separate parts. Each part consists of one cotyledon and a half sphere of the seed coat (Fig. 6).

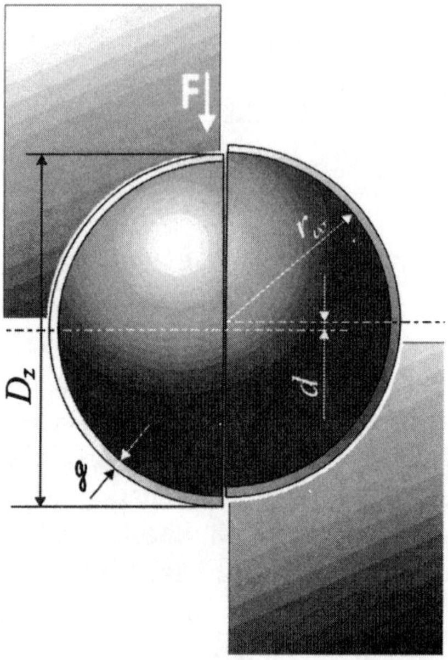

Figure 6. Shearing of the seed coat at cracking of dry seed. D_z – external diameter of the seed before cotyledon's slide, r_w – internal radius of the seed coat, d – axial displacement, g – thickness of the seed coat, F – axial force

For oblong seeds like bean, french bean, and soybean, a cross section area A_s of the seed coat subjected to shearing is determine in a simple way according to the following formula:

$$A_s = \pi(bg - g^2) + 2(a - b)g \quad (6)$$

On the other hand, most spherical seeds like pea seeds are convex with a circular shape of a cross-section. Basing on that, both dimensions a and b are equal to external diameter of the seed D_z, the cross-section area A_s of the seed coat subjected to shearing take the form as follow:

$$A_s = \pi(D_z g - g^2) \quad (7)$$

Determination of the cross section area A_s of the seed coat subjected to shearing allows to estimate approximately stress at cracking:

$$\tau = \frac{F}{A_s} \tag{8}$$

At this process, at the shearing of dry seed coat, the cracking mechanism leads to force-displacement curve, which is rectilinear in a whole range. Basing on trigonometric dependence of triangle of the maximum force F and its respective displacement d, the work of crack L represents the formula (9):

$$L = \frac{Fd}{2} \tag{9}$$

Both values are determine with experimental stand presented on the figure 6. According to the energy conservation law, a kinetic energy of seed impact is equal to cracking work at shearing of the seed coat. After rearranging, presented below formula allows to predict threshold velocity V_p for the grains of known mass m, which did not cracks the seed coat:

$$V_p = \sqrt{\frac{Fd}{m}} \tag{10}$$

Table. 1. Threshold velocity V_p [m/s] for the seeds of different mass and strength of the coat

Species	Seed mass			Seed coat strength	
Cultivars	mean	max.	min.	weak	strong
Faba bean					
Nadw.	13.65	12.49	14.65	11.31	16.68
Bronto	12.80	12.12	13.55	11.26	14.80
Dino	10.09	9.35	11.03	9.12	10.96
Bean					
Blanka	4.61	4.32	5.06	2.59	6.43
Nida	4.95	4.62	5.19	3.32	6.35
Segal	7.62	6.61	8.22	6.10	8.82
Longina	5.30	4.98	5.99	3.66	6.55
Pea seed					
Agat	12.61	12.08	13.46	7.63	15.17
Diament	14.84	14.09	16.40	7.40	20.28
Sześcio.	12.57	11.84	13.25	6.43	17.06
Soybean					
Nawiko	20.92	19.22	21.98	6.00	38.74
Polan	25.64	23.77	28.33	4.27	45.50

Using a quasi-static test, it is possible to estimate the resistance of seed to cracking. Differentiation of speed limits at thrashing for numerous legume seeds is shown in Table 1.

Simple method requires the application of several assumptions and covers some simplifications, however, in this a way a threshold velocity V_p is determine at thrashing process. This method should be useful for prediction of the rotary rate, safe for the seed or the limit of impact at other dynamic processes, frequently meets in practice.

4. COTYLEDON CRACKING - THE MECHANISM OF LEGUME SEED DAMAGE AT COMPRESSION

In the previous study concerning on the seed deformation at compression, Dobrzański, (1998; 2000) observed, that the force-displacement curve is a result of many accidental processes. During compression of dry seed the cracking force is notice at small displacement. It caused an internal crack of the cotyledon, which was frequently invisible. However, the crack observed as a rapid decrease of force on the force-displacement curve, appears as a laud crunch. To clarify the origin of cracks of the seeds of low moisture content, it is necessary to carefully recognise the mechanism of cotyledon cracking separately, except the processes of the seed coat splitting.

The model of crack opening displacement as the mechanism of cotyledon damage at compression of legume seed has been previously reported (Dobrzański, 1998). For this model the following assumptions are use:

- At small deformation and low moisture a hard cotyledon is an elastic body, while at large displacement of the cotyledon the spherical cap is subjected to plastic strain.
- The compression of a half sphere, which consist of legume cotyledon free from the seed coat, being compressed between parallel plates, gave axi-symmetrical strain.
- A large contact area of the cotyledon with a flat plane of support assures a longitudinal top deformation only.

According to the model describing cotyledon cracking mechanism, compression of the cotyledon caused close packing of internal structure of spherical cap. The central part of high density involved a breakage, which appears as the crack opening displacement.

This process caused one invisible crack (Fig. 7a) at elastic range of deformation and one or more (Fig. 7b-7d) crack opening displacements at large deformation ($\sigma_{cod}>R_m$).

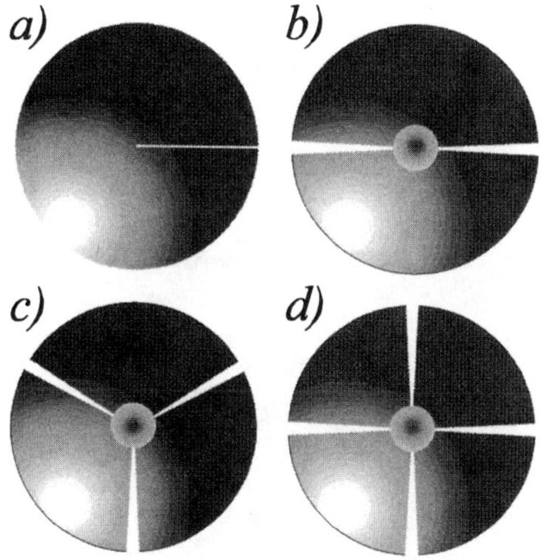

Figure 7. Different cracks at compression of the cotyledon. (a) – crack at small elastic deformation, (b, c, d) – cracks at large deformation

In this case, reuse of the energy conservation law, allows to determine the stress σ_{cod} at cracking of the cotyledon. The work performs at deformation of spherical top cap of the cotyledon L_{dp} is changing into the work of crack opening displacement L_{cod}.

The work of deformation of the cotyledon spherical cap L_{dp} is represented by following formula:

$$L_{dp} = \frac{(F - F_s)d_p}{2} \tag{11}$$

This work depends on the force of upsetting of the cotyledon spherical cap. The close packing of internal core involve a bursting pressure leading the crack opening displacement. According to the figure 8 the work of crack opening displacement L_{cod} is describes by the formula (12):

$$L_{cod} = \frac{\sigma_{cod} A_{cod} d_{cod}}{2} \tag{12}$$

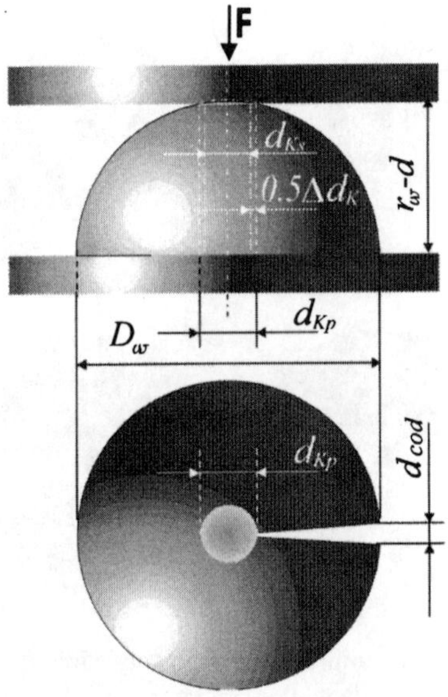

Figure 8. Crack opening displacement d_{cod} at axial compression of legume seed's cotyledon. D_w – diameter of the seed, r_w – radius of the cotyledon, d_{Kp} – diameter of contact area at plastic deformation, d_{Ks} – diameter of contact area at elastic deformation, d – axial displacement, F – axial force

Equating both formulas (11) and (12), and after transformation, the average stress σ_{cod} at cotyledon cracking is represented by:

$$\sigma_{cod} = \frac{(F - F_s)d_p}{A_{cod}d_{cod}} \tag{13}$$

where: A_{cod} – the total area of cracks, D_w – diameter of seed, d_{Kp} – diameter of contact area, $(F\text{-}F_s)$ – force increase, d_{cod} – crack displacement.

Depending on the number of crack's n and area of each crack, the total area of crack surface A_{cod} obtained the following form:

$$A_{cod} = \left(\frac{\pi r_w^2}{4} - \frac{d_{Kp} r_w}{2}\right)n \tag{14}$$

where: r_w=0.5D_w and diameter d_{Kp} is presented by the following formula:

$$d_{Kp} = 2\sqrt{d_p(D_w - d_p)} \tag{15}$$

Finally, unknown total area of crack A_{cod} for axial displacement d_p determines a simple geometrical relation:

$$A_{cod} = \frac{nD_w}{4}\left(\frac{\pi D_w}{4} - 2\sqrt{d_p(D_w - d_p)}\right) \tag{16}$$

On the other hand, crack displacement on the perimeter of cotyledon base d_{cod} (Fig. 8) determines the geometrical proportions:

$$\frac{\pi D_w + d_{cod}}{D_w} = \frac{\pi(d_{Kp} + \Delta d_K)}{d_{Kp}} \tag{17}$$

From above equation after transformation, the crack opening displacement d_{cod} finally represents the following formula:

$$d_{cod} = \pi D_w\left(\frac{d_{Kp} + \Delta d_K}{d_{Kp}} - 1\right) \tag{18}$$

The contact area with a top plate and the increase of its diameter Δd_K is still unknown, however, the approximately value can be estimate using assumption of a constant volume of the cotyledon at deformation. In this case, the upsetting volume of the cotyledon's spherical cup V_{cz}:

$$V_{cz} = \frac{\pi}{2}\left(\frac{d_{Kp}}{2}\right)^2 d_p + \frac{\pi}{6}d_p^3 \tag{19}$$

is equal to the increase of volume of the close packing of the cotyledon internal core, which involve a bursting pressure and crack opening displacement. The volumetric increase of internal core represents V_{cod}:

$$V_{cod} = \pi\left(\frac{D_w}{2} - d_p\right)\left[\left(\frac{d_{Kp} + \Delta d_K}{2}\right)^2 - \left(\frac{d_{Kp}}{2}\right)^2\right] \tag{20}$$

Equating both formulas (19) and (20) the increase of diameter Δd_K of the contact area obtain the following form:

$$\Delta d_K = \sqrt{\frac{2d_p\left[\left(\frac{d_{Kp}}{2}\right)^2 + \frac{1}{3}d_p^2\right]}{\frac{D_w}{2} - d_p} + d_{Kp}^2} - d_{Kp} \tag{21}$$

Substituting into the formula (21) the diameter of contact area d_{Kp} (15), its increase Δd_K depends only on the axial deformation d_p and initial diameter of the seed D_w. Finally, the increase of diameter Δd_K presents the formula (22):

$$\Delta d_K = \sqrt{\frac{2d_p\left[(D_w - d_p)d_p + \frac{1}{3}d_p^2\right]}{0.5D_w - d_p} + 4(D_w - d_p)d_p} - 2\sqrt{(D_w - d_p)d_p} \tag{22}$$

The crack opening displacement d_{cod} at axial deformation d_p allows to determine the increase of close packing core of the cotyledon as horizontal strain ε_l:

$$\varepsilon_1 = \frac{\Delta d_K}{d_{Kp}} \tag{23}$$

and shortening as axial strain ε:

$$\varepsilon = \frac{D_w - 2d_p}{D_w} \tag{24}$$

Dividing the formula (23) by (24), both formulas presented on the previous page, and after rearranging, the Poisson's ratio ν can be determine using below form:

$$\nu = \frac{\Delta d_K D_w}{d_{Kp}(D_w - 2d_p)} \tag{25}$$

After substituting for Δd_K and d_{Kp} formulas (21) and (15) previously presented, the Poisson's ratio ν determined in this way, depends on axial deformation d_p of the seed and its diameter D_w only. Both diameter's d_p and D_w are determine experimentally at compression of the cotyledon.

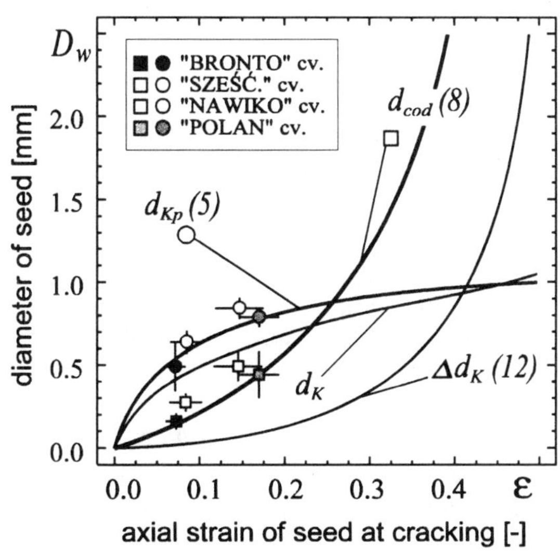

Figure 9. The crack opening displacement – d_{cod}, diameter of contact area – d_{Kp} and its increase – Δd_K determine for legume seed's cotyledon at axial compression at 100 mm/min ("Bronto" fababean, "Sześciotygodniowy" pea seed, "Nawiko" and "Polan" soybean)

According to the formulas (18), (15), and (22) the crack opening displacement d_{cod}, diameter of contact area d_{Kp}, and increase of diameter Δd_K, determine for the cotyledon of legume seeds are shown on the figure 9. The largest value of crack opening displacement d_{cod} is observe at high rate of cross head move (100 mm/min).

The soybean cracks at larger axial deformation of the cotyledon, than other legume seeds. The highest value of strain for the soybean indicates a plastic behaviour of the cotyledon. The behaviour of soybean depends on high level of fats, what explains the plasticity of soybean.

The compression of dry soybean at low rate of crosshead move (5 mm/min) caused a visible crack at axial strain under 0.03. For more elastic cotyledon of pea seed and fababean the cracks appears at higher force, however, at low axial strain close to 0.01.

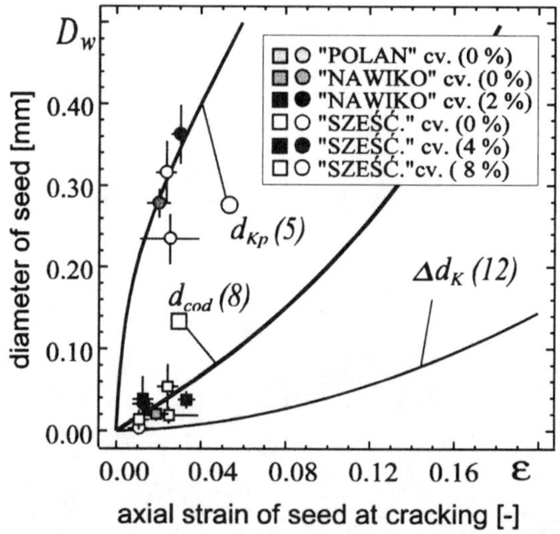

Figure 10. The crack opening displacement – d_{cod}, diameter of contact area – d_{Kp} and its increase – Δd_K determine for legume seed's cotyledon at axial compression at 5 mm/min of cross head move ("Sześciotygodniowy" pea seed, "Nawiko" and "Polan" soybean)

The experimental results being well fitted with theoretical curves, verify the model describing the cracking mechanism of legume seed's cotyledon at compression and clarify the reason of dry seed damage.

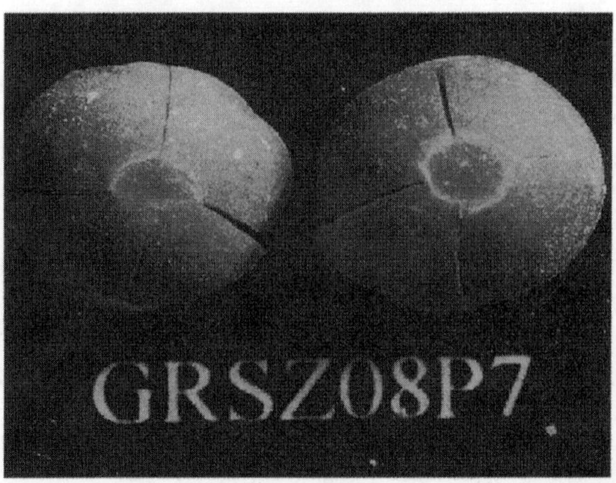

Figure 11. The circular contact area and symmetric cracks around the cotyledon core of "Sześciotygodniowy" pea seeds compressed at 20 mm/min of cross head move

For large displacement, regular cracks around the core of the cotyledon and a circle contact area are observe (Fig. 11). However, deformation at axial compression of partly concave fababean – a not completely convex cotyledon (Fig. 12), causes cracks spaced irregular around the circle of the contact area, which are located to the nearest board of the base.

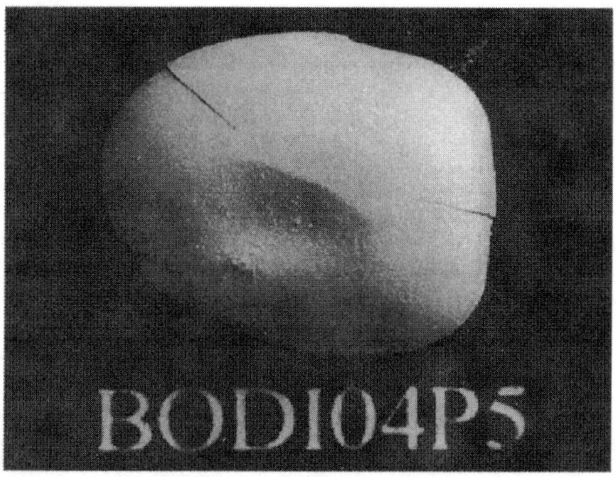

Figure 12. Longitudinal contact area and non-symmetric cracks around the cotyledon core of partly concave "Dino" fababean compressed at 5 mm/min of cross head move

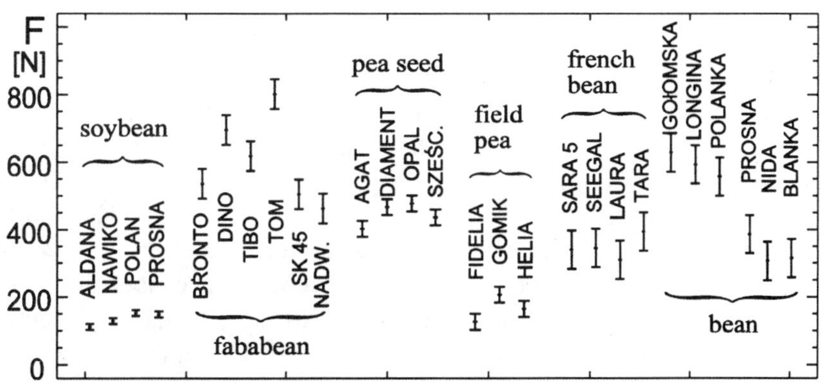

Figure 13. The breakage force of the legume seed's cotyledon at axial compression

The results obtained at compression test indicate different damage force of each cultivar, especially of most fababean. The cotyledons of all studied cultivars of fababean are very diversified. The force noticed at breakage of

"*Tom*" fababean is two times higher than damage force of "*Nadwiślanski*" fababean (Fig. 13). It looks apparent that mechanical strength of the seeds is

also different. On the other hand, mechanical resistance of fababean is from 5 to 8 times higher than the resistance of pea seeds or soybeans.

However, the modulus of elasticity is at the same level and there are insignificant differences at p=0.05 for most cultivars (Fig. 14). The values obtained at compression of dry seeds of most cultivars indicate homogeneity material of the cotyledon. The modulus of elasticity determined at small deformation reveal elastic behaviour of the cotyledon for most seeds.

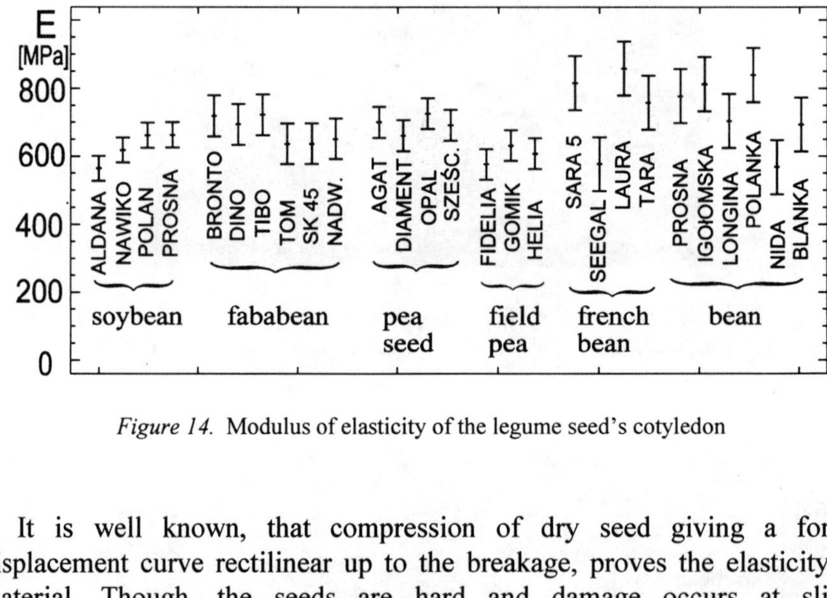

Figure 14. Modulus of elasticity of the legume seed's cotyledon

It is well known, that compression of dry seed giving a force-displacement curve rectilinear up to the breakage, proves the elasticity of material. Though, the seeds are hard and damage occurs at slight deformation in the range of displacement from 0.1 mm to 0.3 mm. All legume seeds at moisture content less 10 % is as elastic body and the modulus of elasticity should be calculated according to the Hertz theory.

For low rate of axial deformation, the compression causes one crack, frequently invisible after unloading. However, for large axial displacement, regular cracks of the cotyledon every 180^0, 90^0, or 60^0 around the core are observe.

A study of the cotyledon breakage during compression clarifies the mechanism of cracking of dry seed, which is describe by the model of crack opening displacement. Generally, the origin of damages of dry grain legume at slight deformation depends on cracking process of the cotyledon.

5. THE SHRINKAGE AND STRESS OF THE SEED COAT DURING DRYING OF GRAIN LEGUME

Temperature and moisture gradients can cause stresses in the seed coat during drying. Overhults *et al.* (1973) observed the physical damage in soybeans during drying, as indentations and cracks. This condition makes the beans susceptible to microbial attack during storage, moreover reduces their germination potential (White *et al.*, 1980). In view of these problems, there is a need, especially in the legume seed industry, to find the optimum combination of drying conditions which will minimise seed coat cracking (Mensah *et al.*, 1984). The major problem in the drying of legume seeds with heated air is the splitting of the seed coat (Liu *et al.*, 1989; 1990). However, in the study of stress cracking due to drying of grains, the moisture content decrease is essential to understand its mechanisms and variation of modulus of elasticity at different levels (Dobrzański and Szot, 1997; Dobrzański, 2000; Herum *et al.*, 1979; Misra et al., 1981; Otten et al., 1984). Looking for the mechanism responsible on the splitting of seed coat, a determination of the elasticity strain limit of the seed coat and its resistance to tension for various legume seeds were studied.

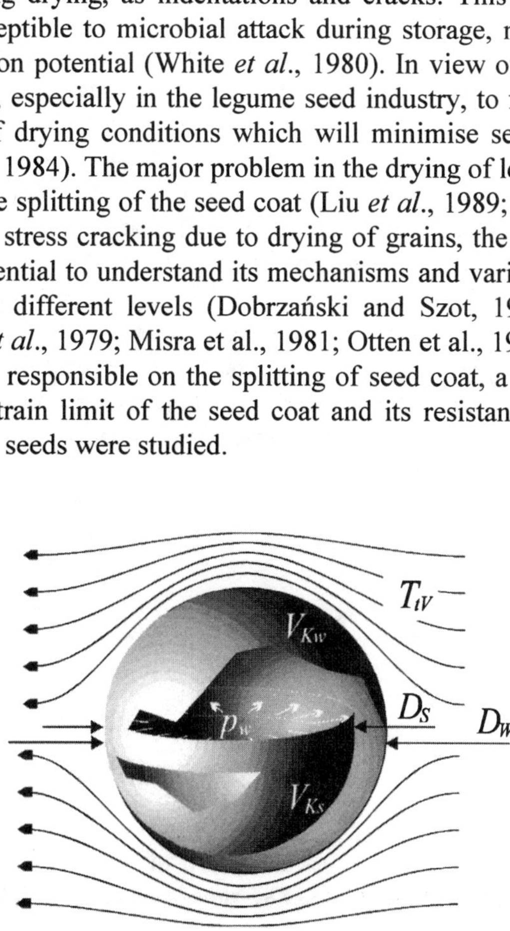

Figure 15. Shrinkage of the seed coat in the stream of air of known temperature and velocity – T_{tV}. D_W – diameter of the seed coat surrounding wet grain, D_S – diameter of the seed coat surrounding dry grain, V_{Kw} – the volume surrounded by wet seed coat, V_{Ks} – the volume surrounded by dry seed coat

On the other hand, the effect of moisture content on expansion volume of soybean shown as an experimental result was previously presented by Dobrzański, (1996). Distinctly effect of moisture content on dimensions of soybean was observed in initial phase of swelling. Further, a rapid change of moisture content of the seed coat during drying caused the shrinkage of coat leading its stress. With an increase of moisture content, the seed coat of legume seeds decreases its strength (Dobrzański and Szot, 1997). All results previously reported by Dobrzański, (1998), made the bases for elaboration of the shrinkage model of the seed coat being drying with a hot air.

The figure 15 presents a spherical seed in the stream of air of known velocity and temperature, specified as T_{tV} stream. The wet seed of volume V_{Kw} is surrounded by the seed coat of diameter D_W. After drying, the seed reduces its volume to the value V_{Ks} represented by a sphere of dry seed coat of diameter D_S. Drying of the cotyledon follow slow volumetric changes, while a rapid shrinkage of the seed coat is observe. Those during drying, the seed coat surrounding wet cotyledons is subjected to the shrinkage. A large volume of wet seed leads a stress of the seed coat, involving an increase of inner pressure p_w of low compressibility cotyledons.

The thickness of the seed coat $g \approx 0.01D_z$ being 5 times less than for shell of thin-walled drum, for which the requirement is $g < 0.05D_z$, allows to assume that the seed coat sphere is as a boiler subjected to inner pressure. In this case, each point of spherical shell is represents a plane stress, where the state of principal stress σ_s of both dimensions is equal:

$$\sigma_s = \frac{D_z p_w}{4g} \leq k_r \tag{26}$$

Rearranging of an equation (26) leads to formula of the inner pressure p_w caused by shrinkage:

$$p_w = \frac{4E_o \varepsilon_s g}{D_z} \tag{27}$$

Taking into account that $D_z = D_W$, the shrinkage strain ε_s depend on a difference of the seed diameter $D_W - D_S$, measured before and after drying. It allows to calculate inner pressure p_w:

$$p_w = \frac{4E_o g(D_W - D_S)}{D_W^2} \tag{28}$$

To determine a criterion of tension strength for the seed coat during drying it is sufficient to estimate circumferential stress σ_o in a cross-section of the seed. According to previous paper (Dobrzański, 1993b) the circumferential stress σ_o is determine with the formula:

$$\sigma_o = \frac{p_w(0.5D_W^2 + D_W g + g^2)}{D_W g + g^2} \tag{29}$$

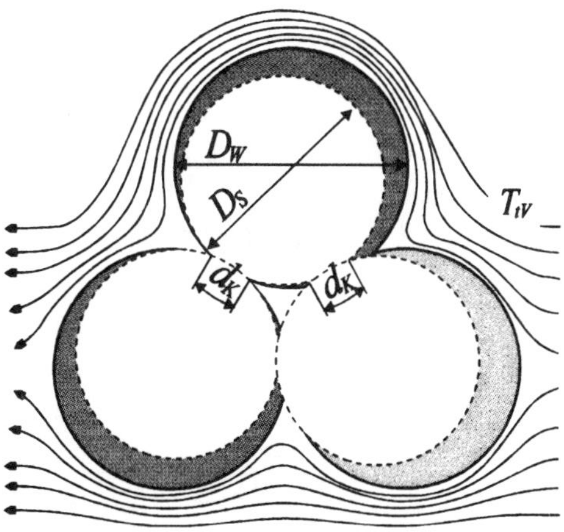

Figure 16. The cross section of seeds on the layer in the stream of air of a known temperature and velocity – T_{tV}. D_W – diameter of wet seed, D_S – diameter of dry seed, d_K – diameter of contact area of seed's pair

To keep safe condition for the seeds at drying, the circumferential shrinkage stress σ_o must fulfil a strength criterion of tension:

$$\sigma_o \leq k_r \tag{30}$$

The figure 16 presents a single seed with whole sphere of the seed coat surrounds a hot stream of air. However, most seeds in real conditions are locate on the layer or the seed being inside the layer is in the contacts to others. It limits the inflow of air to some parts of the seed coat, what keeps a high moisture in those areas. Taking into account, that residual parts of the seed coat being in a contact with hot air, only the parts available for the air flow are subject to the shrinkage.

In this case, the shrinkage of dry part leads to tension of wet part of the seed coat. Diameter of the seed coat decreases from D_W to D_S. However, only a part of the seed coat (D_W - d_K) subjects to the shrinkage cause a tension of wet part of the seed coat (Fig. 16). It allows to determine the shrinkage strain ε_s of dry part of the seed coat:

$$\varepsilon_s = \frac{\pi(D_W - D_S)}{\pi D_W - d_K} \tag{31}$$

The shrinkage leads to the stress σ_s, which depends on the modulus of elasticity of dry seed coat E_{oS} :

$$\sigma_s = E_{oS}\varepsilon_s \tag{32}$$

The shrinkage of dry seed coat causes the elongation ε_o of wet part represented by the segment d_K subjected to tension. Determination of the strain of wet part of the seed coat ε_o and its elasticity modulus E_{oW}, allows to estimate tension strength σ_u and the tension criteria using formula (33):

$$\sigma_u = E_{oW}\varepsilon_o \tag{33}$$

The safe condition of legume seeds drying must fulfil inequality:

$$\sigma_u > \sigma_s \tag{34}$$

Because drying with the air cause the shrinkage of a part available for the stream of air only, a second condition for safe shrinkage ε_s must be fulfil:

$$\varepsilon_s > \frac{2d_K\varepsilon_o}{\pi D_W - 2d_K} \tag{35}$$

For this purpose, the shrinkage strain ε_s, ultimate elongation at rupture ε_o and modulus of elasticity E_o must be determined experimentally at wide range of moisture.

The results obtained at tension of the seed coat showed that the increase of moisture content involves the decrease of force at breakage. Cultivar's differentiation of the seed coat strength in tension test was observed. Frequently, the results obtained in all experiments concerning on the strength of seed coat proved that shrinkage stress σ_s was higher than tension strength σ_u of the seed coat (fig. 17).

The rapid decrease of air temperature during drying causing additional shrinkage of the seed coat for 25 % more, clarifies the origin of the coat cracking after disconnection of the heating set in dryers. This phenomenon was observed for all legume seeds.

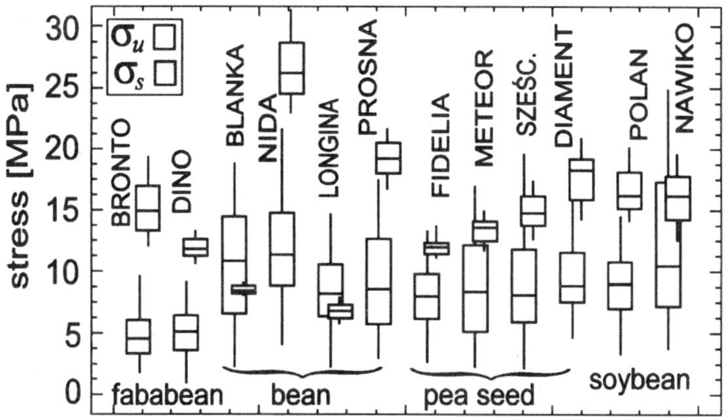

Figure 17. The tension strength σ_u and shrinkage stress σ_s of the seed coat for various legume seeds

The shrinkage of the seed coat during drying was more than 21 % for all studied cultivars (Fig. 18). The highest shrinkage for all cultivars at most severe drying condition was observed. The value of force induced by the shrinkage depended also on drying conditions. Drying of the seed coat at temperature of 62 ^{0}C gave the highest shrinkage stress. Increase of moisture content leads to a decrease of force at breakage. Shrinkage of the seed coat at drying leads to its stress and splitting.

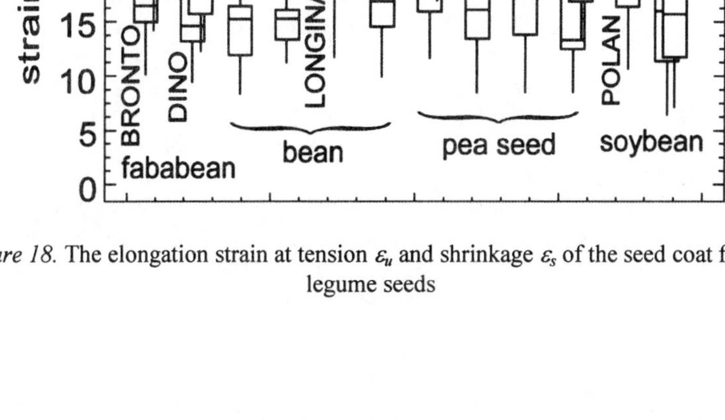

Figure 18. The elongation strain at tension ε_u and shrinkage ε_s of the seed coat for various legume seeds

Theoretical considerations allow to verify the mechanism of seed coat cracking and the model describing geometrical relations during shape deformation and drying. The modulus of elasticity of the seed coat was determined at compression or at shape deformation. These methods enabled the interpretation of the force-deformation curve for the seed of moisture over 12 %. On the other hand, the axial strain of the seed, under 30 %, leads a shape deformation without the seed coat stress. For large deformation of wet seeds mechanical strength is mainly connected with the resistance of seed coat to tension. Sliding of the cotyledons is the most important reason of the seed coat cracking for the seed of moisture content overlarge 20 %.

Shrinkage of the seed coat at drying leads to its stress and splitting. The results obtained in all experiments proved that frequently the stress σ_s was higher than strength of seed coat. Probably, it is the most important reason of the seed coat cracking during drying.

REFERENCES

Anazodo, U.G.N. and Chikwendu, S.C., 1983, Poisson's ratio and elastic modulus of radially compressed biomaterials - I: Small deformation approximation. *Transaction of the ASAE,* **25**(3): 923-929.

Arnold, P.C. and Mohsenin, N.N., 1971, Proposed techniques for axial compression tests on intact agricultural products of convex shape. *Transaction of the ASAE,* **14**(1): 78-84.

Bilanski, W.K., 1988, Mechanical properties of soybeans. *Physical Properties of Agricultural Materials and Products*. Hemisphere Publ. Co., 363-368.

Chikwendu, S.C. and Anazodo, U.G.N., 1984, Poisson's ratio and elastic modulus of radially compressed biomaterials -II: Large deformation approximation. *Transaction of the ASAE,* **26**(6): 1563-1572.

Davison, E., Meiering, A.G., and Middendorf, F.J., 1979, A theoretical stress model of rapeseed. *Canad. Agric. Eng.*, **21**(1): 45-46.

Davison, E., Middendorf, F.J., and Bilanski, W.K., 1975, Mechanical properties of rapeseed. *Canad. Agric. Eng.*, **17**(1): 50-54.

Dobrzański, jr.B. and Szot, B., 1989, Strength of pea seeds during compression. *Proceedings of the 4-ICPP of Agricultural Materials*, Rostock, Germany, vol.**1**: 144-148.

Dobrzański, jr.B., 1991, Theoretical model of large deformation of compressed rapeseed. *Proceedings of 8th GCIRC Rapeseed Congress*, Saskatoon, Canada, vol.**4**: 1261-1266.

Dobrzański, jr.B. and Szot, B., 1992, Mechanical properties of pea seed. *Agricultural Engineering and Rural Development*, A Pergamon-CNPIEC Joint Venture, Beijing, China, II, D, 12-16.

Dobrzański, jr.B., 1993a, Geometrical dependence of seed shell of compressed pea. *Int. Agrophysics*, **7**(4): 259-264.

Dobrzański, jr.B., 1993b, Germination capacity of compressed pea seed. *Zesz. Prob. Post. Nauk Rol.*, **399**: 31-34.

Dobrzański, jr.B., 1996, The influence of moisture content on the shape deformation of soybean. *AgEng'96*, Madrid, Paper 96F-082, 1-7.

Dobrzański, jr.B. and Szot, B., 1997, Strength of pea seed cover. Int. Agrophysics, **11**(4): 301-306.

Dobrzański, jr.B., 1997, Elasticity strain limit for legumineous seeds. *Book of Abstracts of 6th ICA*, vol.**1**: 29-32.

Dobrzański, jr.B., 1998, The analysis of seed cracking mechanisms of grain legume. *Acta Agrophysica*, **13**: 1-98. - (in polish)

Dobrzański, jr.B., 2000, The cracking mechanisms of the legume seeds. CD-ROM of *AgEng'2000*, Warwick, paper: 00-PH-050, 1-10.

Dorell, D.G. and Adams, M.W., 1969, Effect of some seed characteristic on mechanically included seedcoat damage in navy beans. *Agron. Journal*, **61**(5): 672-673.

Fornal, J., Kaczyńska, B., and Gryzowska, A., 1990, Physical properties of faba bean seeds and their significance in particular technological operations. *10th Int. Congress CHISA*, Prague.

Fornal, J., Sadowska, J., and Kaczyńska, B., 1993, Damage of faba bean seeds during drying. *Drying Technology*, **11**(6): 1293-1309.

Herum, F.L., Mensah, J.K., Barre, H.J., and Majidzadech, K., 1979, Viscoelastic behaviour of soybeans due to temperature and moisture content. *Transaction of the ASAE*, **22**(5): 1219-1224.

Hoki, M. and Pickett, L.K., 1973, Factors affecting mechanical damage of navy beans. *Transaction of the ASAE*, **16**(6): 1154-1157.

Liu, M., Haghighi, K. and Stroshine, R.L., 1989, Viscoelastic characterisation of the soybean seedcoat. *Transaction of the ASAE*, **32**(3): 946-952.

Liu, M., Haghighi, K., Stroshine, R.L., and Ting, E.C., 1990, Mechanical properties of the soybean cotyledon and failure strength of soybean kernels. *Transaction of the ASAE*, **33**(2): 559-566.

Mac Donald, jr.M.B., 1985, Physical seed quality of soybean. *Seed Sci. Technol.* **13**: 601-628.

Mensah, J.K., Nelson, G.L., Herum, F.L., and Richard, T.G., 1984, Mechanical properties related to soybean seedcoat cracking during drying. *Transaction of the ASAE*, **27**(2): 550-560.

Misra, R., Young, J.H., and Haman, D.D., 1981, Finite element procedures for estimating shrinkage stresses during soybean drying. *Transaction of the ASAE*, **24**(3): 751-753.

Misra, R. and Young, J.H., 1981, A model for predicting the effect of moisture content on the modulus of elasticity of soybeans. *Transaction of the ASAE*, **24**(5): 1338-1341.

Otten, L., Brown, R., and Reid, W.S., 1984, Drying of white beans-effect of temperature and relative humidity on seed coat damage. *Canad. Agric. Eng.*, **26**: 101-104.

Overhults, D.G., White, H.E., Hamilton, H.E., and Ross, I.J., 1973, Drying soybeans with heated air. *Transaction of the ASAE*, **16**(1): 112-113.

Paulsen, M.R., 1978, Facture resistance of soybeans to compressive loading. *Transaction of the ASAE*, **21**(5): 1210-1216.

Paulsen, M.R., Nave, W.R., and Gray, L.E., 1981a, Soybean seed quality as affected by impact damage. *Transaction of the ASAE*, **24**(6): 1577-1582, 1589.

Paulsen, M.R., Nave, W.R., Mounts, T.L., and Gray, L.E., 1981b, Storability of harvest-damaged soybeans. *Transaction of the ASAE*, **24**(6): 1583-1589.

Perry, J.S. and Hall, C.W., 1966, Evaluating and reducing mechanical-handling damage to pea beans. *Transaction of the ASAE*, **9**(4): 696-701.

Pickett, L.K., 1973, Mechanical damage and processing loss during navy bean harvesting. *Transaction of the ASAE*, **16**(6): 1047-1050.

Rataj, V., 1994, Poisson's constant determination of selected legumes and oil seeds. *Int. Agrophysics*, **8**(2): 295-298.

Rataj, V., Jech, J., and Ponican, J., 1994, The technological properties of pea seeds. *Int. Agrophysics*, **8**(2): 299-303.

White, G.M., Bridges, T.C., Loewer, O.J., and Ross, I.J., 1980, Seed coat damage in thin-layer drying of soybeans. *Transaction of the ASAE*, **23**(1): 224-227.

Properties of Grain for Silo Strength Calculation

JÓZEF HORABIK and MAREK MOLENDA
Institute of Agropysics, Polish Academy of Sciences, Lublin, Poland

1. INTRODUCTION

Handling bulk solid materials is one of the least understood areas associated with solid processing plants (Knowlton et al., 1994). For that reason, mechanics of granular materials have recently gained growing interest among researchers and industrial practitioners. New theoretical approaches and experimental methods have been developed for procedures such as material characterisation and silo design. Silo design based on Janssen's (1895) equation with necessary material parameters such as bulk density, coefficient of friction on the wall and pressure ratio allows for calculation of wall and floor loads with acceptable accuracy. Although one hundred years old, Janssen's equation is still the most widely used analytical solution for the calculation of pressures in silos. A review of 13 design codes by Wilms (1991) showed that all of them use the Janssen formula to calculate pressure distribution in deep vertical wall silo. The Janssen formula predicts vertical and lateral pressure exerted on a silo by a granular material as an exponential function of the depth of fill:

$$\begin{aligned} \sigma_z &= \frac{\rho g D}{4k\mu}\left(1 - e^{-\frac{4k\mu}{D}z}\right) \\ \sigma_x &= k\sigma_z \end{aligned} \qquad (1)$$

where:

D - silo diameter (m);
g - acceleration of gravity (9.81 m/s^2);
k - lateral to vertical pressure ratio (Janssen's constant);

z - vertical co-ordinate, depth of grain in silo (m);
μ - coefficient of wall friction;
ρ - bulk density (kg/m^3);
σ_x - lateral pressure (Pa);
σ_z - vertical pressure (Pa).

Janssen's derivation considered the equilibrium of forces exerted on a horizontal slice of the granular material in the silo. Janssen assumed that the bulk density, ρ, the ratio of lateral to vertical pressure, k, and the angle of wall friction, φ_w, are constant. To obtain a statically determinable system, he also assumed the vertical pressure was uniform across the horizontal section of the silo. A number of investigations have been undertaken to develop a Janssen type equation assuming that the vertical stress is not constant over the horizontal cross section of the silo. A detailed discussion of the implications of various types of stress distribution in a differential slice on the pressure ratio is given by Drescher (1991).

Design codes (Eurocode 1, 1996) give a list of values of material properties, but as an alternative recommend the test methods for determination of the properties. Current practice assumes the lateral to vertical pressure ratio to be about 0.4 or recommends theoretical equations to calculate the ratio as a function of the material's effective friction angle φ (Lohnes, 1993).

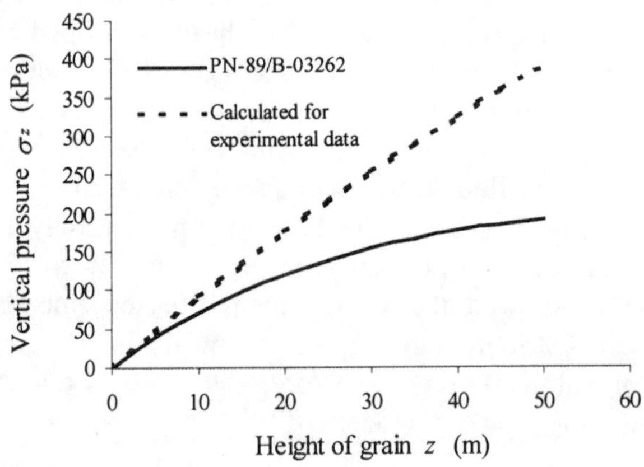

Figure 1. Vertical pressure of rapeseed in the silo as a function of filling height.

Figure 1 illustrates the degree of influence of design parameters on vertical stress σ_z calculated for 20 m in diameter and 50 m high silo. It was assumed that rapeseed is stored in the silo, and following Polish Standard (1989) design parameters were taken such as density $\gamma = 8.5$ kN/m^2, coefficient of wall friction $\mu = 0.35$, and pressure ratio $k = 0.53$. With these parameters vertical pressure reaches value of 193 kPa at the level of silo floor (solid line). However, assuming: $\mu = 0.1$ – low value but possible in condition of wall lubrication, $\gamma = 9.5$ kN/m^2 (Tys, 2001), and pressure ratio $k = 0.41$ – from experiments of Horabik and Molenda (1995), much higher values of vertical pressure are obtained as shown by dashed line in Fig. 1. Maximum value of σ_z reaches 389 kPa that is twice higher than calculated following the code.

The seemingly simple problems of predicting stresses exerted by grain on the silo wall continue to remain a challenge. Detailed discussion of shortcomings in successful design of silos may be found in Sundaresan (2001).

This work illustrates the effects of variability in granular material properties on silo loads. Cereal grain was chosen as a typical granular material of biological origin. The results presented in the paper were selected from several different projects performed by the authors.

2. DENSITY

Bulk density of a granular material (mass divided by the volume) is one of the most important parameters commonly used to predict pressures in silo structures. Whereas in silo calculations the bulk density is generally considered as a constant parameter, for fine granular materials the bulk density cannot be regarded as a constant. In general, it depends on the physical characteristics of the particles such as particle size, size distribution, shape, roughness, moisture content and state of compaction. Porosity, permeability, pressure distribution and flow pattern depend on the packing density and internal structure of bulk solids. Compaction of a granular material in the silo is influenced by filling procedure. Stream filling creates a less compacted structure of the material than sprinkle filling. Bulk density depends on height of filling, fine particle content and filling rate (Britton and Moysey, 1986).

Thompson and Ross (1983) have found that moisture content, compressive pressure and grain properties influence the bulk density of wheat grain. These authors have determined that the bulk density of wheat increased with increases in both pressure and moisture content. The changes in bulk density which occurred for a pressure between 7 and 35 kPa were caused mainly by the rearrangement of grain in bulk. For higher pressures and higher moisture content the elasticity of grains contributes to the compaction. Therefore large changes in the values of the bulk density were observed as a function of pressure at higher moisture contents. Clower et al. (1973) observed similar relationships during compaction of corn.

Due to the wide range of variability of the bulk density and a large number of affecting factors, for different compaction conditions, parameters such as the uncompacted bulk density, the dense packed density and the tapped density are in use (Gu at al., 1992). Eurocode 1 (1996) recommends to determine the bulk density in the uniaxial compression test at a normal stress corresponding to the maximum vertical pressure in a silo.

2.1 Test Procedure

Tests were performed on 0.17 m high and 0.21 m diameter uniaxial compression tester shown schematically in Fig. 2. The tester was used also for determination of the pressure ratio. The wall of the chamber consisted of two semicircular halves cut along the axis and connected with four load cells installed in pairs on the two connection lines. One half of the wall was settled directly on the base and the other was allowed to move in the horizontal direction to follow the deflection of load cells. Bottom and top plates of the chamber transmitted the vertical load through the load cells. The surface of the wall was smooth whereas the surfaces of the top and bottom plates were rough. The experimental set allowed for determination of the mean lateral pressure σ_x, mean vertical pressure on the bottom σ_z, and the mean vertical pressure on the top plate σ_{zo}.

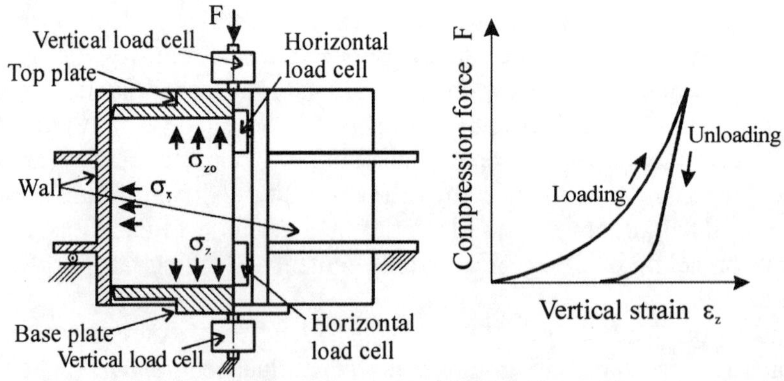

Figure 2. Uniaxial compression tester and typical loading-unloading curve.

The tester chamber was centrally filled through a funnel and the surface was leveled. The specimen was loaded to the reference vertical stress of 100 kPa using a universal loading frame at the constant displacement rate of 0.35 mm/min. The top plate was rotated backwards and forwards three times through an angle of 10° to consolidate the sample. Next the sample was reloaded to the reference vertical stress. Fig. 2 shows a typical graph of vertical force versus vertical strain during loading and unloading cycle.

2.2 Density - Pressure Relationship

The bulk density – compacting stress relationship is a monotonic, nonlinear function. The bulk density increases with applied stress rapidly at low stress and slowly at high stress. The relationship between the bulk density and applied stress during the uniaxial compression test for soft red winter wheat at a moisture content of 10.4% (w.b.) and uncompacted bulk density of 733 kg/m^3 was fitted by the power function:

$$\rho = 738.6 + 6.05\sigma_z^{\,0.5} \tag{2}$$

The initial bulk density (736 kg/m^3) increased to 770 kg/m^3 (approximately 5%) during the compaction under the pressure of 100 kPa. Gu et al. (1992) analysed different empirical functions used for description of bulk density – compacting stress relation. They found that among several bulk density models the power and exponential functions fit the experimental relationships the best.

Table 1. Bulk density of wheat grain (variety Begra) according to three methods: tapped density, uncompacked density and density according to Eurocode 1 procedure

Moisture content (w.b.) (%)	Tapped density (kg/m^3)	Uncompacted density (kg/m^3)	Density according to Eurocode 1 (kg/m^3)
10.0	871 ± 4	773 ± 3	779 ± 2
12.5	861 ± 7	765 ± 3	799 ± 9
15.0	860 ± 10	694 ± 4	784 ± 2
17.5	868 ± 9	704 ± 4	778 ± 5
20.0	823 ± 11	713 ± 5	790 ± 5

Bulk density of wheat grain (variety Begra) at a moisture content of 10-20% determined in the uniaxial compression test under the pressure of 100 kPa was compared with the tapped bulk density and the uncompacted bulk density (Table 1). The tapped density was determined according to the procedure described by Dobrzański et al. (1994). The tapped density was 17% higher than the uncompacted density. The bulk density obtained in the uniaxial compression test was almost independent on the moisture content and significantly lower than the tapped density. Procedure of standardized twisting under the reference stress increased the bulk density about 2%. The tapped density was almost constant for the moisture content in the range from 10 to 17.5% and for the moisture content of 20% significantly decreased. The uncompacted bulk density was the lowest for the moisture content of 15%.

3. PARAMETERS OF INTERNAL FRICTION: c, φ_c AND φ

Jenike shear test is the most popular method for determination of the frictional parameters of bulk solids. For the Jenike (1964) method, the sample is first pre-sheared to a steady state flow and later is sheared to failure at a lower stress level. Pre-shearing to steady-state flow is especially advantageous for failure part of the test, since it will also erase the effect of any preceding deformation and remove influence of all stress history. It gives a very defined starting point for the failure test (Feise, 1998).

Eurocode 1 has simplified the Jenike procedure of a sample preparation. A standardized rotations of the top plate under the reference stress σ_r are recommended for consolidation of the sample. At least two shear tests shall be carried out. One sample shall be sheared when loaded at the reference stress, the other shall be sheared at half the reference stress after pre-loading to the reference stress (Fig. 3).

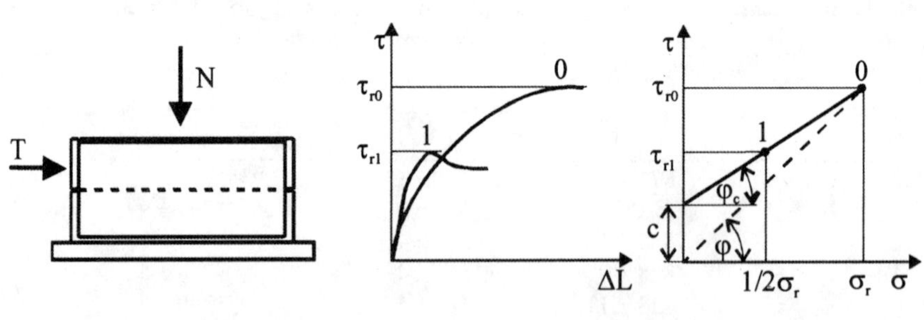

Figure 3. Schematic of Jenike shear tester and procedure of determination of the frictional parameters of granular materials (Eurocode 1, 1996).

In the research reported here, filling the test apparatus and stress history are applied in a way to model conditions that may develop in a bin as an effect of filling method or conditions of operation.

3.1 Equipment and Procedure

Parameters of internal friction (also termed strength parameters): cohesion, c, angle of internal friction, φ_c, and effective angle of internal friction, φ, were determined with a modified direct shear apparatus designed by Ross (Molenda et al., 2000). In this apparatus a sliding tray with the corrugated steel test specimen or vertical blades to determine grain-on-grain friction was pulled horizontally to create a shear zone within the stagnant granular material. The sliding tray was supported on six pairs of roller bearings that allowed it to move freely in the longitudinal direction of the apparatus (Fig. 4). The tray was 0.25 m wide with 0.61 m of its length exposed to grain pressure. A flexible diaphragm mounted on the bottom of the cover of the apparatus was used to exert a known normal pressure on the material located in the grain compartment. The shear apparatus was attached through a chain and gear system to a universal loading frame and shearing speed of 50 mm/min was applied. The shear force was recorded in 0.25 s time intervals with a calibrated electronic load cell having accuracy of ±5 N.

Direct shear tests were performed with the original procedure of Jenike (1964) adopted for the modified direct shear:

1. a prescribed mass of material was placed into the compartment of the apparatus;
2. vertical consolidation reference pressure, σ_r, was applied for 5 minutes;
3. sample was sheared until an asymptotic value of frictional force (steady flow) was approached, thus values of σ_r and τ_{r0} stresses at the terminus of yield locus were determined;
4. steps (1) and (2) were repeated;
5. sample was sheared until 95% of an asymptotic value of frictional force was achieved;
6. vertical consolidation reference pressure, σ_r, was released to zero;
7. vertical pressure, σ_{z1}, was applied for 2 minutes of static holding;
8. sample was sheared until an asymptotic value of frictional force was approached.

Sample was sheared under pressure, σ_{z1}, until the end of the 60 mm shear path. Steps (4) to (8) were repeated for vertical pressures of σ_{z2} and σ_{z3}, and thus three points on the yield locus were obtained.

Seven levels of vertical reference pressure σ_r were used (4.1, 6.2, 8.3, 10.3, 13.8, 17.2 and 20.7 kPa). For each level of reference pressure, σ_r, one sample was sheared at the reference pressure, and three other at lower pressures $\sigma_{z1} < \sigma_{z2} < \sigma_{z3}$ after pre-shearing to the reference stress σ_r. This allowed for estimation of seven yield loci. Soft red winter wheat with a moisture content of 11.2% and uncompacted bulk density of 740 kg/m^3 was used in the experiments.

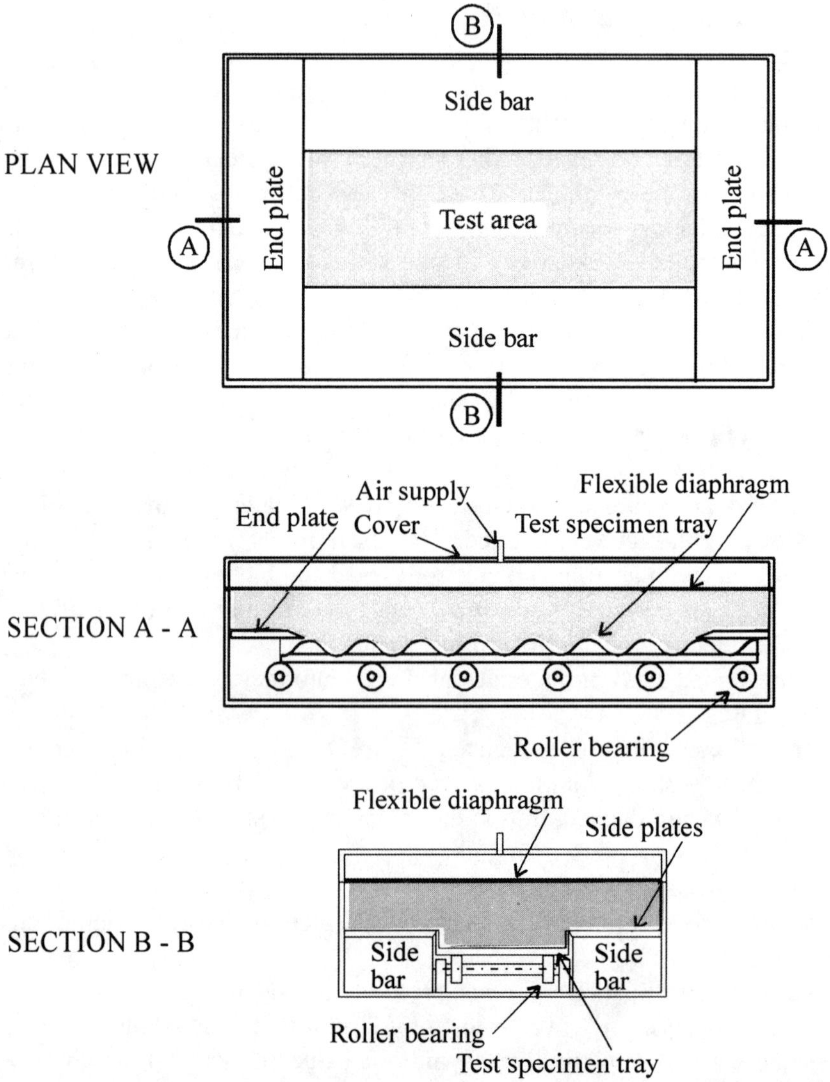

Figure 4. Modified direct shear apparatus.

Friction parameters were calculated following the procedure indicated in figure 3. A straight line was drawn through the origin of co-ordinates (0, 0) and terminus of the yield locus (σ_r, τ_{r0}) and an effective angle of internal friction was calculated from

$$\varphi = \arctan\left(\frac{\tau_{r0}}{\sigma_r}\right) \tag{3}$$

Linear regression was applied to three points (σ_{zi}, τ_{zi}) and parameters c and φ_c were estimated in the regression model:

$$\tau = c + \sigma \tan \varphi_c \tag{4}$$

The effective angle of internal friction was found to be $(26.4 \pm 0.6)°$ and the angle of internal friction $(25.4 \pm 0.5)°$. Relatively low cohesion (0.33 kPa) indicated that dry wheat was a free flowing material. No clear relation was determined between frictional parameters and consolidation pressure.

3.2 Sample Structure

Stress histories have a profound influence on behavior of particulate materials. Ishihara (1978) considered two categories of the stress histories: (1) *precompression*; and, (2) *preshearing*, where a shear stress has been previously applied to the particulate material. He distinguished large and small preshearing. *Small preshearing* influences behavior of a granular material loaded only in the same direction. *Large preshearing* influences all direction of subsequent loading. For the purpose of designing bulk handling equipment for flow, generally, the maximum strength of the bulk material needs to be known. For most granular materials, the way the sample was formed and consolidated influences the yield locus, as it introduces anisotropy in an initially isotropic sample. *Induced anisotropy* is the material anisotropy introduced into the sample by the consolidation procedure while *inherent anisotropy* stems from the shape of the particles and the way the bedding is formed (Wong and Arthur, 1985). The anisotropy introduced during consolidation depends on the type of test procedure used.

Oda (1978) conducted detailed investigations of the influence of the structure of a granular assembly on its mechanical properties. He found that stress-strain behavior depends on the fabric, understood as the spatial arrangement of solid particles and associated voids. This fabric can be described by two parameters: (a) spatial orientation of non-spherical particles and (b) packing density. For describing the spatial orientation of non-spherical particles he used the following two statistics: (a) vector mean direction - which denotes the preferred direction of the long axes of the grain, and (b) frequency of the parallel alignment of the long axes. Oda distinguishes two types of anisotropy produced by the spatial arrangement of granules: (a) anisotropy owing to the preferred orientation of non-spherical particles, and (b) anisotropy owing to the concentration of normal directions at contacts. When a granular assembly is composed of non-spherical particles, the two types of anisotropy are produced.

To investigate the effect of influence of the method of sample deposition on internal friction and friction against corrugated steel the following methods of grain deposition into the test specimen were applied (Molenda et al., 2001):

- standard – the grain was slowly poured into the specimen tray forming a thin layer;
- sprinkle – the grain flowed through the 5 mm openings in the wire mesh located approximately 10 cm above the sample surface and “sprinkled” onto the surface; and
- inclined – the grain was allowed to slide down its plane of natural repose to the front edge of the device.

To investigate the effect of stress history on the friction three methods of the sample consolidation were applied:

- standard – sample was sheared under 20.7 kPa of normal pressure with no pre-treatment;
- compacted – sample was consolidated for a period of 15 minutes under a pressure of 48.2 kPa; and
- pre-sheared – sample was pre-sheared to steady flow and later sheared to failure at lower stress level as recommended by Jenike (1964).

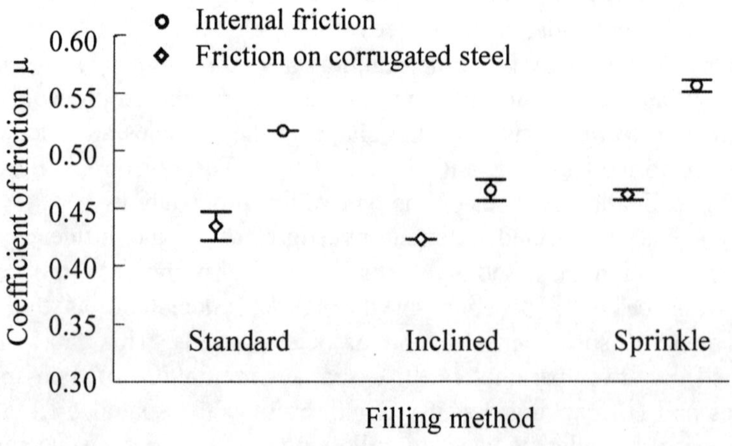

Figure 5. Coefficient of internal friction and friction on corrugated steel (95% C. I.) as influenced by filling method.

The method of sample deposition had a significant influence on both the force-displacement characteristics and grain-on-grain friction coefficient and friction against the corrugated steel in steady flow (Fig. 5). An analysis of variance showed that for all filling methods the coefficient of friction on corrugated steel was significantly lower ($\alpha = 0.05$) than grain-on-grain (approximately 87%). It confirms the formula that friction against a profile sheeting with horizontal ribs is a function of internal friction, friction against the wall and shape of the profile (Australian Standard AS 3774, 1996).

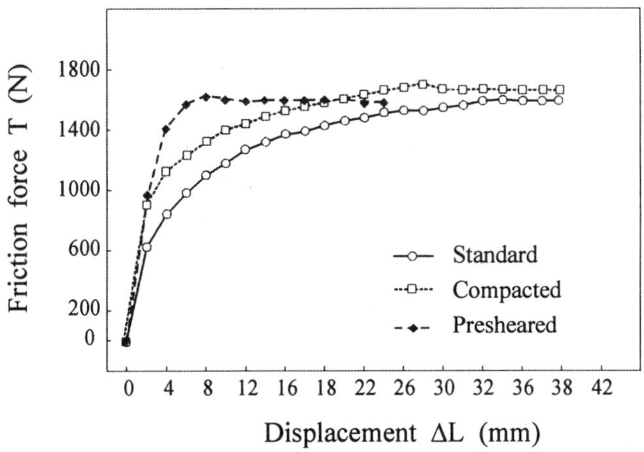

Figure 6. History of loading.

The loading history was found to influence the force-displacement characteristics. Stress history had considerable influence on the rate of force increase (Fig. 6). Friction force for the pre-sheared sample increased the fastest and reached the maximum strength almost linearly. The force-displacement curve shows a form of induced anisotropy typical for this type of deformation. Friction force for the compacted sample reached the sample strength slower and asymptotically. Friction force for the standard sample reached the steady state after the longest displacement.

Similar effect of fabric anisotropy was observed by Szot et al. (1985) during the triaxial compression tests. The angle of internal friction of cereal grains was found to be influenced by the initial orientation of non-spherical grain in the test specimen. The angle of repose and the phenomenon of grains lying nearly along a line of a formed cone was used for producing the desired grain orientation. The angle of internal friction gradually increased as the inclination angle of the preferred orientation of grain to slip plane was increased. The maximum change in the angle of internal friction for wheat grain was about 4° (Fig. 7)

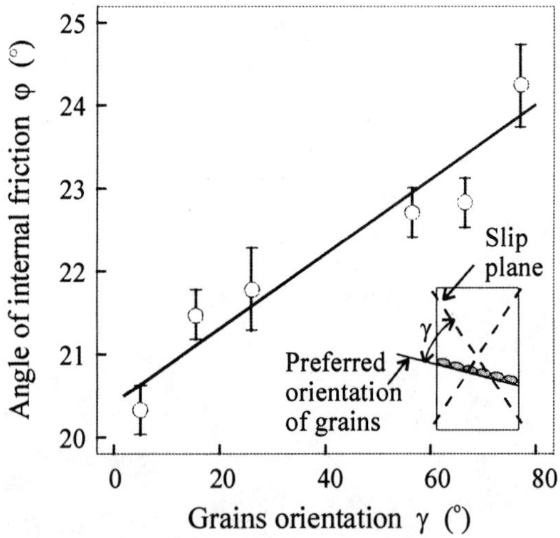

Figure 7. Angle of internal friction φ (95% C. I.) measured in triaxial compression test as affected by the angle γ (Szot et al., 1985).

4. COEFFICIENT OF WALL FRICTION

In the case of wall friction two parameters are recommended for use:

- angle of wall friction φ_w for the evaluation of flow;
- coefficient of wall friction μ for the determination of pressures in silos.

In the first case the reference pressure is kept low to simulate the low pressures occurring during discharge near the silo outlet. In the case of μ the reference pressure should be equal to the horizontal silo pressure in conditions of operation. The tests should be carried out using shear apparatus (for example, Jenike shear tester).

The coefficient of friction between grain and storage structure surfaces varies with surface roughness, moisture content, pressure, velocity and duration of contact between the grain and the surface. What influences friction coefficient to the greatest degree is roughness of the surface. For very smooth surface the adhesive forces may contribute significantly to the friction coefficient. For very high roughness shearing occurs inside grain layer and coefficient of friction corresponds to that of grain-on-grain. Friction against corrugated surface (Fig. 5) is a good example illustrating the contribution of grain-on-grain friction in the resulting friction coefficient.

Coefficient of friction increases with the moisture content increase. Coefficient of friction of wheat grain variety Begra on smooth galvanized steel determined in Jenike shear tester of 0.21 m in diameter as effected by the normal pressure and the moisture content of grain is shown in Fig. 8. Snyder et al. (1967) indicated very strong influence of equilibrium moisture content on friction coefficient of wheat grain. Thompson and Ross (1983) observed an increase in the friction coefficient of wheat grain against a

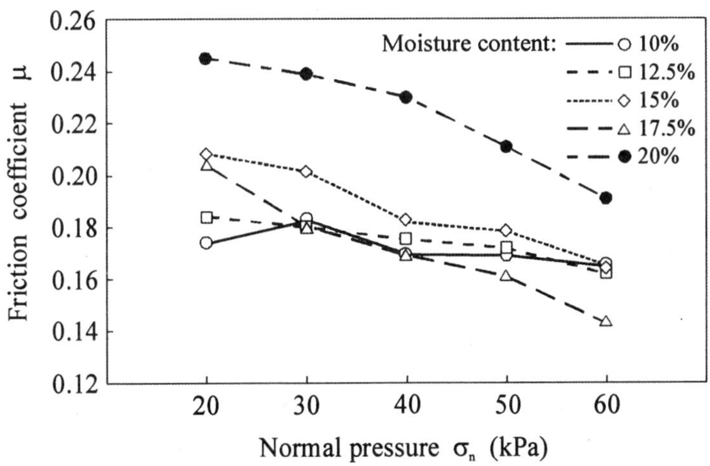

Figure 8. Coefficient of friction of wheat on smooth galvanized steel determined in Jenike shear tester as effected by normal pressure and moisture content of grain.

galvanized steel surface for an increase in the moisture content from 8 to 20%; the friction coefficient decreased for a moisture content of 24%. According to the authors, at this moisture content, grains are very wet and elastic with surface moisture tending to act as a lubricant. The surface water makes the junction between the solids weaker in shear, thus causing the friction coefficient to decrease.

The angle of wall friction φ_w (for flow evaluation) of wheat grain variety Begra against smooth galvanized steel was measured with the tilting table method (Molenda et al., 2000). The table was tilted slowly and the angle at which wheat samples started sliding was measured. The angle of wall friction was found to decrease for an increase in normal pressure and to increase for an increase in moisture content (Fig. 9). Coefficient of friction $\mu = \tan(\varphi_w)$ measured with the tilting table was significantly higher than determined in Jenike shear due to lower value of normal loading (Fig. 8).

Relation between friction coefficient and the normal loading is attributed to less than proportional increase in the normal and tangent stresses at a single contact area of grain and the wall surface with an increase in normal load.

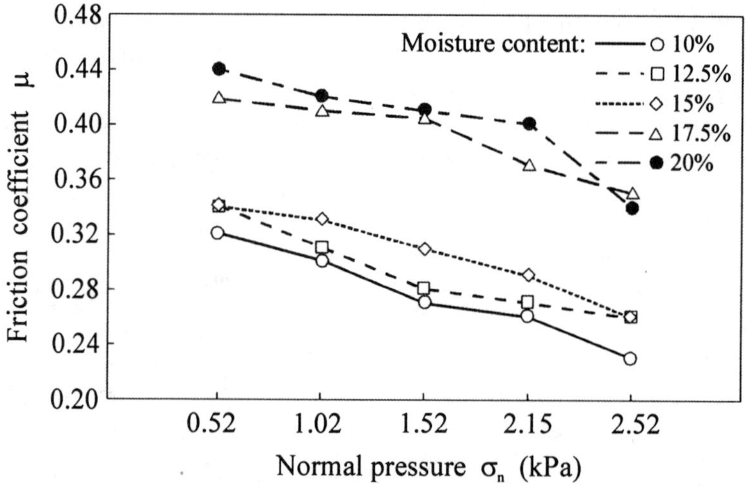

Figure 9. Coefficient of friction of wheat on smooth galvanized steel determined with tilting table as effected by normal pressure and moisture content of grain.

Another factor related to the coefficient of friction is the stick-slip behavior often observed during grain bins discharge. Brubaker and Pos (1965) found that the friction force of grains increases asymptotically to some limit and the slip-stick patterns usually occurs. Stick-slip behavior refers to a jerking action in the flow of grain across a surface at low velocities. This behavior usually disappears at some point as velocity increases (Bucklin et al., 1996).

The influence of displacement velocity on the magnitude of oscillation of friction coefficient of wheat against the corrugated galvanized steel surface obtained by Molenda et al., (2000) is shown in Figure 10. Authors used modified direct shear apparatus described in Chapter 3.1. The normal pressure was 6.9, 27.6 and 48 kPa. The velocity at which the upper and lower limits of coefficients of friction intersected and stick-slip behavior ended was defined as the critical velocity.

The slip-stick behavior exhibited by grain moving over a metal surface can lead to undesirable vibrations and noise in mechanical systems. Ibrahim (1994) summarised the present understanding of friction induced vibration. He stated that the nature of friction induced vibrations depends on several parameters including sliding velocity, time, humidity, overall shapes of the contacting surfaces, surface texture, presence of wear particles, surface films and normal load.

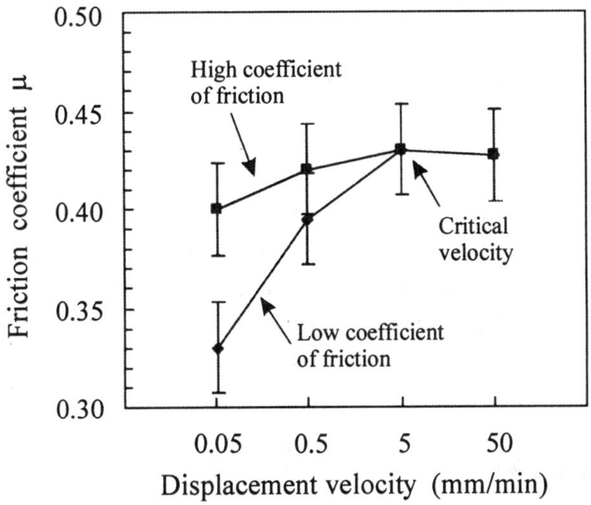

Figure 10. High and low coefficients of friction μ (95% C. I.) for wheat on corrugated galvanized steel surface using the modified direct shear apparatus (Molenda et al., 2000).

5. PRESSURE RATIO

According to Janssen (1895), the pressure ratio is defined as the ratio of the horizontal stress at the silo wall to the mean vertical stress. The pressure ratio is one of the three most important physical properties of bulk solids commonly used for calculation of pressure in a silo (Schwedes, 1996). Since the work of Janssen, several attempts aiming at an expression for pressure ratio, *k*, based on postulating a mechanical model for bulk solids, have been proposed (Moysey, 1979; Drescher, 1991; Lohnes, 1993). Majority of estimations are based on the assumption that the bulk material stored in or discharged from a silo is at a limiting state of stress. Another important assumption concerning location of a region inside the slice of material where the yielding conditions occur involves relations between local stresses and the stresses averaged over the area or the perimeter of a slice.

Jaky (according to Lohnes, 1993) derived a relationship for the pressure ratio at static conditions:

$$k = \frac{(1 - sin\varphi)\left(1 + \frac{2}{3}sin\varphi\right)}{1 + sin\varphi} \qquad (5)$$

Equation (5) was modified and recommended by Eurocode 1 (1996) in the simplified form:

$$k = 1.1(1 - sin\varphi) \qquad (6)$$

The most popular method of experimental determination of the lateral to vertical pressure ratio is the uniaxial compression test (Kwade et al., 1994). Law et al. (1993) and Kwade et al. (1994) have found that the pressure ratio is nearly constant with increasing vertical pressure.

5.1 Experimental Results

Experiments were performed using the uniaxial compression tester described in Chapter 2. The slope of the lateral to the vertical pressure increase was determined after reloading the sample. The pressure ratio k_s appropriate for filling and storing was determined as (Eurocode 1, 1996):

$$k_s = 1.1\Delta\sigma_x/\Delta\sigma_z \qquad (7)$$

at the reference vertical stress $\sigma_z = 100$ kPa (Horabik and Rusinek, 2001).

Values of the pressure ratio k of cereal grains at moisture content in the range from 10 to 20% are presented in Table 2 (experimental - Eq. (7) and calculated according to Eq. (5) and (6)). The angle of internal friction (Tab. 2) was measured using the Jenike shear tester according to the Eurocode 1 procedure. With the moisture content increase the angle of internal friction increases. As a result, smaller part of the vertical loading in the uniaxial tester is transmitted into the lateral direction. Consequently, the lateral to vertical pressure ratio decreases with an increase in moisture content. Equation (5) recommended by Eurocode 1 (1996) to estimate k gave a value considerably higher than the experimental one. Jaky's Eq. (5) provided an estimation for k much closer to the experimental value than Eq. (6). Difference in magnitude between the experimental (Eq. 7) and calculated (Eq. 6) pressure ratios reflect different loading conditions during the uniaxial compression test and in the silo. During the uniaxial compression the horizontal and vertical stresses are approximately principal stresses in the test sample. Pressure ratio corresponds to the calculated one for yielding at centre of the silo. The values calculated according to Eq. (6) are typical for yielding at the rough wall of the silo in the active stress case when principal stresses are not vertical and horizontal (Drescher, 1991).

Table 2. Measured (k_s) and calculated (k_φ) values of the pressure ratio and the angle of internal friction φ for different values of the moisture content of grain (Mean ± St. Dev.)

Grain	Moisture content w.b. (%)	Angle of internal friction φ (°)	Pressure ratio Measured k_s	Calculated k_φ Eurocode 1 (Eq. 6)	Calculated k_φ Jaky (Eq. 5)
	10	27.8 ± 0.4	0.45 ± 0.02	0.59	0.48
	12.5	28.5 ± 0.5	0.47 ± 0.03	0.56	0.47
Barley	15	31.2 ± 0.3	0.43 ± 0.02	0.50	0.43
	17.5	30.6 ± 1.0	0.45 ± 0.03	0.54	0.43
	20	33.2 ± 0.5	0.39 ± 0.03	0.51	0.40
	10	26.7 ± 0.6	0.48 ± 0.04	0.60	0.49
	12.5	31.7 ± 0.5	0.40 ± 0.03	0.52	0.42
Corn	15	32.0 ± 1.4	0.36 ± 0.05	0.51	0.41
	17.5	33.4 ± 0.8	0.34 ± 0.03	0.50	0.40
	20	33.6 ± 1.5	0.30 ± 0.05	0.51	0.39
	10	22.1 ± 1.1	0.49 ± 0.03	0.67	0.57
	12.5	22.4 ± 0.9	0.44 ± 0.04	0.68	0.56
Oat	15	24.0 ± 0.5	0.45 ± 0.03	0.64	0.54
	17.5	23.9 ± 1.0	0.40 ± 0.03	0.66	0.54
	20	26.4 ± 1.7	0.41 ± 0.06	0.63	0.50
	10	25.7 ± 0.3	0.44 ± 0.02	0.62	0.51
	12.5	26.2 ± 0.4	0.38 ± 0.01	0.61	0.50
Wheat	15	27.0 ± 0.5	0.34 ± 0.02	0.60	0.49
	17.5	33.0 ± 1.0	0.31 ± 0.02	0.50	0.40
	20	35.5 ± 0.5	0.35 ± 0.01	0.46	0.37

6. EFFECTS OF VARIABILITY OF GRAIN PROPERTIES ON SILO LOADS

The grain pressures in a silo predicted by Janssen's equation are highly dependent on the assumed values of the wall friction coefficient, μ, the pressure ratio, k, and the bulk density, ρ, as discussed in Chapter 1 (see Fig. 1). Most design codes suggest that these values vary with respect to the type of silo wall and material stored (Bucklin et al., 1989). However, they do not consider changes in these values that may occur with use over the lifetime of the storage structure. Large changes in the coefficient of wall friction may be expected during the initial stages of silo use. An experimental investigation was performed to examine the effects of „wear-in" on a model silo during the initial period of operation (Molenda et al., 1996).

6.1 Experimental Set

A model cylindrical silo 2.44 m in diameter and 7.3 m tall was constructed such that the wall and floor (flat bottom) of the silo are each supported independently on three load cells. The silo was constructed in this manner to isolate the vertical wall loads from the floor loads in the silo. The load cells supporting the silo wall and silo floor were evenly spaced around the circumference of the silo at an angular distance of 120°. This type of experimental configuration allows for the determination of vertical wall and vertical floor loads as well as the calculation of overturning moments on the walls and floor of the silo. Tests were performed in which the model silo was filled with soft red wheat grain and emptied a total of 23 different times (Molenda et al., 1996).

6.2 Results

Tests have shown that repeated sliding contact between stored agricultural materials and silo walls during unloading can result in a decrease in the coefficient of wall friction. The specific objectives were to determine: (a) the extent of changes in wall friction and the mechanical properties of wheat during repeated loading and unloading cycles (LC) of the model bin and (b) the effect of these changes on silo loads.

A single loading cycle (LC) consists of filling and emptying of the model silo one time. As the loading cycle increased, the vertical wall loads decreased along with the coefficient of friction between the stored grain and the silo wall. The vertical-wall-load to total-grain-load ratio after filling was found to decrease from 52.7% for the first cycle to 28.3% for the 10th LC. This decrease in friction is caused by a „wear-in" effect in which the sliding surface is polished by the granular agricultural materials and coated with organic substances originating from the surface of the grain. Vibration, which led to the oscillation of the wall and floor as a result of stick-slip friction, was also observed during the first three LC's. Stick-slip behavior refers to a jerking action as grain flows across a surface at low velocities. The dynamic-to-static wall load ratio (DSR) increased from 1.08 to 1.24 during the first 10 LC's.

Triaxial compression tests were performed on the grain used in these experiments to determine the angle of internal friction of the grain before and after the series of 23 LC's. The angle of internal friction of the grain φ in its initial state was determined to be $(25 \pm 0.8)°$, while the angle internal friction of the grain after the 23rd LC was (22.3 ± 0.9)°. Measurements of the friction coefficient were made on two different plates cut from the same silo wall material used to construct the model silo. One plate was used in its initial or unused state, and the other was smoothed with grain dust collected during the experiments to simulate the condition of the silo wall after sliding contact with the grain. For the untreated plate the coefficient of friction was determined to be of 0.45 ± 0.01, while the treated plate had a coefficient of friction of 0.16 ± 0.01.

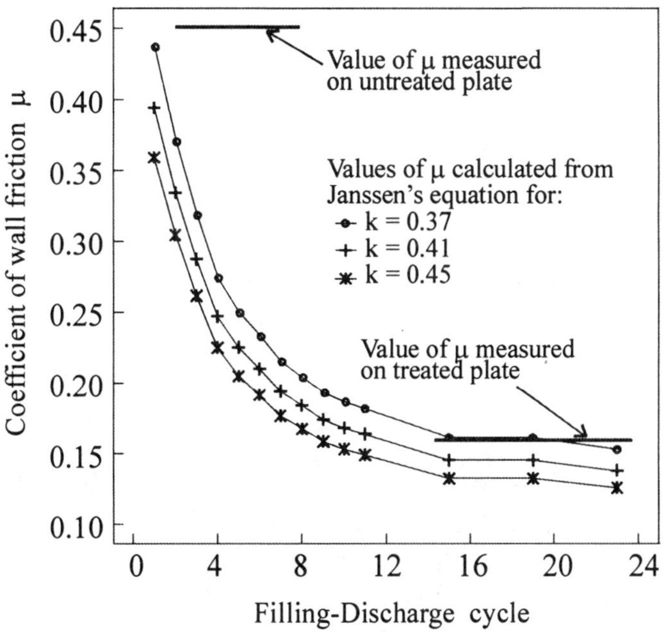

Figure 11. Coefficient of wall friction calculated using Janssen's equation compared to the coefficient measured in friction tests (Molenda et al., 1996).

Values of vertical stress, σ_z, were calculated in the model silo by dividing the measured vertical floor loads by the area of the silo floor. Using the calculated values of σ_z, values of $k\mu$ product were obtained by performing a numerical solution of the Janssen equation. Coefficients of wall friction calculated using k values of 0.45, 0.41 and 0.37 are presented in Figure 11 along with measured values of μ. The values of $k\mu$ product were found to decrease from 0.16 for the first LC to 0.06 for the 23rd LC. This change in $k\mu$ was the result of the combined effects of the decrease in the coefficient of wall friction and the increase in k value resulting from the decrease in internal friction. The ratio of lateral to vertical pressure k calculated using the equation for the active case and yielding at the silo centre (Drescher, 1991):

$$k = (1 - sin\varphi)/(1 + sin\varphi) \quad (8)$$

was found to be 0.41 for φ equal to 25° for the grain used in the first LC and 0.45 for φ equal to 22.3° for grain used in the 23rd LC. Therefore, the observed decrease in the value of $k\mu$ in successive LCs is predominantly influenced by changes in wall friction. The k value of 0.37 provided the closest agreement between the theoretical and measured values of μ.

7. SUMMARY

The first silo standard (DIN 1055-part 6, 1987) for the calculation of loads in silos was published in 1964. Its authors adopted Janssen's equation as the model for predicting the pressure caused by granular materials in storage. Several organizations and design committees have been formed over the last two decades to draft Silo Design Codes. Most of all these codes have also used a Janssen-type pressure distribution to predict silo pressures (Wilms, 1991). A Subcommittee of the Structures Group of ASAE (American Society of Agricultural Engineers), associated with „Loads Due to Bulk Grains and Fertilizers", also prepared and updated their EP433 Engineering Practice (1997). The authors of this standard pointed out that in some areas, engineers' lack the information required to accurately predict the loads caused by granular materials.

The codes published recently recommend Jenike tester as a standard apparatus for determination of parameters of wall friction and internal friction necessary for design. The agreement among specialists has been reached regarding dimensions of the apparatus, procedure of performing tests and method of data elaboration. Standard triaxial compression method is recommended in preference to direct shear method for determining the angles of internal friction φ and φ_c and the cohesion c. Consolidated bulk weight density γ should be determined from the measurements of the weight and volume of material sample consolidated in cylindrical box. The standards recommend that horizontal to vertical pressure ratio k will be determined from the measurements of horizontal and vertical stresses using uniaxial compression method.

Experimental investigations have shown that a number of factors influenced stored material properties. Coefficient of wall friction is under the strongest influence of wall roughness and moisture content of grain. The interaction between the wall material and the bedding of the grain is of major importance for loads exerted by grain on the wall. The filling method produce specific spatial structures of the bedding material that determines the angle of internal friction and pressure ratio. In the case of smooth walled silos, experiments have shown a profound effect of „wear-in" which produced a decrease in the wall coefficient of friction for an increase in grain movement on the walls. This resulted in approximately a twofold decrease in vertical wall load for first ten filling-discharge cycles.

The angle of internal friction of grain is distinctly affected by the moisture content of grain as well as by the bedding structure. As a parameter related to frictional parameters pressure ratio undergoes the same effects as coefficient of friction and angle of internal friction.

Frictional vibrations influenced sample strength and brought additional difficulties in interpretation of test results. Stick-slip may be considered a sequence of compaction-dilation events around the areas of shear zones formed in the material. In industrial practice, stick-slip effects appear in silos as vibrations or shocks (Tejchman and Gudehus, 1993). Currently there is no clear model of frictional vibrations, and the phenomenon needs further investigations.

Wide range of variability of mechanical parameters of grain has to be accepted as a typical feature of the material because a considerable number of factors have influence on these values. Variability of material properties influence the operation of silos, causing flow disturbances, frictional vibrations, asymmetry of loads or just poses a serious problem for the constructor of safe silos (Britton and Moysey, 1986; Johanson, 2000; Molenda et al., 1996).

REFERENCES

ASAE Standards, EP433, 1997, Loads exerted by free-flowing grain on bins. *ASAE, St. Joseph, MI. 49085-9659*, pp. 693-696.

Australian Standard, AS 3774-1990, 1990, Loads on bulk solids containers. *Standard House, 80,* Arthur St, North Sydney NSW.

Britton M.G., Moysey E.B., 1986, Grain properties in the proposed new engineering practice on bin loads. *ASAE Paper No. 86-4502,* St. Joseph, MI.

Brubaker J.E., Pos J., 1965, Determining static coefficients of friction of grain on structural surfaces. *Trans. of the ASAE* **8**(1): 53-55.

Bucklin R.A., Thompson S.A., Ross I.J., Biggs R.H., 1989, Apparent coefficient of friction of wheat on bin wall material. *Transactions of the ASAE* **32**(5): 1769-1773.

Bucklin R.A., Molenda M., Bridges T.C., Ross I.J., 1996, Slip-stick frictional behavior of wheat on galvanized steel. *Transactions of the ASAE* **39**(2): 649-653.

Clower R.E., Ross I.J., White G.M., 1973, Properties of compressible granular materials as related to forces in bulk storage structures. *Transactions of the ASAE* **16**(3): 478-481.

DIN 1055, Part 6, 1987, Design loads for buildings: loads in silo bins. *Deutsche Norm.*

Dobrzański B., Horabik J., Molenda M., 1994, Vibratory method for bulk density estimation. *Powder Handling and Processing* **6**(1): 67-70.

Drescher A., 1991, Analytical methods in bin-load analysis. *Elsevier,* Amsterdam-Oxford-New York-Tokyo.

Eurocode 1, 1996, Basis of design and actions on structures. Part 4. Actions in silos and tanks. *European Committee for Standardization.* Rue de Stassart 36, B-1050 Brussels.

Feise H.J., 1998, A review of induced anisotropy and steady-state flow in powders. *Powder Technology* **98**: 191-200.

Gu Z.H., Arnold P.C., McLean A.G., 1992, Consolidation - related density and permeability models for bulk solids. *Powder Technology* **72**: 39-44.

Horabik J., Molenda M., 1995, Effect of filling method on radial distribution of vertical pressure of rapeseed on bottom of a model silo. *ZPPNR* **427**: 91-97.

Horabik J., Rusinek R., 2001, Pressure ratio of cereal grains determined in uniaxial compression test. *International Conference „Physical Methods in Agriculture", 27-30 August 2001,* Prague, Book of Abstracts, pp. 104-105.

Ibrahim R.A., 1994, Friction-induced vibration, chatter, squeal, and chaos. Part I: Mechanics and friction. *Applied Mechanics Review* **47**(7): 209-226.

Ishihara K., 1978, Stress history effects on the behavior of sand. *Proc. U.S.-Japan Seminar on Continuum-Mechanical and Statistical Approaches in the Mechanics of Granular Materials.* Gakujutsu Bunken Fukyukai, Tokyo, Japan, pp. 71-90.

Janssen H. A., 1895, Versuche über Getreidedruck in Silozellen. *Verein Deutscher Ingenieure*, Zetschrift (Düsseldorf) **39**: 1045-1049.

Jenike A.W., 1964, Storage and flow of solids. *Engineering Experimental Station, University of Utah,* Bulletin No. **123**.

Johanson J.R., 2000, Smooth out Solids Blending Problems. *Chemical Engineering Progress* **96**(4): 21-37.

Knowlton T.M., Carson J.W., Klinzing G.E., Wen-Ching Yang, 1994, The Importance of Storage, Transfer and Collection. *Chemical Engineering Progress* **90**(4): 44-54.

Kwade A., Schulze D. Schwedes J., 1994, Determination of the stress ratio in uniaxial compression tests-Part 1. *Powder Handling and Processing* **6**(1): 61-65.

Law G.J., Negi S.C., Jofriet J.C., 1993, A method for measurement of horizontal to vertical pressure ratios of wheat and barley in a circular bin. *Canadian Agric. Eng.* **35**(1): 45-49.

Lohnes R.A., 1993, Lateral stress ratios for particulate materials. *Powder Handling and Processing* **5**(4): 331-335.

Molenda M., Horabik J., Ross I.J., 1996, Wear-in effects on loads and flow in a smooth-wall bin. *Transactions of the ASAE* **39**(1): 225-231.

Molenda M., Thompson S.A., Ross I.J., 2000, Friction of Wheat on Corrugated and Smooth Galvanized Steel Surfaces. *Journal of Agricultural Engineering Research* **77**(2): 209-219.

Molenda M., Horabik J., Ross I.J., 2001, Sample structure and strength characteristics of wheat in direct shear test. *International Conference „Physical Methods in Agriculture", 27-30 August 2001,* Prague, Book of Abstracts, pp. 162-163.

Moysey E.B., 1979, Active and passive pressures in deep grain bins. *Transactions of the ASAE* **22**(6): 1409-1413.

Oda M., 1978, Significance of fabric in granular mechanics. *Proc. U.S.-Japan Seminar on Continuum-Mechanical and Statistical Approaches in the Mechanics of Granular Materials..* Gakujutsu Bunken Fukyukai, Tokyo, Japan, pp. 7-26.

Polish Standard, PN-89/B-03262, 1989, Concrete bins for storing granular materials and silage. *Design rules* (in Polish).

Schwedes J., 1996, Measurement of flow properties of bulk solids. *Powder Technology* **88**: 285-290.

Snyder L.H., Roller W.L., Hall G.E., 1967, Coefficients of kinetic friction of wheat on various metal surfaces. *Transactions of the ASAE* **10**(3): 411-419.

Sundaresan S., 2001, Some outstanding questions in handling of cohesionless particles. *Powder Technology* **115**: 2-7.

Szot B., Horabik J., Molenda M., 1985, The effect of the spatial orientation of cereal grain in bulk on its mechanical behaviour. *Collection of papers, Faculty of Mechanization, College of Agriculture, Agricultural University in Prague*, pp. 873-878.

Tejchman J., Gudehus G., 1993, Silo-music and silo-quake experiments and a numerical Cosserat approach. *Powder Technology* **76**: 201-212.

Thompson S.A., Ross I.J., 1983, Compressibility and Frictional Coefficients of Wheat. *Transactions of the ASAE* **26**(4): 1171-1180.

Tys J., 2001, Influence of conditions and time of storage on viscoelastic properties of rapeseed. *Internal report. Institute of Agrophysics* (in Polish).

Wilms H., 1991, Criteria for evaluation of silo design codes. *Bulk Solids Handling* **11**(1): 55–61.

Wong R.K.S., Arthur J.R.F., 1985, Induced and inherent anisotropy in sand. *Géotechnique* **35**(4): 471-481.

Digital Image Analysis – Essence and Application in Cereal Science

TOMASZ JELIŃSKI, JADWIGA SADOWSKA and JÓZEF FORNAL
Institute of Animal Reproduction and Food Research, Polish Academy of Sciences, Olsztyn, Poland

Sight is the most important sense of a human allowing him to perceive the surrounding world. Size, shape, structure and colour of examined objects are easily recognised by human eye and brain and simultaneously verify whether they correspond to the standard. Interpretation of images registered solely by the sight is somehow limited by the observer's perception. Increased number of objects and their mobility resulted in difficulties of proper interpretation of human observations. Moreover, the eyes become tired quite fast, thus their ability to differentiate selected features of the object decreases.

Along with dynamic development of electronic engineering, optics and computer techniques, it became possible to replace the eye with digital systems. Development of new systems of registration and processing of different signals is most likely visible in audio-visual applications. Application of the appropriate converter of the signal enables processing of images created by visible, infrared and ultraviolet radiation, ultrasounds, X-rays and electron beams. Processing of such type of images is commonly called "image analysis" and obtains the processes connected with registration and processing of any information describing object under observation (Wojnar, 1994).

The set used for image analysis comprises three modules responsible for: a\ acquisition, b\ processing and analysis of image, and c\ analysis of the obtained results. Image acquisition proceeds by a module responsible for conversion of an analogue signal into digital form with the help of a converter and a acquisition card. Processing and analysis of images proceeds most often in a program module frequently supported by sophisticated software for improvement of the image quality and extraction of the image details especially important in further analysis. The final analysis of the results is most often conducted with external statistical-mathematical software. Many modules of image analysis and processing are capable of carrying out such tasks, at least at the basic level.

Incessant development of digital image analysis creates quite new possibilities of application of visual computer systems for the estimation of agricultural materials including these of the food industry (Lewicki, 1995; Wiwart, 1999). A fine example of such application may be classification of grains or seeds based on detection of the object's edge (kernels, seeds, impurities), segmentation or shape analysis and look of object surface. Application of visual computer systems are forecast to be used in the food industry for monitoring and steering of technological processes in cereal grain processing, i.e. determining the degree of dehulling, mechanical or thermal damage, granulation of milling fraction and swelling of particles.

1. DIA IN KERNEL RECOGNITION

The first application of digital image analysis for the recognition of wheat varieties was described by Zayas et al. (1985). They proved that the discriminant analysis of the few shape features may be used for the recognition of six samples of American durum wheat var. Arkan from four samples of morphologically similar soft wheat var. Arthur. The same authors used successive morphological features of a kernel for the discrimination of various classes and varieties of other American wheat (Zayas et al., 1986) and reached discrimination accuracy of 77 % and 85 % for varieties and classes of wheat, respectively. Also Neuman et al. (1987) used digital characteristics of kernel morphology features for objective recognition of Canadian wheat varieties. Unstable discrimination accuracy (from 4 to 85 %) for varieties within classes disabled proper classification. Using the same method, Sapirstein et al. (1987) tried to distinct durum wheat (HRS), barley, oat and rye. For the sample of 580 grains, classification error reached ca. 1 % only, but for a more numerous sample and randomly selected grains, the percentage of properly distinguished grains was, however, less satisfactory (78.3-98.4 %). After completing the model based on morphological parameters with the surface quality expressed as mean reflectance value, they achieved a great progress in kernel recognition, i.e. from 95.3 to 99.2 % of properly classified cases. Keefe (1992) used digital image analysis for testing of 20 wheat varieties for which 69 parameters characterising kernel size and shape were measured. He noted that morphological features of the examined British wheat overlaid and permeate impairing identification of varieties. While investigating morphological features of kernels of Australian wheat, Myers and Edsall (1989) stated that geometric characteristics of kernel end projection differentiate varieties to a smaller extent than the features of view of kernel directed crease-downwards.

In the works on classification or recognition of cereals, most of authors use the view of crease-downwards kernel. Symons and Fulcher (1988b) estimated morphological difference among nine varieties of three Canadian wheat classes: hard red winter (HRW), soft white winter (SWW) and hard red spring (HRS). The model of four-class linear discrimination allowed proper discrimination of over 80 % HRS and HRW wheat and almost 100 % of SWW wheat. A less accurate discrimination of HRS and HRW wheat noted also by Neuman et al. (1987) may indicate higher morphological diversity of these varieties of wheat. Then, Symons and Fulcher (1988b) carried out the analysis of variability of kernel morphology supplemented with characteristics of its cross-section among six varieties of SWW wheat. They stated that consideration of geometrical parameters of the kernel, crease characteristics, thickness of the seed coat and size of aleurone cells enable precise recognition of the examined SWW wheat varieties. Mean share of properly classified kernels accounted for 89 %. This work was a great milestone that indicated limited possibilities of wheat varieties recognition while making use of geometric features of kernel view only. In order to distinguish wheat varieties, Chen et al. (1989), using camera contour extractor and laser range finder, developed therefore a system of image analysis with the capability to acquire and combine three-dimensional laser range data and two-dimensional camera contour extracted images.

Colour is an important visual attribute of kernels used for estimation and classification of wheat. The use of colour in image analysis makes it more advantageous to that which uses grey levels despite the necessity of using of more expensive equipment, more complicated illumination (Neuman et al., 1989a) as well as more effective software, resulting from the necessity of collecting a larger number of data, the storage and processing (with images and numbers), somehow limits popularisation of that technique.

Types and varieties of cereals are generally characterised by different colours of grains and grain degradation express its discolourisation. In one of the first papers using the colour analysis of image, kernels of ten wheat varieties representing six wheat classes cultivated in West Canada were investigated (Neuman et al., 1989a,b). It was stated that red, white and amber-coloured wheat varieties were well distinguished, while errors occurred only between some types of red kernels. Almost 88 % and 34-90 % of cases were properly classified among wheat classes and varieties, respectively.

Image analysis is often used for recognition of wheat grain impurity. Draper and Travis (1984) presented a simple system of image analysis and software for the analysis and recognition of the shape of wheat and barley kernels (and other seeds as well) in which area, perimeter, length, width and shape coefficient of the largest kernel projection were measured. While continuing their studies on 49 weedy samples of grain, these authors found that most of cases may be distinguished using shape coefficient and length of objects (Travis and Draper, 1985). Westerlind (1985) described a computerised separator for the estimation of the grain purity according to changes in colour, size, shape and texture of objects. Zayas et al. (1989) used digital image analysis for recognition and identification of wheat impurity. They suggested a 3-step method of mixture identification consisting in: a\ discrimination of wheat grains, b\ separation of grains using fixed parameters of mechanic device, and c\ use of multi-factor discriminant analysis for identification and recognition of non-wheat components. The results of Kubiak and Usajewicz (1993) indicate usefulness of the analysis of faba bean seed. coat for quick estimation of quality of seed batches. Calculation of basic statistic measure and determination grey level frequency distribution for seed coat allowed to select mouldy, wrinkled and badly-developed seeds.

Digital image analysis can be applied also for estimation of microbiological quality as well as the characterisation of the degree of thermal and mechanical micro-damages. Thomson and Pomeranz (1991) used digital image analysis for identification of damaged kernels in wheat samples. Using some morphological features 89 % germinated and 83 % non-germinated wheat kernels were properly classified. Classification of damages is often based on the evaluation of the surface quality expressed by changes in its colour (Wiwart, 2000). The analysis of colour images was used for identification of six types of damages of Canadian red winter wheat (CWRS) grains (Luo et al., 2000). The colour features were only helpful in successful recognition of healthy and damaged wheat kernels but analysis of colour changes supplemented with morphological features of the kernels highly improved effectiveness of identification. Non-parametric classification with twenty four selected colour features and four morphological features allowed to obtain identification accuracy from 90 to 100 % for examined types of damages.

The most of the known systems of digital image analysis needed time-consuming positioning of kernels to measure their morphological features. In order to apply an automatic system of grain selection and classification for the industry an algorithm for segmentation of connected objects should be developed. Such algorithm was presented by Shatadal et al. (1995a) who tested it on grain of different wheat, barley, oat and rye varieties (Shatadal et al., 1995b). Recognition accuracy ranged from 79 to 95 %. The studies on cereal grain recognition are so advanced that complete automatic systems of different computer techniques are currently developed (Chen et al., 1995; Fang et al., 1998; Kubiak et al., 1998).

2. DIA IN MILLING YIELD PREDICTION

Digital analysis of photographic or microscopic pictures provides potential help in monitoring of the milling process, as the milling yield could be significantly increased through available wide and direct monitoring of changing parameters of products obtained during milling (Dexter, 1997). Witworth (1994) described a technique of image analysis allowing to quantitatively estimate degree of pollution with bran and to measure continuously the content of specks in the obtained flour. A device installed within eight places of the technological line allows to analyse the final or semi-products of milling. The device makes it possible to detect problems in the milling process early enough to find the defects before large amount of low-quality flour have been produced.

Berman et al. (1996) used DIA for the investigation of relationships between milling yield i.e., one of the parameters of quality evaluation and selected geometric parameters of kernels (area, major axis, minor axis and ellipsoidal capacity of an object). Using multi-factor variance and regression analysis of collected data, they obtained empirical equation explaining about 70 % of the estimated variability. Zayas et al. (1996) also established a DIA method for measuring of fine and coarse milling fractions of soft and hard wheat. The study of Kubiak and Fornal (1994) on geometric features of grain fractions of 12 Polish wheat varieties indicated statistically significant diversity in width, perimeter and area of kernel images. There was also stated a significant correlation between kernel geometric features and its mass and protein content.

3. DIA IN FLOUR QUALITY EVALUATION

Systems of image analysis can be also applied for the estimation of the dark spots content in semolina from durum wheat – an important indicator of the semolina quality on the market. Symons et al. (1996) presented an instrumental method allowing to estimate the specks content in semolina with the use of digital image analysis irrespectively of their size and colour. Kim and Flores (1999) used DIA for the estimation of flour contamination with bran and noted statistically significant correlation between ash content and contamination range. A great advantage of these methods is direct measurement of the content of non-endosperm components of flour being much faster than traditional estimation of the ash content.

Other potential applications of image analysis are related to quick quantitative determination of kernel structural elements in wheat flour on the basis of fluorescence phenomenon (Fulcher, 1992; Jensen, 1982; Wetzel, 1993). The elements of both aleurone layer and seed coat can be measured quantitatively in flour due to the estimation of energy emitted by them. Harrigan (1995, 1998) described the possibilities of practical application of microscopic fluorescence method and digital image analysis for the evaluation of flour quality. Symons and Dexter (1991, 1996) indicated that correlation between the ash content in flour and fluorescence of seed coat is very high for flours from all of the investigated wheat classes milled in different mills. Attention should be paid to the fact that such combination of fluorescence method and DIA may be also advantageous in the estimation of other quality features of flour, including the degree of starch damage and the flour granulation. Novales et al. (1996) already presented the multispectral fluorescence imaging for identification of four products: maize, pea, soybean and wheat.

4. DIA IN PASTA AND NOODLE QUALITY

Digital image analysis can be also applied for the estimation of noodle and pasta quality. Fardet et al. (1998) analysed changes of protein network of different pasta during cooking. The samples were analysed with a microscope using the fluorescence technique and images obtained in a CLSM microscope were stored as TIFF files with 256 grey levels. The image analysis was done in two steps: a\ extraction of texture features and transformation into numeric values and b\ statistical analysis of the obtained data. Hatcher et al. (1999) used DIA for measurements of discoloured spots on noodles. The images were obtained using a colour camera and analysed with KS400 program (Carl Zeiss Version). Then, the authors used the developed technique for the evaluation of the effect of different factors on the occurrence of undesired discolourisation.

5. DIA AND BREAD STRUCTURE

The main problem in determination of bread structure (porosity) is the proper acquisition of the true slice image and the determination of the border between pores and crumb. As technology developed, video cameras (Zayas et al., 1989; Sapirstein et al., 1994), scanners (Bertrand et al., 1992; Rogers et al., 1995; Noll and Kuhn, 1997), and different types of microscopes (Harrigan, 1997; Harrigan and Bussmann, 1998), were used to obtain images of bread slices. They of all paid attention for the problem of proper sample illumination being a requisite of achieving precise and repetitive results. While for the analysis of processed images, firstly the authors used software adapted by themselves (Zayas et al., 1989; Sapirstein et al., 1984; Noll and Kuhn, 1997), in the meantime some commercial software has appeared.

One of the first authors describing the use of digital image analysis for the estimation of baking goods was Smolarz (1989). He presented a method of acquisition and processing of images as well as extracting of contours of pores and cross-sections. He determined six parameters of structure, including five which characterised pores and one which characterised the biscuit. He noted that structural parameters connected with general evaluation of texture can be found more reliable for distinction of extrudates than such describing individual pores. In the study of a Japanese team (Gohtani et al., 1992), DIA was used for the estimation of traditional ink imprint of the "sudachi" baking product on a sheet of paper which showed distribution of pores. Accurate correlation between geometrical parameters of spots on the paper measured with traditional and computer method was presented. Next, the mathematical methods of expressing image texture have been used. Bertrand et al. (1992) applied two-dimensional Haar transform that gives a vector characterising the texture. They obtained about 80 % accuracy of the investigated breads recognition. Zayas (1993) developed the algorithm of multivariate discriminant analysis using patterns of the image texture features for bread crumb grain evaluation. Bread structure was also evaluated through Fourier transform analysis (Miura et al., 1995; Rogers et al., 1995). Roger et al. (1995) used a scanner for acquisition of bread slice image, which finally resolved the problem of the proper illumination of the sample. Contrary approach to that problem was presented by Sapirstein et al. (1994), who characterised particular elements of bread structure, i.e. crumb brightness and area, density, wall thickness and distribution of pore sizes, on the transformed binary image of a slice. Image thresholding, a significant element of image analysis, was conducted by introduction of K-mean algorithm, what in the opinion of the authors made the measurements fully objective. The results obtained were also used for the estimation of other qualitative parameters of crumb. The accurate correlation between pore characteristics and density of bread crumb was found and expressed by adequate equation (Zghal et al., 1999). While investigating the effect of additives and parameters of the production process on the wheat bread quality, Crowley et al. (2000) used a simple but efficient system capable of objective, quantitative characteristics of pores. They applied a different thresholding system consisting in finding a characteristic point of inflexion on the curve of grey level frequency which corresponded to the border between bright crumb and dark pore. Kvaal et al. (1998) presented few algorithms of image processing without pre-treatment. The authors showed that mathematical transformation of images into vectors would enable the use of multi-factor techniques and probably would simultaneously form many features of the image. Different methods of extracting texture features from images were compared: the angle measure technique (ATM), the singular value decompression (SVD), the autocorrelation and autovariance function (ACF), and the so called size and distance distribution (SDD) method. It was proved that, although all the methods are proper for description of sensory-evaluated porosity, the ATM method is dominant. The same methods were used by Baadseth et al. (2000), who investigated the structure of French baguette. Jeliński et al. (1998) presented a method of acquisition, transforming and calculation of object features on crumb images which promises a simple quantitative measure of crumb porosity. By using the scanner to obtain the image from a bread slice, the problem of proper illumination of sample is eliminated. A "Well" filter, used for image segmentation by detecting and emphasising the points that are darker than background, allows to omit the stage of image thresholding. Attention should be paid also to the use of digital analysis of colour image of bread crumb obtained by microscopic fluorescence for the estimation of the share of seed coat and aleurone layer particles (Autio et al., 1997; Parkkonen et al., 1997).

6. DIA IN FOOD MICROSTRUCTURE

The microscopic methods are widely used in the food science for characterisation of the structure of raw materials and products. In practice, the knowledge of fundamental cell microstructure enables, for example, the identification of raw material used as a source of the examined product. The knowledge of relationship between the microstructure characteristics and physical, physicochemical and biological properties of raw materials is applied to predict the reactions of materials during a mechanical or thermal process. The examination of changes in the microstructure during these processes allows to establish their parameters and the quality of the final product. Studies of microstructure grow in importance in sensory analysis (determination of the correlation between different sensoric attributes of food) and biotechnology (observation of the structural changes in the genetically manipulated food). A conventional microscopy generates data in the form of images which are visually analysed and compared. In modern microscopy micropictures obtained are transformed into numerical form and stored as a files on hard or removable discs of computers. The objectivisation and quantification of microscopic pictures is the most important scientific problem (Kalab, 1995).

6.1 Methodical Aspects of Digital Analysis of Wheat Endosperm Micropictures

Mealy and vitreous kernels of Polish, Italian, and Spanish varieties of *T.durum* and *T.aestivum* were examined. Selected varieties were characterised by different hardness and quality of grain (Tab. 1).

Table 1. Selected physical and mechanical properties of wheat grain

Variety	Kernel fracture resistance [N]			Vitreosity	Protein content
	Random	Vitreous	Mealy	[%]	[% dm]
		Triticum durum			
Appio A	164.1	219.1	138.9	40	12.1
Appio B	185.6	226.1	149.3	92	15.2
Simeto A	231.9	270.9	154.0	91	12.1
Simeto B	235.8	289.2	147.0	95	15.5
Vitron	174.0	194.5	125.6	88	13.8
		Triticum aestivum			
Alkora	142.8	154.0	92.1	80	15.3
Sigma	164.7	176.2	103.8	89	13.4
Alba	72.0	100.8	60.7	66	12.2
Rosa	102.4	128.9	80.4	54	12.8
Alcotan	110.4	131.6	101.5	29	11.2

In previous studies on the microstructure of wheat endosperm (Fornal, 1999; Sadowska, 1999) we have distinguished few typical SEM micrographs: a\ large, lens-shaped, as well as small, pretty spherical starch granules loosely packed in a delicate protein matrix, b\ structure similar to previously described structure but appearing more dense and compact, and a part of starch granules is covered with protein material, c\ large and small starch granules are recognisable, although strong adhesion of protein to starch granules was observed, d\ structure similar to structure previously described but larger number of starch granules broken during the sample preparation, indicated stronger protein-starch linkages, which are ordered according to increased mechanical resistance.

All of these types of endosperm microstructure were found in kernels of examined varieties. Microstructure of very hard kernels (*T.durum* - Vitron) is characterised by compact protein matrix and starch granules covered often by protein layer (Fig. 1). On the micropicture of very soft wheat endosperm (*T.aestivum* - Alcotan) big oval starch granules between many small granules connected loosely to protein matrix and air spaces between cells are shown (Fig. 2). Differences of endosperm pictures of medium-hardness wheat changed accordingly to their hardness. Typical picture of Alkora (*T.aestivum*) (Fig. 3) showed empty places small starch granules confirming connection to protein matrix weaker then in hard wheat. On starch granule surface fragments of adhered protein (absent on picture of Alcotan endosperm) are found. Special attention should be paid to the fact that microstructure of mealy kernels contains always higher number of elements which are characteristic for soft wheat.

The method of acquisition, transformation and analysis of images was as follow: the endosperm images (magnification 1000x) sized 592/395 pixels were stored as TIFF files with 256 grey levels. After introducing to the IPP program, the image was negativised and filtered with Sobel filter. Then, the area of interest (AOI) was limited to the size of 588/391 pixels (smaller by 2 pixels on both sides) to eliminate outlines produced by the filter. After image transformation, the elements of the grey level ranging from 135 to 255 were analysed by measuring the area, A, and number of objects, N, while discriminating the objects smaller than 14 pixels.

During microscopic observation significant problem appeared - are more accurate results reached if the pictures of higher number of cells (at lower magnification of micropicture) or higher magnification pictures which more accurately show details of microstructure are used (Figs 1-3).

It seemed that the values of average area and number of objects were different but more effective magnification was not distinguished in discriminant analysis in which the grouping variable contained three, *a prori* defined, groups i.e., soft, hard and durum wheat and the independent variables were average area and number of objects on transformed endosperm picture (Tab. 2).

Wilks' coefficients and percentage of properly defined cases confirmed almost the same efficiency of discrimination of wheat in both magnification groups. Separated microscopic studies of vitreous, mealy and random kernels were done, then similar discriminant analyses in respective groups were calculated. Better discrimination was obtained because of lower data variability in mentioned groups (Tab. 2). Thus, it was decided to use magnification 1000×, which, on the other hand, was accepted as standard in our previous microscopic investigation.

Table 2. Accuracy of wheat type discrimination

Magnification		Wilks' λ	Percentage correct
1000×	Vitreous	0.1043	89.3
	Mealy	0.1193	96.3
	All	0.2183	81.8
500×	All	0.1901	80.4

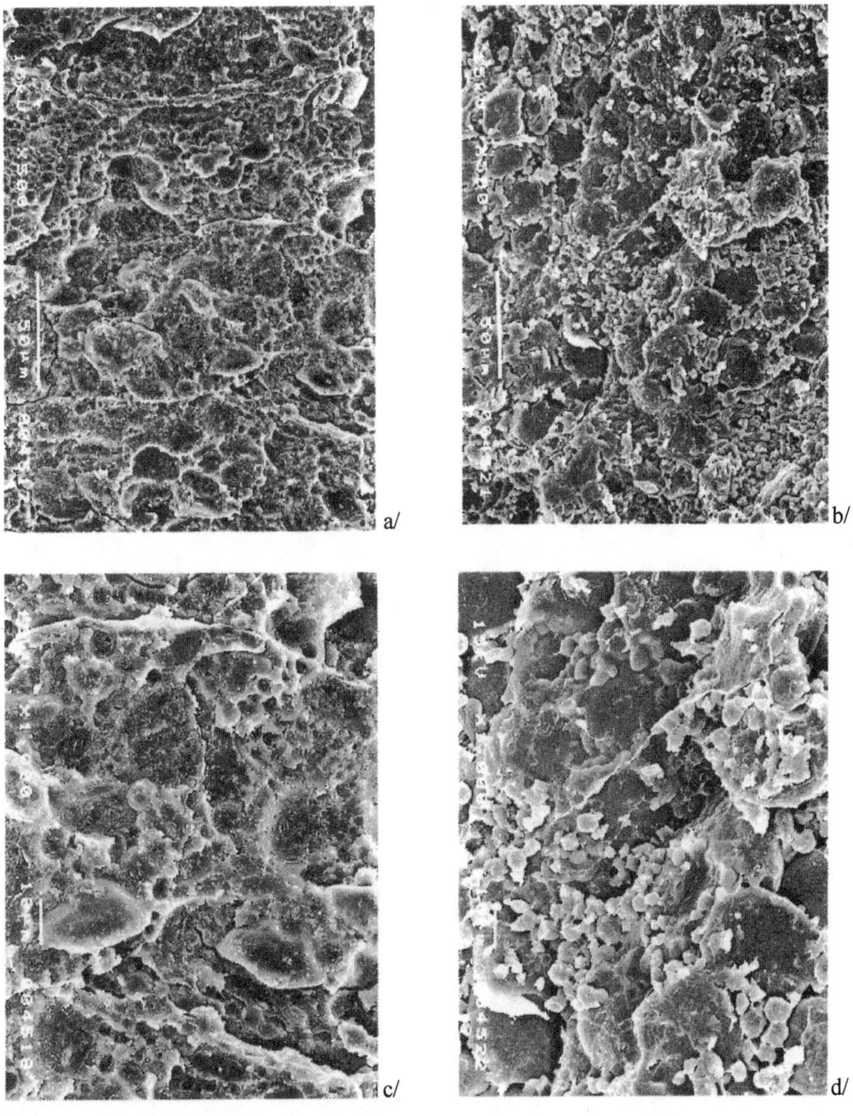

Figure 1. SEM pictures of endosperm microstructure of Vitron (*T.durum*)
a/ (500×), c/ (1000×) - vitreous kernel; b/ (500×), d/ (1000×) - mealy kernel

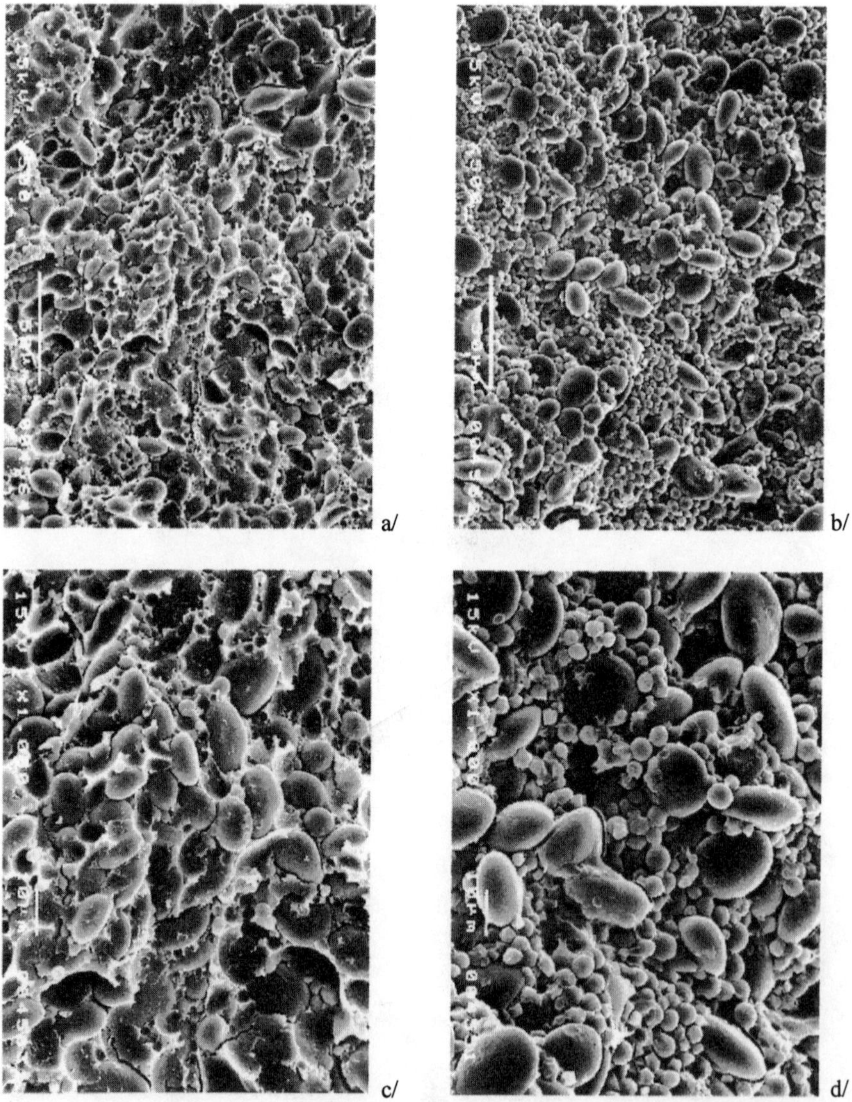

Figure 2. SEM pictures of endosperm microstructure of Alcotan (*T.aestivum*)
a/ (500×), c/ (1000×) - vitreous kernel; b/ (500×), d/ (1000×) - mealy kernel

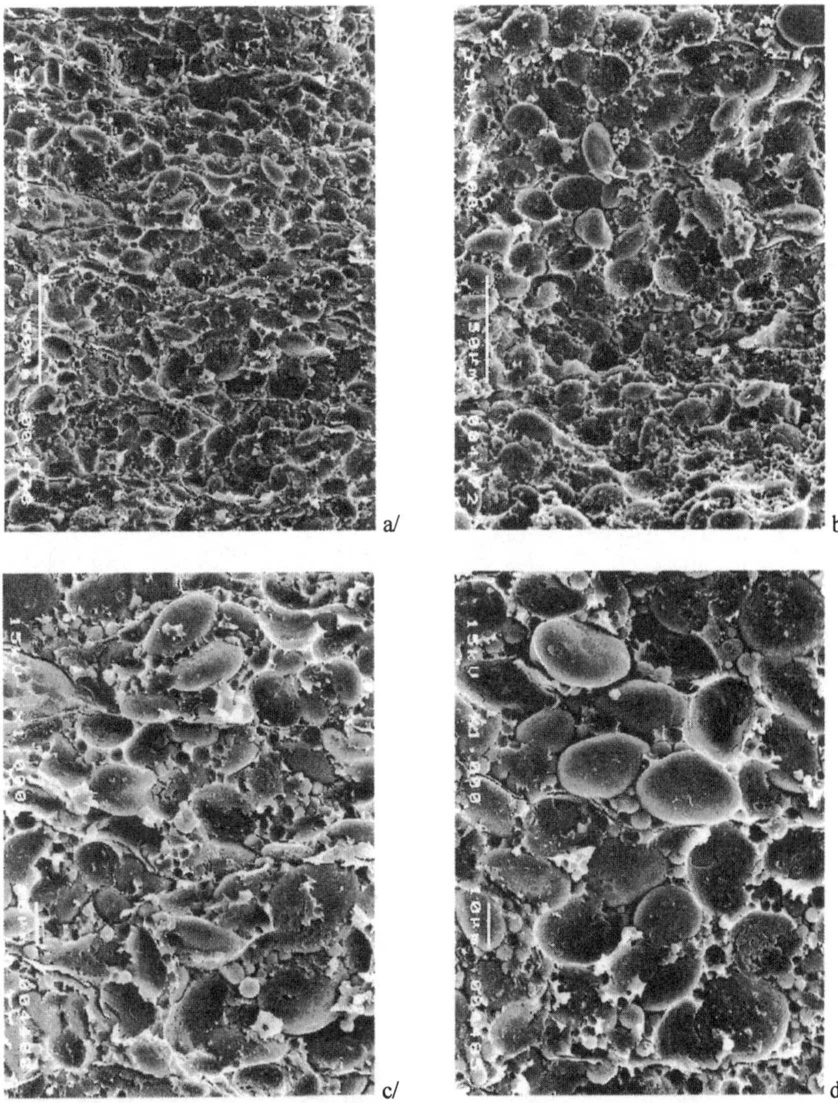

Figure 3. SEM pictures of endosperm microstructure of Alkora (*T.aestivum*) a/ (500×), c/ (1000×) - vitreous kernel; b/ (500×), d/ (1000×) - mealy kernel

The influence of location of sampling area on the cross-section surface of kernel and kernel type on accuracy of endosperm microstructure evaluation were determined using digital parameters of microstructure, which are presented in Table 3. The number of objects was higher in hard than in soft wheat while average area of object was larger in soft wheat.

Table 3. Geometrical characteristics of object on endosperm image

Variety		Vitreous kernels		Mealy kernels		Random kernels	
		$A_{ave.}$	$N_{obj.}$	$A_{ave.}$	$N_{obj.}$	$A_{ave.}$	$N_{obj.}$
			Triticum durum				
Appio A	Mean	267.8	320	330.2	261	299.0	290
	Cv	26.7	27.6	23.5	18.0	9.9	9.4
Appio B	Mean	307.3	278	361.7	257	320.9	273
	Cv	19.4	17.6	24.9	13.5	18.4	13.9
Simeto A	Mean	229.8	345	796.2	138	513.0	237
	Cv	15.4	12.7	15.9	13.0	16.7	11.9
Simeto B	Mean	238.9	333	325.5	262	260.6	315
	Cv	21.6	11.3	24.4	28.1	23.9	28.5
Vitron	Mean	348.5	294	440.2	185	394.4	239
	Cv	12.6	11.1	24.6	22.9	16.5	27.0
			Triticum aestivum				
Alkora	Mean	434.2	208	422.2	202	428.2	204
	Cv	23.9	21.1	17.6	14.9	11.2	11.9
Sigma	Mean	365.6	234	474.4	189	447.2	201
	Cv	12.2	11.5	17.8	14.0	14.8	14.0
Alba	Mean	424.9	193	587.5	145	506.4	169
	Cv	20.1	16.4	19.4	21.3	20.8	17.3
Rosa	Mean	488.4	189	567.5	161	527.8	175
	Cv	24.1	17.3	20.5	20.2	12.2	12.5
Alcotan	Mean	588.3	143	910.4	107	749.3	125
	Cv	24.7	25.8	276.3	22.3	28.1	23.1

cv – coefficient of variance

Figures 4 and 5 show microstructure on all sampling locations of extremely different kernels, i.e., vitreous kernel of hard variety Vitron (*T.durum*) and mealy kernel of very soft variety Alcotan (*T.aestivum*). For both varieties, microstructure of endosperm from five sampling locations looked different, especially for mealy kernels of mechanically weak variety Alcotan. Also micropictures from opposite to crease position seemed different compared to pictures from other positions. Results of variance analysis (ANOVA) for separate varietal groups confirmed, however, statistically significant differences in DIA indices for particular kernels but did not allow to indicate location of sampling area which assure proper selection of image typical for kernel type and wheat variety.

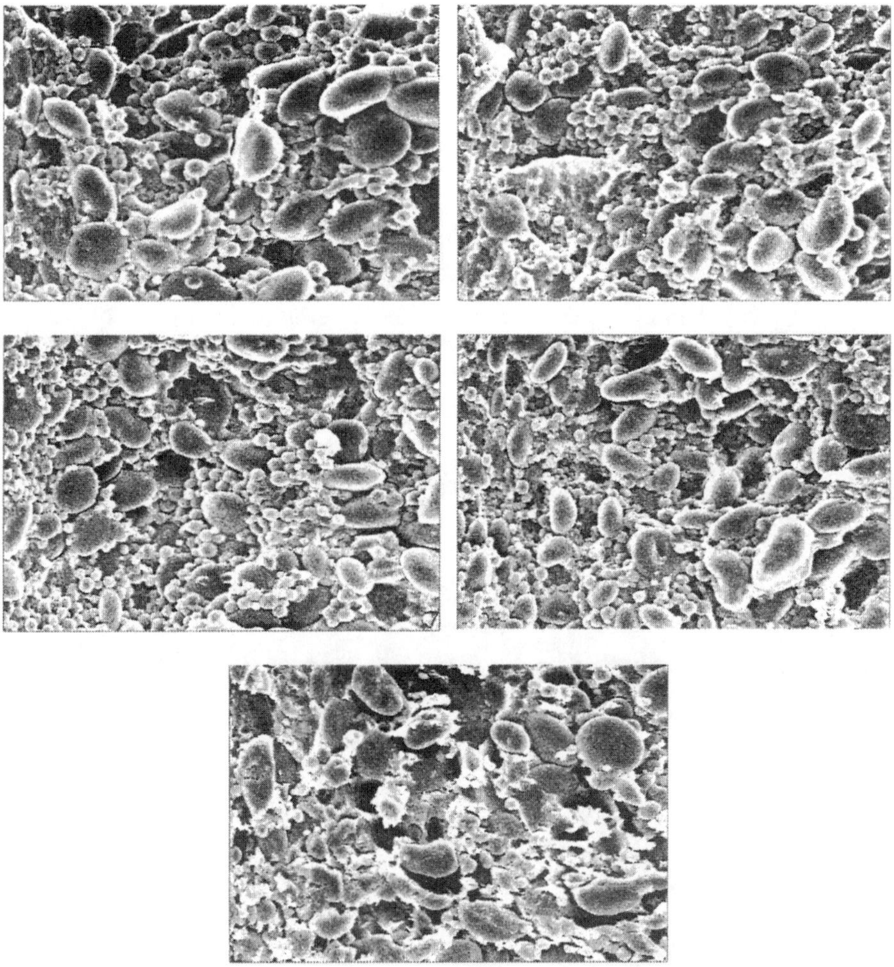

Figure 4. SEM pictures of endosperm microstructure of Alcotan (*T.aestivum*) from sampling places on the cross-section surface of mealy kernels.

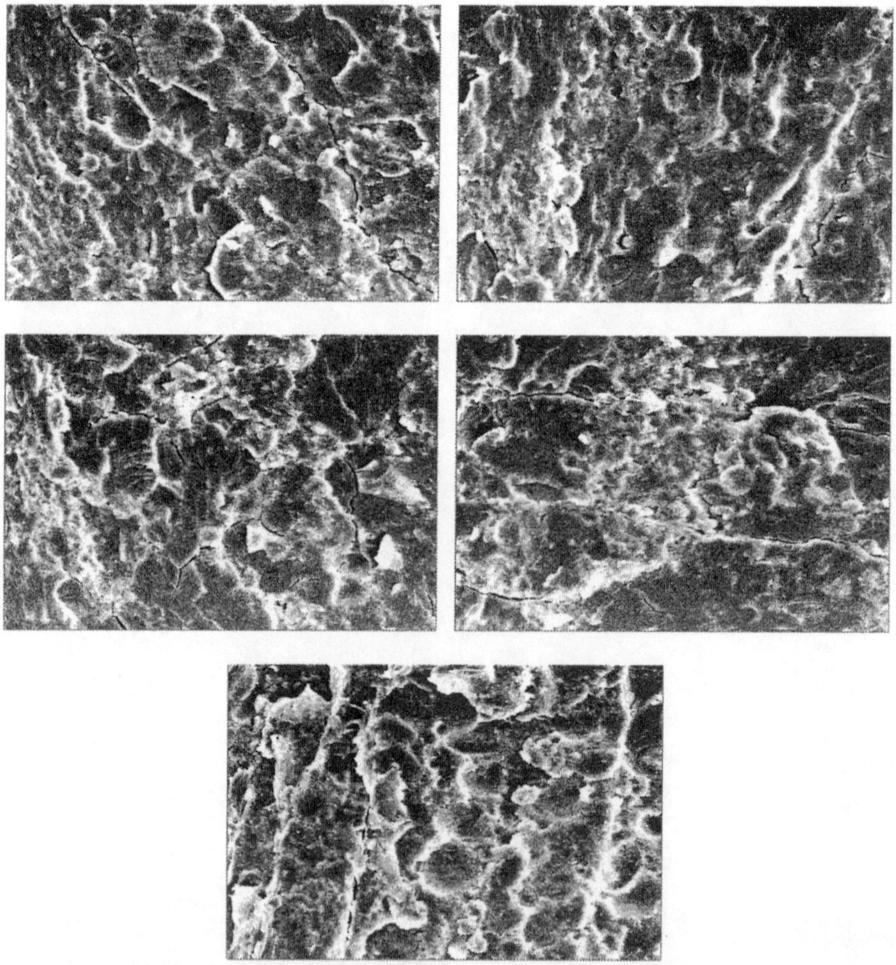

Figure 5. SEM pictures of endosperm microstructure of Vitron (*T.durum*) from sampling places on the cross-section surface of vitreous kernels.

The influence of sampling location was not confirmed by the results of multivariance analysis (MANOVA) for all varieties while influences of varieties and kernel type were statistically significant (Table 4). These results showed that differences in endosperm microstructure in single kernel were lower, despite high data variability, than those caused by kernel type and wheat variety. Thus, the results obtained for all positions should be considered in the calculations.

Table 4. Geometrical characteristics of object on endosperm image

	Wilks' λ	Rao coefficient	df-1	df-2	P
Variety (1)	0.1005	28.0045	22	286	0.0000*
Kernel (2)	0.5268	18.0042	6		0.0000*
Sampling place (3)	0.9766	0.8512	4		0.4937
Interactions:					
(1)×(2)	0.1607	6.4771	66		0.0000*
(1)×(3)	0.7786	0.8661	44		0.7120
(2)×(3)	0.9071	1.1900	12		0.2895
(1)×(2)×(3)	0.4201	1.1760	132		0.1321

* - statistically significant

For estimation of discrimination efficiency of groups related to mechanical properties and technological hardness of wheat varieties statistical analyses were done separately for vitreous, mealy and random kernels (Table 5). Discriminant function analysis is used to determine which variables discriminate between two or more naturally occurring (or *a priori* defined) groups. Basic idea of underlying this analysis is to determine whether groups differ with regard to the mean of the variable, and to use that variable to predict group membership. Computationally, discriminant analysis is similar to analysis of variance and the standard statistic that is used to denote the statistical significance of the current model is Wilks' coefficient. Its value ranges from 1.0 (no discriminatory power) to 0.0 (perfect discriminatory power).

Table 5. Geometrical characteristics of object on endosperm image

	Wilks' λ	Percentage correct
For all repetitions in group of kernels		
Vitreous	0.3384	74.5
Mealy	0.3039	80.5
All	0.3876	71.2
For means	0.0071	100.0

Discrimination in all cases was good - Wilks' coefficients for groups of vitreous, mealy, and random kernels were 0.3384, 0.3039 and 0.3876, respectively. Real classification of hardness groups showed good discrimination of durum, while discrimination of soft and hard wheat was not satisfactory (Fig. 6). Better discrimination of wheat types on analysing the mean values for particular varieties was obtained - Wilks' coefficient was extremely low i.e., 0.0071 (Fig. 7). The difference in discrimination efficiency was probably caused by high variability in data for particular kernels and their types. It should be said, that the differences of microstructure of mealy and vitreous kernels within the variety strongly influenced mean values of digital structure parameters. Thus, percentage of vitreous and mealy kernels according to vitreosity of particular varieties should be kept during preparation of specimens for microscopic analysis.

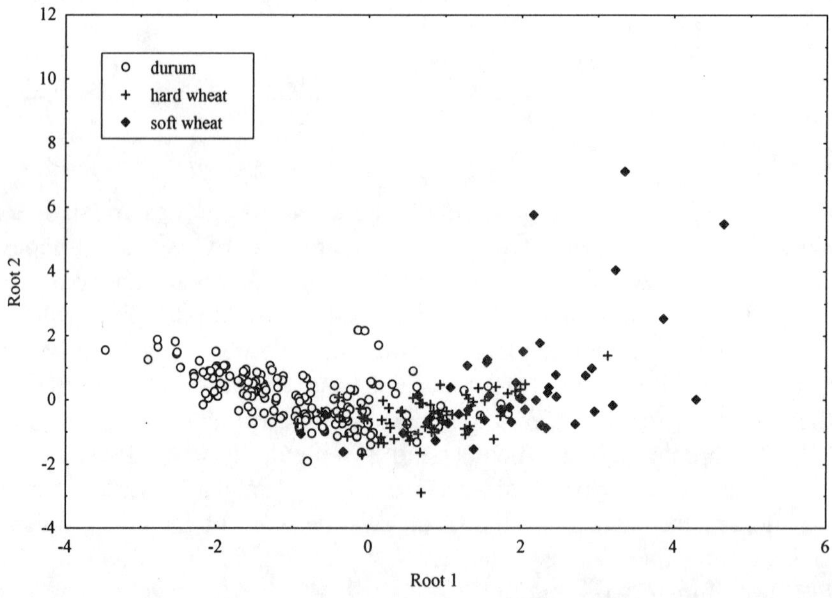

Figure 6. Discrimination of hardness group among wheat varieties using microstructure digital parameters as independent variables (all repetitions)

Concluding, it should be said that the algorithm developed for transformation and digitisation of micropictures is useful for estimation of changes in endosperm microstructure.

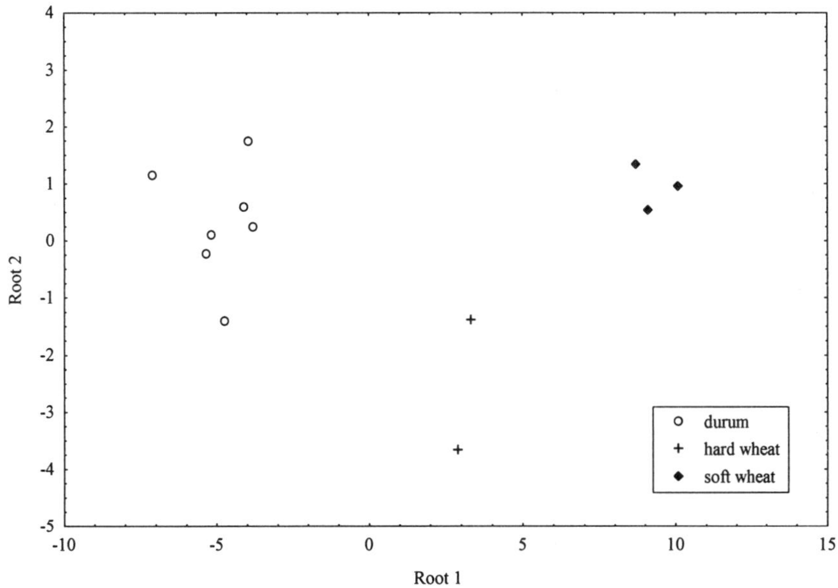

Figure 7. Discrimination of hardness group among wheat varieties using microstructure digital parameters as independent variables (means)

7. CONCLUSIONS

Digital image analysis is a quick and precise method of recognition and identification of different-shape objects but actually it will not be able to provide absolute perfection in discrimination and identification of cereals or grain quality. Versatile development of DIA techniques for this application has been noted because of precise and objective grain characteristics compared with traditional subjective classification and evaluation.

DIA techniques used for objective quality evaluation of cereal by-products and products (i.e., flour contamination, bread porosity, pasta or noodle discolourisation) enable automation of industrial processes and standardisation of final products.

REFERENCES

Autio K., Parkkonen T. and Fabritius M., 1997, Observing structural differences in wheat and rye breads. *Cereal Foods World* **42**: 702-705.

Baardseth P., Kvaal K., Lea P., Ellekjær M.R. and Færgestad E.M., 2000 The effect of bread making process and wheat quality of french baguettes. *Journal of Cereal Science* **32**: 73-87.

Berman M., Bason M.L., Elison F., Peden G. and Wrigley C.W., 1996, Image analysis of whole grains to screen for flour milling yield in wheat breeding. *Cereal Chemistry* **73**: 323-327.

Bertrand D., LeGuernevé C., Marion D., Devaux M.F. and Robert P., 1992, Description of the textural appearance of bread crumb by video image analysis. *Cereal Chemistry* **69**: 257-261.

Chen C., Chiang Y.P. and Pomeranz Y., 1989, Image analysis and characterisation of cereal grains with a laser range finder and camera contour extractor. *Cereal Chemistry* **66**: 166-470.

Chen Y.R., Delwiche S.R. and Hruschka W.R., 1995, Classification of hard red wheat by feedforward backpropagation neural networks. *Cereal Chemistry* **72**: 317-319.

Crowley P., Grau H. and Arendt E.K., 2000, Influence of additives and mixing time on crumb grain characteristics of wheat bread. *Cereal Chemistry* **77**: 370-375.

Dexter J.E. 1997, Application of research results in wheat milling. Part 2. Optimisation of milling process – regulation of mills (in Polish). *Przegląd Zbożowo-Młynarski* **41(2)**: 16-19.

Draper S.R. and Travis A.J., 1984, Preliminary observation with a computer based system for analysis of the shape of seeds and vegetative structures. *Journal of National Institute Agriculture Botany* **16**: 387-395.

Fang Q., Biby G., Haque E., Hanna M.A. and Spillman Ch.K., 1998, Neural network modelling of physical properties of ground wheat. *Cereal Chemistry* **75**: 251-253.

Fardet A., Baldwin P.M., Bertrand D., Bouchet B., Gallant D.J. and Barry J-L., 1998, Textural images analysis of pasta protein networks to determine influence of technological processes. *Cereal Chemistry* **75**: 699-704.

Fornal J., T. Jeliński, J. Sadowska J. and Quattrucci E., 1999, Comparison of endosperm microstructure of hard and soft wheats using digital image analysis, *International Agrophysics* **13**: 215-220.

Fulcher F.G., 1982 Fluorescence microscopy in cereals. *Food Microstructure* **1**: 167-175.

Gohtani S., Ariuchi N., Kawasome S. and Yamano Y., 1992, Computerized image analysis of sudachi (gas cell distribution) of baked cereal products. *Nippon Shokuhin Kogyo Gakkaishi* **39(9)**: 749-754.

Harrigan K.A. and Bussman S., 1998, Digital speck counting of semolina using automated image analysis. *Cereal Foods World* **43**: 11-16.

Harrigan K., 1995, Flour power: microscopic image analysis in the food industry. *Cereal Foods World* **40**: 11-14.

Hatcher D.W., Symons S.J. and Kruger J.E., 1999, Measurement of the time-dependent appearance of discolored spots in alkaline noodles by image analysis. *Cereal Chemistry* **76**: 189-194.

Hatcher D.W. and Symons S.J., 2000, Image analysis of Asian noodle appearance: Impact of hexaploid wheat with a red seed coat. *Cereal Chemistry* **77**: 388-391.

Jeliński T., Sadowska J. and Fornal J., 1998, Evaluation of bread crumb structure by digital image analysis method. *Polish Journal of Food and Nutrition Sciences* **7/48(2S)**: 137-142.

Jensen S.A., Munck L. and Martens H., 1982, The botanical consistuents of wheat and wheat milling fractions. Quantification by autofluorescence. *Cereal Chemistry* **59**: 477-484.

Kalab M., Allan-Wojtas P. and Miller S.S., 1995, Microscopy and other imaging techniques in food structure analysis. *Trends in Food Science & Technology* **6**: 177-186.

Keefe P.D., 1992, A dedicated wheat grain image analyzer. *Plant Var. Seeds* **5**: 27-33.

Kim Y.S. and Flores R.A., 1999, Determination of bran contamination in wheat flour using ash content, color, and bran speck counts. *Cereal Chemistry* **76:** 957-961.

Kubiak A. and Usajewicz I., 1993, Computer image analysis for the estimation of kernel quality in faba bean (Vicia Faba L. minor), *Polish Journal of Food and Nutrition Sciences* **2/43**: 51-57.

Kubiak A. and Fornal Ł., 1994, Interaction between geometrical features and technological quality of Polish wheat grains. *Polish Journal of Food and Nutrition Sciences* **3/44**: 75-85.

Kubiak A., Fornal L. and Ojczyk T., 1998, Polish wheat varieties recognition by artificial intelligent system. *Polish Journal of Food and Nutrition Sciences* **7/48(3S)**: 160-167.

Kvaal K., Wold J.P., Indahl U.G., Baardeseth P. and Næs T., 1998, Multivariate feature extraction from textural images of bread. *Chemometrics and intelligent laboratory systems* **42**: 141-158.

Lewicki P.P., 1995, Application of computer image analysis in food technology (in Polish). *Przemysł Spożywczy* **5**: 155-157.

Luo X., Jayas D.S. and Symons S.J., 1999, Identification of damaged kernel in wheat using a colour machine vision system. *Journal of Cereal Science* **30**: 49-59.

Miura M., Yuki A., Nishihara E. and Matsumoto S., 1993, Computer aided image analysis of crumb grain of commercial white pan breads. *Cereal Food World* **38**: 621.

Myers D.G. and Edsall K.J., 1989, The application of image processing techniques to the identification of Australian wheat varieties. *Plant-Varieties & Seeds* **2**: 109-116.

Neuman M.R., Sapirstein H.D., Shwede K.F. and Bushuk W., 1987, Discrimination of wheat classes and variety by image analysis of whole wheat samples. *Journal of Cereal Science* **6**: 125-132.

Neuman M.R., Sapristein H.D., Shwedyk E. and Bushuk W., 1989a, Wheat grain color analysis by digital image processing. I. Metodology. *Journal of Cereal Science* **10**: 175-182.

Neuman M.R., Sapristein H.D., Shwedyk E. and Bushuk W., 1989b, Wheat grain color analysis by digital image processing. II. Wheat class discrimination. *Journal of Cereal Science* **10**: 183-188.

Noll B. and Kuhn M., 1997, Bildanalyse zur Strukturbeschreibung von Lebensmitteln am Beispiel von Brotkrumen und Blatterteiggebeck. *Getraide Mehl und Brot* **51**: 98-103.

Novales R., Bertrand D., Devaux M.F., Robert P. and Sire A., 1996, Multispectral fluorescence imaging for the identification of food products. *Journal of the Science of Food and Agriculture* **71**: 376-382.

Parkkonen T., Heinonen R. and Autio K., 1997, A new method for determining the area of cell walls in rye doughs based on fluorescence microscopy and computer assisted image analysis. *Food Science and Technology/Lebensmittel-Wissenschaft und Technologie* **30**: 734-747.

Rogers D.E., Day D.D. and Olewnik M.C., 1995, Development of an objective crumb-grain measurement. *Cereal Foods World* **40**: 498-501.

Sadowska J., Jeliński T. and Fornal J., 1999, Comparison of microstructure of vitreous and mealy kernels of hard and soft wheat. *Polish Journal of Food and Nutrition Sciences* **8/49**: 3-15.

Sapirstein H.D., Neuman M.R., Wright E.H., Shwedyk E. and Bushuk W., 1987, An instrumental system for cereal grain classification using digital image analysis. *Journal of Cereal Science* **6**: 3-14

Sapirstein H.D., Roller R. and Bushuk W., 1994, Instrumental measurement of bread crumb grain by digital image analysis. *Cereal Chemistry* **71**: 383-391.

Shatadal P., Jayas D.S. and Bulley N.R., 1995a, Digital image analysis for software separation and classification of touching grains. I. Disconnect algorithm. *Transaction of the ASAE* **38**: 635-643.

Shatadal P., Jayas DS. and Bulley NR., 1995b, Digital image analysis for software separation and classification of touching brains. II. Classification. *Transaction of the ASAE* **38**, 645-649.

Smolarz A., van Hecke E. and Bouvier J.M., 1989, Computerized image analysis and texture of extruded biscuits. *Journal of Texture Studies* **20**: 223-234.

Symons S.J. and Fulcher R.G., 1988a, Determination of wheat kernel morphological variation by digital image analysis. I. Variation in Eastern Canadian milling quality wheats. *Journal of Cereal Science* **8:** 211-218.

Symons S.J. and Fulcher R.G., 1988b, Determination of wheat kernel morphological variation by digital image analysis. II. Variation in cultivars of soft white winter wheats. *Journal of Cereal Science* **8**: 219-229.

Symons S.J. and Dexter J.E., 1991, Computer analysis of fluorescence for measurement of flour refinement as determined by flour ash content, flour grade and tristimulus color. *Cereal Chemistry* **68**: 454-460.

Symons S.J., Dexter J.E., Matsuo R.R. and Marchylo B.A., 1996, Semolina speck counting using an automated imaging system. *Cereal Chemistry* **73**: 561-566.

Symons S.J. and Dexter J.E., 1996, Aleurone and pericarp fluorescence as estimators of mill stream refinement for various Canadian wheat classes. *Journal of Cereal Science* **23**: 73-83.

Thomson W.H. and Pomeranz Y., 1991, Classification of wheat kernels using three-dimensional image analysis. *Cereal Chemistry* **68**: 357-361.

Travis A.J. and Draper S.R., 1985, A computer based system for the recognition of seed shape. *Seed Science and Technology* **13**: 813-820.

Westerlind E., 1985, Computerized separator as a mechanical aid for determination by number of other seed in cereal species. *Medd. Statens Utsadeskon* **59**: 83-98.

Wetzel D.L. 1993, Pixel counting as routine quantitative fluorescence quality control in food processing. In *Proceedings of Food flavors, ingredients and composition*, ed. G.Charalambous, Elsevier Science Publishers B.V., pp.1059-1079.

Whitworth M., 1994, Under the spotlight. *Milling* **188**: 10-13.

Wiwart M., 1999, Digital image analysis. - modern research tool in agriculture science (in Polish). *Postępy Nauk Rolniczych* **5**: 3-15.

Wiwart M., 2000, Digital image analysis of Triticale grains obtained from heads infected with Fussarium Culmorum sacc. (in Polish). *Folia Universitatis Agriculturae Stetinensis* **206**: 325-330.

Wojnar L. and Majorek j., 1994, *Computer image analysis*. Warszawa:Computer Scanning Systems Ltd..

Zayas I., Pomeranz Y. and Lai F.S., 1985, Note: Discrimination between Arthur and Arkan wheats by image analysis. *Cereal Chemistry* **62**: 478-480.

Zayas I.Y., Lai F.S. and Pomeranz Y., 1986, Discrimination between wheat classes and varieties by image analysis. *Cereal Chemistry* **63**: 52-56.

Zayas I., Pomeranz Y. and Lai F.S., 1989, Discrimination of wheat and non-wheat components in grain samples by image analysis. *Cereal Chemistry* 1989; **66**: 233-237.

Zayas I.Y., 1993, Digital image texture analysis for bread crumb grain evaluation. *Cereal Foods World* **38**: 760-766.

Zayas I.Y. and Steele J.L., 1996, Image texture analysis for discrimination of mill fraction of hard and soft wheat. *Cereal Chemistry* **73**: 136-142.

Zghal M.C., Scanlon M.G. and Sapirstein H.D., 1999, Prediction of bread crumb density by digital image analysis. *Cereal Chemistry* **76**: 734-742

Thermal Conductivity of Macaroni Products

A Model for Agricultural Materials in Granules

SEBNEM TAVMAN[1] and PERIHAN YOLCI[2]
[1]*Ege University, Bornova – Izmir, Turkey*
[2] *TUBITAK-MRC, Food Science and Technology Research Institute, Gebze, Turkey*

1. INTRODUCTION

Thermal properties of agricultural products are important to understand their thermal behaviour. They are also used to model and develop processes, for example drying, sterilization, and freezing, in that mass and energy are exchanged. Proper storage of grains without loosing their quality requires the prediction of their temperature in the stock-pile. The success of the prediction of this temperature also depends on the knowledge of the thermal properties of the agricultural products. The primary thermal properties of agricultural products are thermal conductivity, specific heat and thermal diffusivity. Although specific heat is the only thermal property needed for estimating the amount of energy required to change the temperature of a product, very often the rate of heat transfer must be known for efficient equipment and process design. Thermal diffusivity is also required for this case. For agricultural products it is often more simple to calculate thermal diffusivity from specific heat, thermal conductivity and mass density than to measure thermal diffusivity directly. Therefore, thermal conductivity data is most frequently used (Wallapan and Sweat, 1982; Chuma *et al.*, 1981).

Several methods have been developed for determining the thermal transport properties of agricultural products by analysing the heat conduction equation. The suitability of a particular technique is based on physical characteristics of agricultural products, type of application and required accuracy (Voudouris and Hayakawa, 1994). The conventional methods for measuring thermal conductivity are steady state methods. That require the simultaneous measurement of the steady-state heat flux and temperature gradient through test samples in the form of slabs, cylinders, or spheres, for which solutions to the differential equation of heat conduction are known. Since no perfect thermal insulation exists, complicated guarding systems are required to minimize edge or end effects. A further drawback common to all steady-state methods is the relatively long time required to attain thermal equilibrium. The transient methods are rapid and therefore less likely to produce nonuniform saturation distributions (Tavman and Tavman, 1994).

Zuritz *et al.* (1989), reported that thermal conductivity of individual agricultural products is difficult to measure because of physical constraints. Studies in literature on thermal conductivity of small agricultural materials are mostly on bulk thermal conductivity. Bulk thermal conductivity refers to a combined property of materials and air in the pore spaces. It may be suitable for calculating the amount of heat transfer in some processes such as batch drying and cooling that involve stationary materials. Furthermore, the thermal conductivity of individual particles is necessary for calculating heat transfer when materials are either in motion or surrounded by the heating or cooling medium. Zuritz *et al.* (1989), also reported that bulk thermal conductivity is not appropriate in evaluating heat transfer in processes like continuous-flow drying, soaking, and blanching. The measurement and modelling of bulk thermal conductivity of granular agricultural products is a challenge as it depends not only on moisture along with composition, but also on the structure and physical arrangement of the sample (Sweat, 1985). The structural characteristics, which affect conductivity, include the number and size of voids, heterogeneity, impurities, particle-to-particle contact and fibre orientation.

Kazarian and Hall (1965) measured the thermal conductivity of soft white wheat by a line heat source method. A linear increase of thermal conductivity with water content expressed as percent by weight has been deduced from the experimental results:

$$k = 0.1169 + 0.001131\,X_W \tag{1}$$

There are also other studies in literature to reveal us that thermal conductivity of granular food increases linearly with moisture content (Bilanski and Fisher, 1976; Tavman and Tavman, 1998; Sharma and Thompson, 1973; Kusterman *et al.*, 1981; Chuma *et al.*, 1981; Zuritz *et al.*, 1989; Andrieu *et al.*, 1989) and/or with bulk density (Chang, 1986; Tavman and Tavman, 1999; Tavman and Yolci, 1998).

Data on thermophysical properties of processed agricultural products are scarce compared with fruits, vegetables, and other agricultural material. A study by Tavman and Tavman (1998) included bulgur, which is a wheat product similar to macaroni, and two varieties of *Triticum durum* wheat. The moisture contents of the samples studied ranged from 9.17 to 38.65 percent wet basis and the bulk densities ranged from 675 to 827 kg/m^3. Under these conditions, the measured effective thermal conductivities ranged from 0.159 to 0.201 W/m.K. The effective thermal conductivity is found to be increasing linearly with moisture content. Andrieu *et. al.* (1989) studied the thermophysical properties of disc shaped macrospagetti (d = 30mm l = 7-10mm) with different temperatures and moisture contents. Nicolai *et al.* (1995) applied on finite element analysis of heat conduction in Lasagne during thermal processing.

Since macaroni products for soup are granular-shaped materials made from wheat, they can easily represent the behaviour of other granular agricultural products in terms of thermal conductivity. Therefore, they were chosen as a model system in this study. The objective of this study was to determine experimentally the thermal conductivity values of three different macaroni products made for soup in bulk, to relate thermal conductivity with water content and bulk density of the samples, to compare this relation with the other most suitable theoretical models for each sample and to ascertain that macaroni products have characteristics similar to the other granular agricultural products in terms of thermal conductivity.

2. CONDUCTION MODELS

Powdered, porous and granular materials, as two phase systems, consist of a solid phase and a gas phase (mostly air). Since each phase contributes to the thermal conductivity, then any measured conductivity is called as effective thermal conductivity (Wallapan and Sweat, 1982). All the models predict the effective bulk thermal conductivity (k_e) of porous materials as a function of thermal conductivity of the solid phase (k_s) and fluid phase (k_f) and also of porosity (ε), which is defined as the ratio of the fluid volume to the total volume. Some of the models take further into consideration such parameters as the shape of the solid particles, the orientation and distribution of the particles and the contact resistance between the particles.

In some literature, various models and formulae have been proposed for the prediction of effective thermal conductivity of porous systems (Hahne *et al.*, 1991; Tavman and Tavman, 1994; Murakami and Okos, 1990). Some equations for the prediction of the effective thermal conductivity areas follow;

The maximum value of effective thermal conductivity is estimated by the model in which the solid and fluid phases are arranged in alternate layers parallel to the direction of heat flow. In this model, the effective thermal conductivity is given by the weighed arithmetic mean of the thermal conductivities of the two phases:

$$k_e = \varepsilon k_f + (1 - \varepsilon) k_s \tag{2}$$

The minimum value of effective thermal conductivity occurs for the series distribution in which the solid and the fluid phases are in layers normal to the direction of heat flow. In this case the effective thermal conductivity is given by the weighed harmonic mean of the solid and fluid conductivities:

$$k_e = \frac{k_s k_f}{\varepsilon k_s + (1 - \varepsilon) k_f} \tag{3}$$

Maxwell using potential theory for electrical conduction through heterogeneous media obtained an exact solution for the conductivity of randomly distributed and non-interacting homogeneous solid spheres in a homogeneous continuous medium (Hahne *et al.,* 1991):

$$k_e = k_f \frac{2 \varepsilon k_f + (3 - 2 \varepsilon) k_s}{(3 - \varepsilon) k_f + \varepsilon k_s} \tag{4}$$

The model proposed by Woodside and Messmer (1961) is a combination of the series and parallel distributions. They use the electrical conductivity analogy of an aggregate of conductive particles saturated with conducting electrolyte, in order to derive a modified resistor model equation to predict the effective thermal conductivity of porous media:

$$k_e = \frac{a k_s k_f}{k_s (1 - d) + d k_f} + c k_f \tag{5}$$

where $c = \varepsilon - 0.03$, $a = 1 - c$, $d = \frac{1 - \varepsilon}{a}$

Krupiczka (1967) derived a numerical solution for effective thermal conductivity of materials using a model made up of long cylinders ($\varepsilon = 0.476$). In view of the complicated character of the formulas obtained, the two solutions are approximated by a general correlation, which takes into account the effect of porosity;

$$\frac{k_e}{k_f} = \left(\frac{k_s}{k_f}\right)^{A+B\log\frac{k_s}{k_f}} \tag{6}$$

where $A = 0.280\text{-}0.757 \log \varepsilon$, $B = -0.057$

The correlation formula obtained is valid for $0.251 \leq \varepsilon \geq 0.476$, but it can also be used without too much error for porosity near the region under considerations.

Rahman (1992) developed a general form of thermal conductivity correlation,

$$\left(\frac{k_e}{k_o}\right)\left(\frac{1}{1-\varepsilon_a}\right) = \left[1.82 - 1.66\exp\left(-0.85\frac{X_w}{X_{wo}}\right)\right] \tag{7}$$

where X_{wo} is initial water content and k_o is the initial thermal conductivity. This model applied to five food materials (apple, pear, squid, beef, and potato) as a function of water content and porosity. Rahman and Chen (1995) presented that this correlation was developed for the water content between 5-88 % (wet basis), temperature 20-25 °C and porosity 0-0.5. Rahman *et al.* (1997) has explained the disadvantages of this correlation. The first one is that the left side of the equation (7) becomes infinity, which is physically incorrect, and the other is that it does not include any temperature effects on thermal conductivity and it also needs the conductivity values of the fresh foods.

Rahman and Chen (1995) proposed another model having a parameter α, which accounts for residual effect of temperature and structure of a food material as,

$$\frac{k_e - \varepsilon_a k_a}{(1 - \varepsilon_a - \varepsilon_w)k_s + \varepsilon_w k_w} = \alpha \tag{8}$$

where k_a is thermal conductivity of air, k_w is thermal conductivity of water, ε_a is volume fraction of air, and ε_w is volume fraction of water. It is reported that the values of α is applicable for apple for the porosity between 0.482-0.512, water content (wet basis) 0.20-0.60 and temperature 5-45 °C. It is also reported that the values of α varied within a range from 0.371 to 0.575.

Rahman *et al.* (1997) developed a general power law correlation as given as follows:

$$\frac{\alpha}{1 - \varepsilon_a + \frac{k_a}{(k_w)_r}} = 0.996\left[T/T_r\right]^{0.713} X_w^{\ 0.285} \qquad (\alpha = 0.12) \tag{9}$$

where "$_r$" is reference temperature (0 °C) and X_w is the water content (wet basis). It is reported that this correlation is developed by considering the thermal conductivity of apple, pear, raisin, cornstarch and potato as a function of water content, porosity and temperature. The water content is between 14 to 88 % (w.b.), porosity 0-0.56 and the temperature 5-100 °C.

3. MATERIALS AND METHODS

3.1 Materials

In this study, thermal conductivity of three kinds of uncooked macaroni products made for soup was studied. These were (1)- Melon seed shaped macaroni (Risi) with 2.7 mm width, 1.5 mm thickness and 8 mm length, (2)- Small disk shaped macaroni with 3.9 mm average diameter, 1.6 mm thickness and (3)- flat noodles with 14 mm length, 5 mm width and 2.6 mm thickness. They were measured in bulk at different moisture contents and bulk densities. Samples of various moisture contents were obtained by soaking them into boiling water for predetermined periods of time and were placed in airtight plastic bags for subsequent thermal conductivity measurements. Chemical composition of each group of the material determined by laboratory experiments carried out according to AOAC (Anonymous, 1990) is given in Table 1.

Table 1. Chemical composition of the macaroni samples.

Food	Moisture Content (% w.b.)	Ash (% d.b.)	Fat (% d.b.)	Protein (% d.b.)	Carbohydrate (% d.b.)
Melon shaped macaroni	10.36	0.792	0.31	11.35	67.90
Small disk shaped macaroni	10.27	1.49	1.50	11.75	75.50
Flat noodles	10.51	1.25	1.49	11.75	74.90

d.b. refers to dry basis, and w.b. refers to wet basis.

3.2 The Theory of Thermal Conductivity Measurement

A modified hot wire method selected to measure the effective thermal conductivity of sample was used in this study. This method has been used for many food materials (Tavman and Yolci, 1998; Tavman and Tavman, 1997). A line source, i.e., a thin straight wire, through which a constant electric current passed by generating constant heat, was embedded in the sample under test. When the sample and the wire was at uniform and constant temperature, constant power was supplied to the heater element and the temperature rise of the heating wire was measured by a thermocouple and recorded with respect to time during a short heating interval (see Fig. 1).

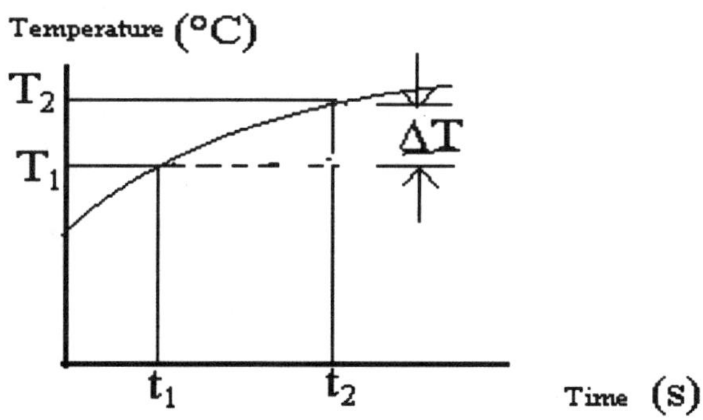

Figure 1. Time- temperature relationship

The thermal conductivity (k) of the sample was obtained from the temperature-time record and power input as;

$$k = \frac{Q \ln(t_2 / t_1)}{4\pi (T_2 - T_1)} \tag{10}$$

where T_1 and T_2 are temperatures at times t_1 and t_2, Q is the heat flow per unit time, per unit length of the heating wire. The values $d(\Delta T)/d(\ln t)$ were obtained from the straight line portion of a plot of ΔT versus ln t obtained from a test run recording of ΔT versus t, (see Fig. 2). Q was calculated from the voltage drop (V) and the current (I) in the resistance wire;

$$q = 0.86\, R I^2 = 0.86\, V(I/L) \tag{11}$$

where R is the resistance per unit length of heating wire in (Ω/m), and L the length of the heating wire in m. Equation (10) gives results with error of 1 % or less when the condition $r/(2at) < 0.17$ is satisfied ("a" is the thermal diffusivity in m^2/s, r is the radius of the heating wire in m)

Using Eqs. (10) and (11), thermal conductivity was given by the equation:

$$k = 0.00277\, R\, I^2 \frac{\ln(t_2/t_1)}{V_2 - V_1} \tag{12}$$

where

I : electric current passing through heating wire (A),

t_1, t_2 : sampling time (s),

V_1, V_2 : electrical potential of K-type thermocouple (mV) at time t_1 and t_2,

V_1, V_2 : ηT_1, ηT_2

η : thermoelectric power of K-type thermocouple(mV/°C) ($\eta = 0.0405$ mV/K, -10 °C < T < 200 °C)

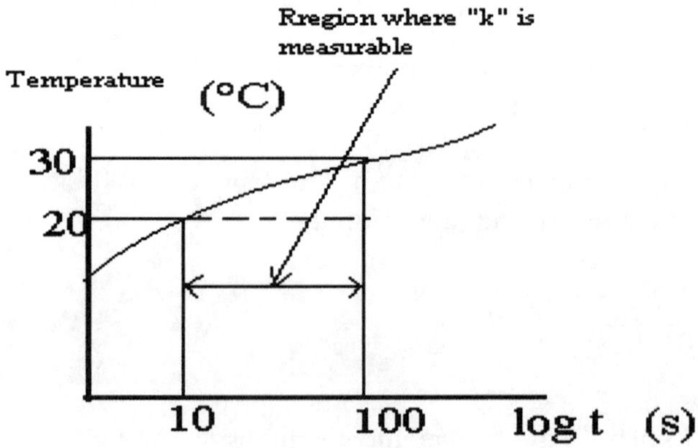

Figure 2. Temperature plotted against log t

Tavman and Tavman (1994) reported several possible error sources of this method:

1. Approximation character of Eq. (10): This error is minimised by using a thin wire, by localising the thermocouple close to the heater, and by discounting the early period of the temperature-time record.
2. The finite length to diameter ratio of the real heating wire in contrary to theory with the ideal linear heat source of infinite length to diameter ratio: Blackwell (1956) has shown theoretically that probes with a length to diameter ratio of 30 or more will have negligible errors in this respect.
3. Finite volume of the tested sample in contrary to the theory based on the infinite sample. However, if the testing intervals are limited to such cases, in which no increase of temperature at the sample surface is observed, the samples of practical dimensions behave as though they are of infinite size. Healy (1976) has proved that for $b^2/(a\ t) > 5.768$, the effect of an outer boundary is below ±0.01 % and can be neglected, (b is the outside radius of the medium).
4. Contact resistance between the probe and the surrounding sample. This error can be minimised by exclusion short heating times.

Some additional error can be caused by changes of the wire resistance with changing temperature (e.g. nonconstant power input). This error is negligible if the heater wires of low temperature coefficient of resistance are used (Tavman and Tavman, 1997).

3.3 Experimental Apparatus

In our experiments, a modified hot wire technique was used. The heating wire was placed between two rectangular shaped materials. The first one was an insulating material of known thermal properties which was a part of the measuring probe, and the second one was the granular sample placed in a rectangular shaped sample holder, placed at the top side of the heating wire. The dimensions of the sample holder were 10 cm in length, and 3 cm in width and 4 cm in height (see Fig. 3). Granular material was placed into the holder loosely. The heater wire with a low temperature coefficient of resistance was used. Time interval between 10 and 90 seconds of heating was used for measurement. The thermal conductivity of sample is given by the equation, which is similar to Eq. (12);

$$k = F\frac{I^2 \ln(t_2/t_1)}{(T_2 - T_1)} - H \qquad (13)$$

where F and H are specific constants of the probe to be determined with materials of known thermal conductivities. The slope of the curve ΔV versus $\ln t$ was measured in the linear part and thermal conductivity (k) was measured by using this slope and current passing through the wire (see Fig. 2). The temperature increase of the heating wire was less than 20 °C.

By this method, the thermal conductivity is measured with an accuracy of ± 5% and reproducibility of ± 2%.

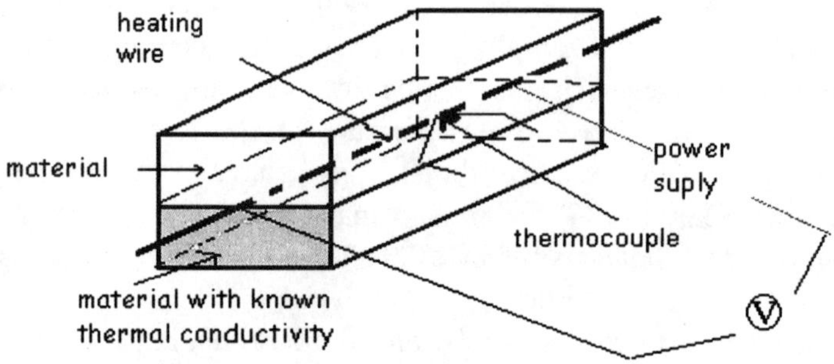

Figure 3. Hot wire method used for measuring thermal conductivity

3.4 Modelling of Thermal Conductivity

Theoretical thermal conductivity data were obtained by using Eqs.(2)-(6). In the present study the solid phase was the grain and its thermal conductivity k_S may be estimated from the thermal conductivity and mass percent of the constituents by the equation proposed by (Sweat, 1985):

$$k_s = 0.60X_w + 0.20X_p + 0.245X_c + 0.18X_{fat} \tag{14}$$

where the subscripts stand for various components: w – water, p- protein, c-carbohydrate, *fat* - fat, a – ash. 0.61, 0.20, 0.205, 0.175, 0.135 are respectively the thermal conductivities of water, protein, carbohydrate, fat and ash in W/m.K , at room temperature (25°C) (Sweat, 1985). The fluid phase is the air between solid grains and has thermal conductivity of k_f = 0.026 W/m.K at 25°C. Porosity is found from the following equation (Sweat, 1985);

$$\varepsilon = 1 - \frac{\rho_{bulk}}{\rho_{solids}} \tag{15}$$

The bulk density of each sample in the sample holder was determined by measuring the volume of a known weight. The numerical values of k_s for individual products can be seen in Table 3.

Two different thermal conductivity models of experimental values were obtained by application of linear and multiple regression analysis (AXUM, Version 3.0). They are expressed as:

$$k_e = a + b\, X_w \tag{16}$$

$$k_e = a + b\, X_w + c\rho_{bulk} \tag{17}$$

where a, b, c are the empirical constants, given in Table 2 .

4. RESULTS AND DISCUSSION

Thermal conductivities of three kinds of macaroni product were measured as a function of moisture content (w.b.) and bulk density. All measurements were performed at room temperature, at about 25°C. Each measurement was repeated six times and the mean values were reported. Thermal conductivity in this study refer to bulk conductivity, i.e. thermal conductivity of the sample plus the entrained air. The results of thermal conductivity of the samples are shown in Figure 4.

Within the ranges of density, thermal conductivities reached 0.124-0.324 W/(m.K) for melon seed shaped macaroni (M.C. w.b. 10.23-44.78 %), 0.163-0.283 W/(m.K) for small disk shaped macaroni (M.C. w.b.10.27-39.01 %), and 0.138-0.256 W/(m.K) for flat noodles (M.C. w.b. 10.51-37.97 %). Chang (1986) measured thermal conductivity of hard red winter wheat with a M.C. ranged from 11 to 18 % (w.b.), and it is presented that experimental values ranged from 0.181 to 0.202 W/m.K. Tavman and Tavman (1998) also measured the effective thermal conductivity of *Triticum durum* wheat and of bulgur which is a product of wheat. The values obtained ranged from 0.159 to 0.201 W/m.K. As macaroni products are made from wheat, these values correspond reasonably our experimental results. Thermal conductivity of granular agricultural products was also studied by other authors (Bilanski and Fisher, 1976; Tavman and Tavman, 1998; Sharma and Thompson, 1973; Kusterman et al., 1981; Chuma et al., 1981; Zuritz et al., 1989; Chang, 1986; Tavman and Tavman, 1999; Tavman and Yolci, 1998). The measured effective thermal conductivity of all samples linearly increased with moisture content and bulk density and these results are in agreement with the results given in the above literature.

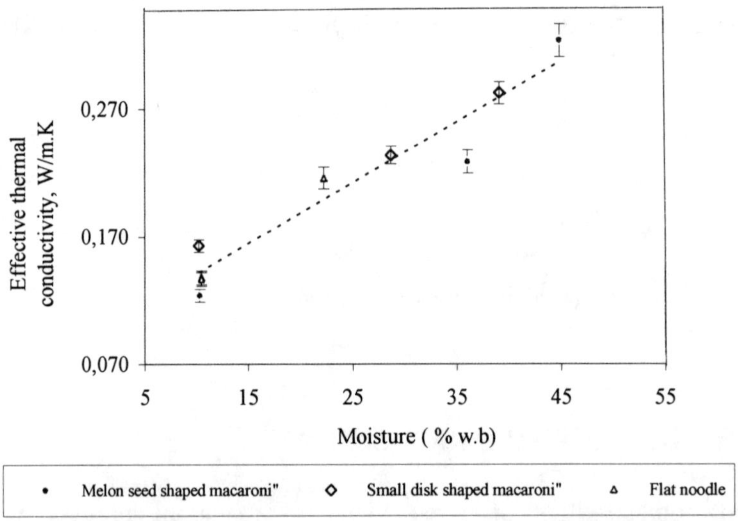

Figure 4. Graph for thermal conductivity vs. moisture content of macaroni products. (Vertical lines show 95% confidence intervals, $R^2 = 0.93$, $k_e = 0.0049.X_W + 0.00849$)

For all of the macaroni samples used in this study, two linear equations were obtained: The first one depends on the moisture content and the other depends on bulk density besides moisture content. The regression equations are given in Table 2.

Table 2. Thermal conductivity results and equations of different kinds of macaroni products used in this study

FOOD	X_w (%)	ρ_{bulk} kg/m^3	k_e W/m.K	Equations	R^2
Melon seed shaped	10.3-44.8	621-800	0.124 -0.324	$k_e = 0.0054.X_W + 0.0602$	0.941
Small disk shaped	10.2-39.0	733-873	0.163 -0.283	$k_e = 0.0051.X_W + 0.0866$	0.999
Flat noodles	10.5-38.0	560-707	0.138 -0.256	$k_e = 0.0042.X_W + 0.1048$	0.926
Macaroni samples	10.3-44.8	560-324	0.124 -0.324	$k_e = 0.0049.X_W + 0.0085$	0.936
Macaroni samples	10.3-44.8	560-873	0.124 -0.324	$k_e = 0.0052.X_W + 0.0787.\rho_{bulk} + 0.023$	0.927

Table 3. % difference between the experimental thermal conductivities and the models (where k_s is thermal conductivity of individual sample, and $k_1, k_2, k_3, k_4, k_5, k_6$, are thermal conductivity obtained from Perpendicular, Maxwell, Woodside, Krupizcka, Geometric, Parallel models respectively).

Food	ε	k_s	k_1	k_2	k_3	k_4	k_5	k_6
Melon seed shaped macaroni	0.310	0.2267	46.4	17.2	5.2	21.4	6.8	32.6
	0.261	0.2267	41.9	11.2	8.4	14.5	0.2	35.1
Small disc shaped macaroni	0.254	0.2445	51.8	25.3	8.3	27.3	14.9	16.0
	0.200	0.2445	46.3	18.7	5.8	17.1	8.1	18.1
Flat noodles	0.589	0.2456	70.1	56.4	39.5	53.8	52.5	15.7
	0.560	0.2456	69.0	54.2	36.0	52.4	49.6	11.7

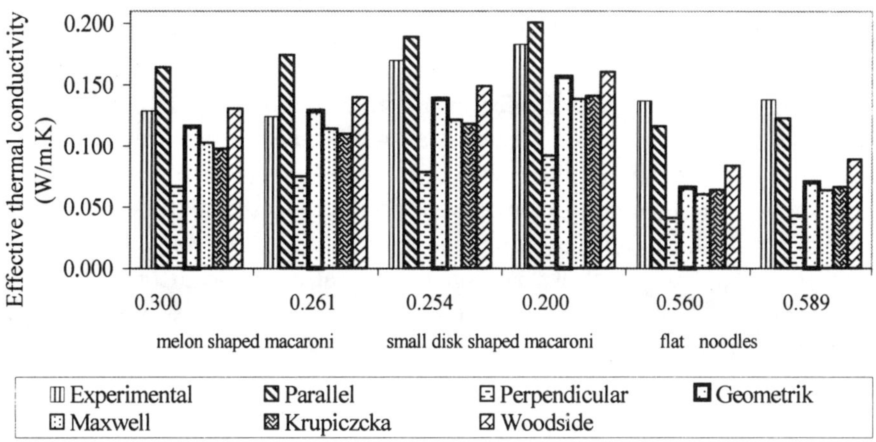

Figure 5. Bulk thermal conductivity – experimental data and model values. The numbers at the horizontal axis express product porosity.

For the samples used in this study (porosity between $\varepsilon = 0.20$ and 0.6), the effective thermal conductivity was calculated from geometric, Maxwell, Woodside and Messmer models, as well as the maximum and minimum values calculated by the parallel and series phase distribution models. The results obtained were compared with the experimental data. The results of the calculations are given in Fig. 5 and Table 3. In respect to experimental material, the ratio $k_s/k_f \approx 10$ was used. Table 3 shows that Geometric model predicted the best values for Melon seed shaped macaroni. Similar conclusions can be made for Woodside and Messmer's model at small disk shaped macaroni, and for Parallel model at flat noodles (see Table 3).

5. CONCLUSIONS

In this study, the variation of thermal conductivity of macaroni products is investigated in bulk as a function of moisture content. Bulk thermal conductivity is a linear function of the moisture content and bulk density. As thermal conductivities of macaroni products for soup increase with increasing moisture content, it is concluded that they have the same characteristics with the other granular agricultural products. The measured thermal conductivity of macaroni products is of the same order of magnitude as thermal conductivity of hard wheat.

Comparing standard deviations between experimental data and the data predicted by the mathematical models, the Parallel model, the Geometric model, and Woodside & Messmer's model were found to be the best ones explaining the dependence of effective thermal conductivity on porosity. It is observed that the most suitable theoretical models also depend on the kind of macaroni product used in this study. The theoretical models are found to be dependent on the geometry of macaroni product. Since only one model can not explain the heat conduction phenomenon in the whole range of k_s/k_f and for different porosities, it can be concluded that it is also difficult to apply one generalized model for thermal conductivity calculations of other granular agricultural products.

REFERENCES

Andrieu, J., Gonnet, E., and Laurent, M., 1989. Thermal conductivity and diffusivity of extruded durum wheat pasta. *Lebensm.-Wiss.u.Technol.* **22**: 6-10.

Anonymous, 1990, *Official Methods of Analysis of the Association of Analytical Chemist.*, AOAC Incorporation, Virginia.

Bilanski, W.K., and Fisher, D.R., 1976, Thermal conductivity of rapeseed. *Trans. ASAE* **19**: 788-791.

Blackwell, J. H., 1956, The axial flow error in the thermal conductivity probe. *Can. J. Phys.* **34**: 412-417.

Chang C.S., 1986, Thermal conductivity of wheat, corn, and grain sorghum as affected by bulk density and moisture content. *Trans. ASAE* **29:** 1447-1450.

Chuma, Y., Uchida, S., and Shemsanga, K.H.H., 1981, Bulk physical and thermal properties of cereal grains as affected by moisture content. *J. Fac.Agr. Kyushu Univ.* **26:** 57-70.

Hahne, E., Song, Y.W., and Gross, U., 1991, Measurement of thermal conductivity in porous media. In *Convective Heat and Mass Transfer in Porous Media (*S. Kakac, B. B.Kelkes, F. A. Kulacki and F. Arın, eds.), Kluwer Academic Publishers, Dordrecht, pp. 523-538.

Healy, J. J., de Grot, J. J., and Kestin, J., 1976, The theory of the transient hot-wire method for measuring thermal conductivity. *Physica* C **82**: 391-408.

Kazarian, E. A., and Hall., C. W., 1965, Thermal properties of grain. *Trans. ASAE* **8:** 33-38.

Krupiczka, R., 1967, Analysis of thermal conductivity of packed beds. *Journal of Chem. Engng.* **7**: 122-144.

Kustermann, M., Scherer, R., and Kutzbach, H.D., 1981, Thermal conductivity and diffusivity of shelled corn and grain. *J. of Food Process Eng* **4**: 137-153.

Murakami, E. G. Okos, and M. R., 1990, Measurement and prediction of thermal properties of food. In: *Food Properties and Computer-Aided Engineering of Food Processing Systems* (R.P. Singh and A.G. Medina, eds.), Kluwer Academic Publishers, London, pp. 105-112.

Nicolai, B. M., Vanden Broek, P., Schellenkens, M., De Roeck, G., Martens, T., and De Baerdemaeker, J., 1995, Finite element analysis of heat conduction in lasagne during thermal processing. *International Journal of Food Science and Technology* **30**: 347-363.

Rahman, M. S., 1992, Thermal conductivity of four food materials as a single function of porosity and water content. *J. Food Eng.* **15**: 261-268.

Rahman, M. S., and Chen, X. D., 1995, A general form of thermal conductivity equation as applied to an apple: effects of moisture, temperature and porosity. *Drying Technol.* **13:** 1-18.

Rahman, M. S., Chen, X. D., and Perera, C. O., 1997, An improved thermal conductivity prediction model for fruits and vegetables as a function of temperature, water content and porosity. *Journal of Food Engineering* **31**: 163-170.

Sharma D.K., and Thompson T.L., 1973, Specific heat and thermal conductivity of sorghum. *Trans. of the ASEA* **19**: 114-117.

Sweat, V.E., 1985, Thermal properties of foods. In *Engineering Properties of Foods (*Rao M.A., and Rizvi S.S.H, eds), Marcel Dekker Inc., Newyork., pp 120- 285.

Tavman, I. H., and Tavman, S., 1994, Thermal conductivity of granular porous media. In *Proceedings of the 1994 Engineering Systems Design and Analysis Conference,* ASME **64**: 165-169.

Tavman, S., and Tavman, I.H., 1997, Thermal conductivity of granular food materials. In *Modelling of Thermal Properties and Behavior of Foods during Production, Storage and Distribution* (Z. Mayer and P. Nesvadba, eds.), Food Research Institute Prague, pp. 223-228.

Tavman, S., and Tavman, I.H., 1998, Measurement of effective thermal conductivity of wheat as a function of moisture content. *Int. Comm. in Heat and Mass Transfer* **25**: 733-741.

Tavman, S., and Tavman, I.H., 1999, Thermal conductivity of dairy products. *Journal of Food Engineering* **41**: 109-114.

Tavman, S., and Yolcı, P., 1998, Thermal conductivity measurements of granular and powdered foods. In *Proceedings of 3rd Karlsruhe Nutrition Symposium: European Research Towards Safer and Better Food, Review and Transfer Congress,* Karlsruhe, pp. 451-460.

Voudouris, N., and Hayakawa, K., 1994, Simultaneous determination of thermal conductivity and diffusivity of foods using a point heat source probe: A theoretical analysis. *Lebensm.-Wiss. U. Technol.* **27**: 522-532.

Wallapan, K., and Sweat, V. E., 1982, Thermal conductivity of defatted soy flour. *Transactions of the ASAE* **25**: 1440-1444.

Woodside, W., and Messmer, J. H., 1961, Thermal conductivity of porous media, I- unconsolidated sands. *Journal of Applied Physics* **32**: 1688-1699.

Zuritz C.A., Sastry S.K., McCoy S.C., Murakami E.G., and Blaisdell J.L., 1989, A modified fitch device for measuring the thermal conductivity of small food particles. *Trans. ASAE* **32:** 711-718.

Transgenosis for Changes of Physical Properties of Plants and Plant Products

MILOŠ ONDŘEJ
Institute of Plant Molecular Biology., Academy of Sciences of Czech Republic

1. INTRODUCTION

Integration of cloned genes into the plant genome - plant transgenosis became reality after the finding of Chilton et al. (1977) that the soil bacterium *Agrobacterium tumefaciens* introduces part of its DNA (called T-DNA) into the plant genome. The first transgenic commercionaly available plant varieties appeared in 1994 and the area of growing transgenic plant varieties in 2001 already reached approximately 500 000 km^2, which is probably the greatest breeding success in the history.

Although the change of physical markers accompanies transgenosis from the very begining as several T-DNA genes acting singly lead to morphological changes. Up to now, there was no systematic research of the possibilities of changes of physiological properties of plants by transgenosis and this is probably the first systematic review of this topic. This article will be just an introduction to the problem, not exhaustive review of it.

1. At least two main types of transgenes of potential plant breeding use should be distinguished:
2. Transgenes which carry new traits absent in the plant genome. These transgenes do not have any homologous sequences in given plant genome. The coding sequences can originate from any genome (plant, animal, bacterial, virus) but the regulatory sequences (promotor, terminator) should be either from the plant genome, plant virus genome or T-DNA genome. In some instances, the mRNA could be unstable or

the efficiency of translation can be low. In such cases, the gene expression can be increased by remodelling the primary coding sequence to fulfill better the needs of the translation machinery of plant cells. Such coding DNA sequences can be synthesized *in vitro*, introduced into the insert construction of the vector plasmid and integrated into the plant genome.

3. Transgenes which inhibit the action of already existing plant gene.
The transgene code for antisense RNA, which is complementary to mRNA and which, together with corresponding mRNA, forms *in vivo* unstable double-stranded RNA structure which can not be translated and is degraded by specific nucleases. Gene expression can also be inhibited sometimes very efficiently by introduction of another copy of already existing gene in *sense* orientation. These types of transgenic plants correspond to deletion or inactivation mutations.

Although there is enormous number of types of transgenic plants produced in scientific research, the overall number of types of transgenic agriculture plants which have already been commercially released is still very limited. There are herbicide resistant plant varieties, virus resistant varieties, male sterile ones and combination of these characters. Of minor importance are transgenic tomatoes with prolonged shelf-life, transgenic rape with changed spectra of fatty acids in their seed oils, blue-flowering carnations or carnations with prolonged vase-life of inflorescences.

The last mentioned types concern physically measurable markers (timing, colour). Hunderds of other types of plants are being prepared for the process of release of new commercial varieties and part of them code for traits which can be characterized also by physical parameters. Changes of physical properties due by transgenosis concerns either transgenic plants as such (their architecture, life cycle time and other characters) or physical properties of their products (lignin, starch, proteins, oils or entirely new products). Although there are many possibilities of genetic manipulation of traits of transgenic plants described in scientific literature, the details of some perspective transgenes are unknown as they were developed by biotechnological companies and are the subjects of commercial secret.

Before new types of transgenic plants with changes in physical properties can be released into the environment, they must be considered from the point of view of safety to the environment, biodiversity and human health as far as concerns not only immediate, but also long-term possible detrimental effects.

Before considering the application of individual types of transgenic plants, the process of plant transgenosis should be described in brief. The natural genes involved have special plant developmental regulation effects. Transgenosis in plants can be achieved by two ways: Transformation by bacteria of the genus *Agrobacterium* or direct gene transfer. Bacteria of the genus *Agrobacterium* carry large plasmids called Ti in *A. tumefaciens*, *A. rubi* and *A. vitis* and Ri in *A. rhizogenes*. These Ti (tumor inducing) plasmids have the ability to induce plant tumors by inserting constant segment of their large plasmids (T-DNA or transferred DNA) into plant chromosomes. T-DNA carries specific set of genes which trigger plant cell divisions. These genes have been studied and they are known to cause specific developmental changes of plant cells and whole regenerated plants. For the use of T-DNA as vector, all the original T-DNA genes can be removed leaving only 24 base pair borderline sequences, and substituted by any genes which the experimenter desires to introduce into the plant genome. The direct DNA transfer uses either microparticle bombardment (biolistic transformation), microinjections, electroporation or other means. Vector plasmids amplified in *E. coli*, which do not need any T-DNA border sequences are used in direct transformation method either in original circular or in linearized forms.

2. GENETIC MANIPULATION OF PHYSICAL TRAITS OF PLANTS

2.1 Developmental Changes Induced by T-DNA Auxin Synthesizing Genes

Two of the Ti-plasmid T-DNA genes have been shown to encode a biosynthetic pathway for IAA production is absent from plants. The *iaaM* gene encodes an enzyme converting tryptophan to indolacetamide (Thomashow et al. 1986, Van Onckelen et al. 1986), which is subsequently hydrolyzed to IAA by the action of the *iaaH* gene product, enzyme indolacetic acid hydrolase (Schröder et al. 1984, Thomashow et al. 1984). The auxin ß-indolacetic acid (IAA) is produced by this (bacterial) auxin biosynthesis pathway, different from plant one. Thus, IAA is overproduced in transgenic tissues expressing feedback regulatory changes in auxin degradation as well as on the synthesis and degradation of all the other plant hormones. Different vector plasmids which transfer to the plant genome individual T-DNA morphoregulatory genes and their combinations have been constructed (see Vlasák and Ondřej 1992).

The production of auxin by transgenic plant tissues carrying *iaaM* and *iaaH* genes was shown to be up to ten fold increased in comparison with controls, which has developmental and morphogenetic consequences (Klee et al. 1987, Sitborn et al. 1991, 1992a, 1992b, Romano et al. 1983). General features of transgenic tobacco plants overproducing IAA are decreased growth, increased apical dominance, reduced branching of stems and formation of roots in lower parts of the stems. The leaves are smaller and narrower than in controls and they show curling. Stems produce more secondary xylem and phloem cells than untransformed controls. Most of cell types in leaf tissue are larger than in control leaves. The increased cell and intracellular space in transgenic plants result in leaves thicker and paler than in controls. The leaf curling is the result of greater tissue volume on the adaxial halves of leaves relative to the abaxial halves.

Of practical importance is the breeding of transgenic poplar (*Populus tremula* x *P. tremuloides*) which overexpress auxin synthesis genes. The trees showed alternations in height and stem diameter length, internode elongation, leaf enlargement and degree of apical dominance. Changes in wood anatomy, most notably reduction in the vessel size, an increase in vessel density and changes in ray development have been observed (Tuominen et al. 1995). These results showed the possibility of manipulating wood properties in forest trees through controlled changes of IAA concentration and distribution.

The auxin level can also be decreased by transgenosis by formation of IAA-amino acid conjugates, for instance by transgenosis with the *Pseudomonas sevastanoi* indoleacetic acid -lysine synthetase gene (Romano et al. 1991). IAA-amino acid conjugates are inactive, or less active forms of IAA and their relative level has been increased and the IAA level concomitantly decreased by transgenosis (Spena et al. 1991). Leaves of such transgenic tobacco and potato plants showed epinasty (upcurvature of the leaves), increased internode length and decreased root growth.

2.2 Developmental Changes Induced by T-DNA Cytokinin Synthesis Gene

The *ipt* gene of *A. tumefaciens* T-DNA codes for isopentenyltransferase, the first enzyme in the cytokinin biosynthetic pathway (Akiyoshi et al. 1983, Barry et al. 1984). There are tens of different natural plant and bacterial cytokinins and their pathway in transgenic plants is not new, except for the rate-limiting enzymatic step which is enforced by isopentenyltransferase gene. The increase of the level of cytokinins in transgenic tissues can be up to hunderd fold. Overexpression of *ipt* by by strong constitutive promotors induces shooty teratomas, but at the same time it blocks almost fully root growth and regeneration of whole plants (Smigocki and Owens 1998).

General symptoms of cytokinin overproduction are the decrease of apical dominance, excess branching of the stems, decrease of the chlorophyll content (Ondřej et al. 1991), changes of leaf structure and morphology and reduction of senescence (Smart et al 1999, Hewelt et al. 1994, Smigocki et al. 1995) and osmotic stress symptoms (Thomas et al. 1995). Endogenously increased cytokinins not only influence different growth parameters, but have the potential to alter differentiation pattern (Hewelt et al. 1994).

Smigocki (1995) has constructed chimeric wound-inducible cytokinin biosynthesis transgene by replacing the original T-DNA promotor with proteinase inhibitor IKK promotor. Heat-shock induced chimeric *ipt* transgene (Schműlling et al. 1985) and promotorless transgene (Hewelt et al. 1994) have been also used. Gan and Amasino (1995) have developed a senescence inhibiting transgenosis system, where the production of cytokinins is specifically targeted to senescent leaves as the senescence-specific SAG2 promotor was used to drive the expression of the *ipt* gene. This senescence-inhibition system is autoregulating because the prevention of senescence attenuates the promotor activity and the expression of the gene to prevent the overproduction of cytokinins. Zhang et al. (2000) obtained flooding-tolerant *Arabidopsis* carrying this chimeric transgene.

2.3 Developmental Changes Induced by Integration of *rol* T-DNA Genes of *Agrobacterium rhizogenes*

Integration of the group of so called *rol* T-DNA genes of *A. rhizogens*, which consist of four genes *rolA*, *rolB*, *rolC* and *rolD* leads to the induction of hairy roots. These transformed roots in agar culture have increased ability to regenerate plants. Plant regeneration can sometimes be induced even directly after transformation, without the stage of hairy root formation. In the early experiments, the effect of integration of all four *rol* T-DNA genes have been studied and only later, when individual clonned *rol* genes became available, the effects of individual genes have been studied. There are two main groups of *A. rhizogenes* strains: mannopine strains, the T-DNA of whose consists of single region an it carries mainly *rol* genes and genes coding for the synthesis of plant tumor specific substances - opines and agrocinopines. The T-DNA of agropine group of *A. rhizogenes* strains consists of two regions: the left one (left or T_L-DNA), which is similar to that in the mannopine group and the second one (right or T_R-DNA) , which carries genes coding for auxin synthesis similar to genes *iaaM* and *iaaH* of *A. tumefaciens*. Agropine strains generally have higher potential to root induction and plant regeneration than mannopine strains. In this outlook we shall deal mainly with plant cells carrying TL-DNA genes of the mannopine strains of *A. rhizogenes*. The T_L-DNA induces 18 potentiaal genes. (Slightom et al. 1986) the four *rol* genes of which are best studied. The transformation by both whole *A. rhizogenes* T-DNA and its individual genes usually leads to characteristic phenotype changes: wrinkled leaves, shortened internodes, rapid root growth, protruded pistils and other flower abnormalities resulting sometimes in male sterility, reduced apical dominance and others (Oono et al. 1990). There are no characteristic changes in the auxin levels in transgenic plants (Ondřej et al. 1989).

The characteristic altered phenotype induced by *A. rhizogenes* is usually regarded as undesirable, but the altered phenotype features can have potential applications in the plant improvement in special cases. For example, in the horticultural industry some morphological alternations like dwarfing, altered flowering or increased branching can be useful in ornamental plants (Chrisley et al. 1997). In snapdragon, flower number was increased dramatically by increased branching induced by *A. rhizogenes* transformation (Handa 1992). There are also changes in the timing of the plant. Damiani and Arcioni (1991) showed that some perennial forage legumes became annual and other lost the capacity to flower. Inhibition of flowering (Tanaka and Matsumoto 1993) as well as accelerated flowering (Suginuma and Akihama 1995) have been reported after *A. rhizogenes* transformation.

Chrisley (1997) has reviewed the results of transformation of different plant species by whole *A. rhizogenes* T-DNA and individual *rol* genes and reports the regeneration of transformed plants of 53 species, mainly belonging to crop plants. He has shown that transformation by *A. rhizogenes* can stimulate rooting of cuttings from some recalcitrant crops, especially woody species. In some species (olive) transformed trees grew faster and produced significantly more flowers, fruits and oil per tree (Strobel et al. 1988). Vartanian et al. (1989) have shown that *A. rhizogenes* transformation of *Brassica napus* have increased its drought resistance.

Concerning individual T-DNA genes, the *rolA* gene causes most of the morphological changes caused by *A. rhizogenes* T-DNA (Martin-Tanguy et al. 1996). The *rolB* gene drastically increases the sensitivity of plant cells to auxin (Maurel et al. 1991) and accelerates senescence (Schműlling et al. 1988). Single *rolA* and *rolB* loci have the ability to trigger root formation in excised leaves or stems of tobacco (Cardarelli et al. 1987), while the *rolC* postpones senescence and accelerates flowering of tobacco (Oono et al 1990). This gene is responsible for the release of cytokinins from glucoside conjugates (Estruch et al 1991a). Concerning protein coded by *rolB*, similar activity concerning auxin conjugates has been proposed (Estruch et al. 1991b), but it has been disproved. The *rolD* gene seems to be without any morphological effects. Concerning hormonal status of transgenic plants carrying individual *rol* genes, there are no clear-cut differences from untransformed controls (Dehio et al. 1983, Nilsson et al. 1993, Schműlling et al. 1993).

2.4 Other Developmental Changes Induced by Transgenes

Phytochromes are important factors which direct the plant development which perceive and transmit light signals. The light is perceived by a complex system of different photoreceptors, which detect different light qualities over a wide spectrum range. These photorecetors include phytochromes which are responsible for the detection of far/red and red light and akin cryptochromes sense also blue and UV-A light. Phytochromes are pigments consisting of polypeptide of about 125 kDa carrying chromophore moiety, the phytochromobilin (Batschauer et al. 1998). Sharrock and Quail (1990) and Clack et al. (1994) described five phytochromes as derived from *Arabidopsis* cDNA sequences and designated them PHYA, B, C, D and E. Phytochromes regulate so called shade avoidance syndrome. When plants grow in proximity, for instance in forests or in herbaceous communities, there is a competition for light. It is manifested by a rapid increase in the extension growth of stems and petioles at the expense of leaf growth. In fields the shade avoidance decreases photosynthesis and yield. Multiple phytochromes are involved in this phenomenon. Increase of the synthesis of PHYA by transgenosis in tobacco suppress shade avoidance, causing proximity-conditioned dwarfing, with concomitant increase of the harvest index, or abolishes the shading effect without any difference in the growth (McCormick et al. 1999, Robson et al. 1996). There are several other ways of modifying plant development by transgenosis. Dwarfism is often the result of a mutation blocking the gibberellin synthesis pathway. Similar effect was obtained also by PHYA gene transgenosis (Fray et al. 1995).

As an instance of the opposite, beneficial effects of transgenosis, Goto et al. (2000) have described interesting developmental changes of lettuce after introduction of the soybean iron storage protein ferritin into the lettuce genome. They observed enhanced growth at the early developmental changes resulting in weight up to 42% greater than those in transgenic control plants. The transgenic lettuce showed photosynthesis rates superior to controls and grew larger and faster as compared with controls.

2.5 Reporter Genes that Change the Physical Properties of Plant Tissues

The inserted T-DNA of transgenic plants usually contains two or three transgenes. A selectable transgene is necessary immediately after transgenosis to select the transgenic cells in the overhelming majority of nontransgenic cells. The main selectable transgenes code for antibiotic (usualy kanamycin or hygromycin) resistance or herbicide (phosphinothricin) resistance. Second important class are reporter genes, the activity of whose can be easily scored and usually also measured. They are very useful in science and they are often included also in insert constructions together with transgenes of importance in plant breeding to facilitate the detection of transgenic plants.

Three main types of reporter transgenes are used:

1. **β-glucuronidase (GUS)** gene coding for bacterial (*E. coli*) enzyme (Jefferson et al. 1987) which can be clearly identified histochemicaly or biochemicaly, but unfortunately it is destructive test, which can not be used universaly.
2. **Luciferase transgene** which activates added substrate luciferin to emit light (Ow et al. 1986). There are two types of luciferase transgenes: bacterial transgene from *Vibrio harvei* and the beetle *Photinus pyralis* transgene. This type of reporter transgenes is now used only for special purposes.
3. **Transgene coding for green fluorescent protein (GFP)** from sea urchin *Aequora victoria* (Chalfie et al. 1994). This transgene was first used mainly in animals and its use in plant was limited because of different codon usage different from that in the plant genome and because of cryptic intron operating in the plant genome. New mutated, gene shuffled and synthetic variants are, however, well suited for the plant genome and now of wide use. GFP expression can be scored easily using a long-wave UV lamp in the level of whole plants or by epifluorescence microscopy using blue light. Blue laser can also be used. There are also spectral variants of GFP (Heim et al. 1994, 1995, Heim and Tsien 1996). The advantage of this system is that subcellular localization of the protein can be traced histochemicaly. There are variants of the transgene which localize the GFP protein in the endoplasmic reticulum of plant cells (Haseloff et al. 1997), in the cell nuclei (Galbraith et al. 1995), in the

mitochondria (Köhler et al. 1997a), chloroplasts (Köhler et al. 1997b, Jang et al. 1999), or without specific location. The method of delivery of GFP marker and following selection chloroplast DNA is also available (Sidorov et al. 1999).

2.6 Tolerance to External Factors

Plants are exposed to several physical stress conditions: draught, salt cold, frost, overflooding, heat. There are several ways to enforce the stress defense reaction by transgenosis and some of them will be mentioned in this brief review.

Salt has detrimental effect on plants because it causes a water deficit which results in osmotic stress. Plants which thrive in hypersensitive environments posses specific mechanisms to adjust their internal osmotic status. One of such mechanisms is the ability to accumulate low-molecular-weight organic compatible solutes such as sugars, some amino acids and quarternaly ammonium compounds, which are essential for adaptability of plant cells to high salinity (Bohnert et al. 1995). Synthesis of glycinbetaine caused by introduced bacterial choline oxidase gene has been shown to increase the tolerance to salt in tobacco (Lilius et al. 1996) and cold in rice (Sakamoto and Murata 1998). Another mechanism of adaptation to high salinity are the exclusion of Na^+ ions from the sodium-sensitive cells, which has been proposed to be a function of a Na/H antiporter and Na ATPase. Expression of compatible solutes and heterologous sodium efflux transporters are useful approaches to improve the salt tolerance of photosynthetic organisms. The increase of salt tolerance or water stress tolerance of plants transformed with genes for the synthesis of compatible solutes was demonstrated (Tarczynski et al. 1993, Bohnert et al. 1995, Takebe et al. 1998). One of the possibilities to increase salt tolerance is transgenosis by yeast halotolerance gene HAL2 as demonstrated in tomato by Arrilage et al. (1998) and another is the overexpression of chloroplast glutamine synthetase transgene (Tanaka et al. 2000). Also Zhang and Blumwald (2001) have shown that is possible to generate crop plants that can tolerate high salinity levels with modification of single gene. Salt stress causes also the induction of oxidative stress and thus the enhancement, by transgenosis, of enzyme activities involved in active oxygen scavenging system may be a potent strategy to increase salt tolerance (Fover et al. 1994, Gupta et al. 1993, McKersie et al. 1996). Photorespiration has been shown to have protection effect against salt stress. When rice has been transformed with rice chloroplastic glutamine synthetase, plants with increased photorespiration capacity showed enhanced tolerance to salt stress (Hoshida et al. 2000). Stress conditioned by hypersaline soil, frost and draught are accompanied by generation of reactive oxygen intermediates. They lead to the damage of cell membranes and of mitochondria, which leads to oxygen stress. The plant defence mechanism can be supported by introduction into the genome of superoxid dismutase gene (McKersie et al. 1996).

Jaglo-Ottonsen et al. (1998) showed that the overexpression of the low temperature transcription activator *CBF1* reintroduced into *Arabidopsis thaliana* genome by transgenosis can induce the expression of *COR* (cold-regulated) genes in *Arabidopsis*. Antifreeze proteins were identified in marine teleosts where they increase their ability to survive in ice-laden seawater at temperatures below the expected freezing point in their blood sera. The appropriate gene has been cloned and transferred to tobacco (Kenward et al. 1999), where it was expressed in apoplast. Curiously, however, freeze resistance has not been increased.

2.7 Fruit Maturation

Several ways of modifying the process of fruit maturation by gene engineering have been applied in the tomato. As already mentioned in this volume by Sakurai et al., the inhibition of pectin degradation as the result of transgenosis in the tomato with the insert carrying antisense construction of the polygalacturonidase gene results in the increase of shelf/life of tomato fruits (Smith et al. 1988, 1990). This increase was not due to the inhibition of fruit softening as was originally expected. Although pectin degradation in the fruit cell walls was inhibited, only negligible changes in the softening of the transgenic fruits were observed (Grierson and Schuch 1993), which demonstrates that the polygalacturonase activity, which was inhibited here, can not be the major determinant of the fruit softening. The fruits show other positive changes which lead to the prolongation of shelf life. The fruits are resistant to cracking and mechanical damage, which allows to harvest the fruits in later stages of ripening and results in more flavoursome product. The low polygalacturonase fruit extracts have considerably higher viscosity and increased soluble solids, both valuable traits for production of tomato pastes. It has been the first plant transgenic plant variety released to the marked under the name "Flavr Savr" in early nineties.

Another way to increase the shelf life of (not only) tomato fruits is the manipulation of ethylene biosynthesis genes (or ethylen receptor genes).

Ethylene is plant gas hormone which triggers fruit ripening and other processes. The last step of ethylene biosynthesis involves 1-aminocycylopropane-1-carboxyzlic acid oxidase (ACC oxidase). Transformation of tomato by antisense construction to ACC oxidase (Hamilton et al. 1990, Oeller et al. 1991) showed drastic reduction of ethylene biosynthesis. The colour changes of their fruits were initiated at about ordinary time, but the extent of geting red was reduced. When stored for several weeks at room temperature, antisense transgenic fruits were more resistant to overripening and shrivelling than untransformed controls. Similar effects were obtained also by antisense inhibition of the gene controlling the preceding step of the ethylene biosynthesis, S-adenosylmethionine hydrolase (Good et al. 1994). On the other hand, transgenosis by antisense construction of another ripening-related gene E8 causes increased ethylene production in tomato fruits and speeds-up fruit ripening (Pennarubia et. al 1992). Tomato fruit ripening can also be influenced by transgenosis with antisense construction to another tomato-ripening related gene, TOM5 (Bird et al. 1991), which is crucial for carotenoid synthesis. The fruits of transgenic tomatoes have yellow colour.

2.8 Genetic Manipulation for Inhibition of Potato Tuber Browning

The development of brown discoloration in a wide range of fruits and vegetables reduces consumer acceptability and is thus of significant genomic importance. In potato tubers, injury during mechanical harvesting and subsequent handling causes areas of the tuber to develop discoloured patches (black spots) which extends from the site of impact. The discoloration of the damaged tissues results from the enzymatic production of complex polyphenols. In plants, the coloured oxidation products of the activity of polyphenoloxidase can result from polymerization of a wide variety of different phenolic compounds. Bachem et al. (1997) transferred into potato genome antisense construction of the polyphenoloxidase gene, which led to the prevention of browning of tuber tissues after mechanical damage.

3. GENETIC MANIPULATION OF PHYSICAL PROPERTIES OF PLANT PRODUCTS

3.1 Genetic Manipulation of Starch Quality and Quantity

Starch is the main carbohydrate reserve in most of plants. It constitutes a staple diet of much of the world human population. In many countries, starch is used in food and beverage industries as a thickener or sweetener. It improves uniformity, stability and texture of food. Amylopectin also imparts better freeze/thaw ability of frozen industry. It is increasingly used as a renewable and biodegradable polymer as foamed starch (Fergason 1994). High amylose starches have numerous industrial applications: in fried snack products, where they hamper the penetration of cooking oils, which leads to beneficial reduction of fat intake by consumer. They are used as thickeners and gelling agents and in the production of gum candies. Many types of photofilms also have a starch component. On the other hand, starches with higher levels of amylopectin are useful in food industry, paper industry, adhesive industry, livestock and food industry (Slater et al. 2000).

Production of various types of starch directly in the plants would reduce the postharvest modification s currently being used. As starch is the principal constituent of the harvestable parts of many agronomic plants, genetic engineering for the increase of its synthesis would increase also crop yields.

Reserve starch is composed of two different glucan chains: short, branched chains of amylopectin and long, unbranched chains of amylose. Amylose forms 11%-37% of starch mass. It might be of value to produce polymers with features that are intermediate between amylose and amylopectin, or those that are more highly branched or have a higher molecular weight (Műller-Rőber amd Kossman 1994, Slater et al.2000). Starch quality is influenced also by the presence of lipids, proteins and phosphorus. It is possible to manipulate all these traits. As new starches are produced, new applications will be found. Starch debranching enzymes affect amylopectin synthesis (Kubo et al 1999) and their genes are also potential subjects of manipulation and transgenosis.

The first step in the starch biosynthesis is moderated by ADP-glucose pyrophosphorylase (AGPase) and the next enzyme is moderated by a gene coding antisense RNA to AGPase mRNA decrease starch synthesis in tubers and leads to the formation of tubers of reduced size than store sugars (Műller-Rőber et al. 1992). On the other hand, the addition of *E. coli*

AGPase transgene increased starch production by 30% (Stark et al. 1992). AGPase has a profound effect on the starch synthesis not only in potato but also in other plant species. (Sweetlove et al. 1999) and the responsible gene has been subjected to metabolic engineering (gene farming) (Greene et al. 1998). By manipulation of the gene coding for the second important enzyme of starch biosynthesis, starch synthase, by alternating its substrate specificity or changing its binding specificity for starch glucan chains it could be possible too generate new starches with changed physico-chemical properties. This possibility has been experimentally proven by Edwards et al. (1999). The granule-bound starch synthase (GBSS) catalyses the synthesis of unbranched glucose polymer amylose in transgenic plants. The corresponding gene have been cloned by Visser etal (1989) and its function unequivocaly proven by introduction of the cloned gene into amylose-fee starch mutant of potato. The GBSS transgene restored the amylose synthesis (Flipse et al. 1994). Amylose synthesis can also be inhibited by antisense construction of GBSS. There are amylose-free commercial potato varieties obtained by transgenosis.

The starch branching enzyme (SBE) occurs in two isoforms: SBE A is present in low levels in potato tubers, but it has the major impact on starch structure, while SBE B is the major isoform with little effect on starch structure. The SBE A gene has already been cloned (Jobling 1999), which is the prerequisite of its manipulation and transgenosis. Unexpected result of transgenosis of potato by *E. coli* pyrophosphate was the increase of sucrose degradation and starch biosynthesis. The tubers accumulated much more starch than untransformed controls.

3.2 Accumulation of Fructose Polymers by Transgenic Plants

Fructose has became an interesting popular sweetener in many food products, primarily at the expense of sucrose. One of the important reasons is the availability of bulk quantities of corn stark in low cost. Hydrolyzed corn starch is enzymaticaly converted to fructose on industrial scale. An alternative source of fructose would be fructans, the storage carbohydrates of many plants. Fructans are fructose polymers which can be enzymaticaly or chemicaly hydrolyzed to yield fructose. There are no enzymes to degrade fructans in human digestive truct and therefore fructans are low caloric food ingradients. Low degree polymerization oligofructose molecules are sweet tasting and this property has been explored for the production of natural low cost sweeteners.

Polyfructan accumulating transgenic plants could have therefore two practical uses:

1. As source of fructose, alternative cheap sweetener
2. As source of organic polymers.

In this context, the second use will be considered. About 12% of *Angiospermae* produce polyfructans, but their chain length is only 5-60 units. Fructans are produced also by bacteria and such fructans consist on up to 100 000 units (*Bacillus*, *Pseudomonas*, *Streptococcus*). After transgenosis by appropriate transgenes, polyfructans of this length can be synthesized also by plants. Fructans can not be decomposed in plant tissues. Bacterial glucosyl transferase converts sucrose to bacterial fructan called levan. The levansucrase gene of *Bacillus subtilis* has been introduced into tobacco genome and transgenic plants accumulated polyfructans with chain length up to 25 000 units. Transgenosis by this bacterial transgene created new sink for sucrose in transgenic plants. Polyfructan accumulating tobacco plants did not show any phenotype differences to controls (Ebskamp et al. 1994).

In potato, starch-deficient transgenic clone expressing antisense RNA to ADP-glucose pyrophosphorylase, which showed reduction in tuber formation, have been transformed by *Erwinia carotovora* levan sucrase synthase gene. Accumulation of fructans in the apoplast did not restore the tuber forming ability. On the contrary, further decrease of tuber yield was observed (Rőber et al. 1996).

3.3 Seed Storage Proteins

The practical goal of most of the experimental attitudes to transgenosis with seed storage protein genes is to increase the amount of certain essential amino acids, which are in lack in seed storage proteins to make the legume and grain seed storage proteins to make the seed storage proteins more useful as food and feed. Most of the introduced types of seed storage proteins do not cause any changes of physical properties of seed products, except some cereals.

The processing properties of wheat flour are strongly influenced by the glutein protein fraction. This fraction of the glutein storage proteins form a network in dough, conferring visco-elasticity that is essential for the production of bread pasta and many other foods. For bread making, highly elastic doughs are required. The wheat glutein is a complex mixture of proteins which can be divided into two groups: the gliadins are monomeric proteins while the gluteins comprise individual subunits assembled into high Mr (above 1 x 10^6) polymers stabilized by interchain hydrogen bonds. The gliadin monomers and glutenin polymers interact by non-covalent forces, resulting in a continuous gluten network of the dough. The high molecular weight group of glutenin proteins is particularily important in determining gluten viscosity. The bread wheat *Triticum aestivum* is hexaploid (2n = 42) and consists of three distinct genomes derived from wheat ancestors: A, B and D. The D genome mostly contributes to high molecular weight glutein storage protein genes, which improve s its function properties (Barro et al. 1997). Tetraploid *Triticum turgidum* var. *durum* lacks the D genome and most of the fraction of high molecular weight proteins in the grain (Branland et al. 1989). It is produced mostly for production of the pasta. Some *durum* wheats are used also for bread making and the potential for increasing viscoelastic properties can be improved by transgenosis by high molecular weight storage protein coding genes (He et al. 1999).

3.4 Seed Oils

Vegetative oils are of economic importance as human food and animal feed and in many industrial applications. The fatty acid composition of most vegetative oils are not optimal for these uses (Hamada et al. 1996). Rape and other vegetable oils are composed predominately of unsaturated 18-carbon fatty acids: the monounsaturate oleic acid (18:1), polyunsaturated linoleic acid (18:2), and linolenic acid (18:3) and small amounts of the saturated palmitic acid (16:0) and stearic acid (18:0) (Knudzon et al. 1996). A shift of the carboxylic acid for instance to lower molecular weight (enrichment by lauric acid 12:1) changes the utilisation of rape oils and possibility of their use in cosmetic industry. A shift to higher molecular weight fatty acids, like erucic acid (22:1) make the plant oils more useful to industrial use (for instance bio oil and lubricants). There are several ways to induce these changes. The first of them was transgenosis of tobacco by rat ώ-9 desaturase gene (Grayburn et al. 1992). Antisense expression of the steraoyl-acyl carrier protein desaturase gene resulted in decrease of oleic acid (18:1) and in increase of stearic acid (18:0) in *Brassica napus* (Knudzon et al. 1992). Another interesting aim is to induce the production of γ-linoleic acid. Consumption of vegetable oils containing γ-linoleic acid is thought to alleviate hypercholesterolemia and other related clinical disorders that correlate with susceptibility to coronary heart disease. The *Synechocystis* γ-desaturase gene has been cloned and introduced into the tobacco genome. (Reddy and Tomas 1996), where it has been expressed in whole plants. Of course, the ultimate goal is the expression in the seeds.

3.5 Lignin Synthesis and Wood Quality

Lignin is formed in cell walls and between cells of woody tissue by polymerization of monomeric precursors, such as coniferyl and synapyl alcohols, which are secreted by these cells after biosynthesis *via* several general phenylpropanoid and lignin - specific enzymes. In woody tissues, lignin acts as cellulose fibre "encrusting material" which impact rigidity to the plants. In three species, lignin constitutes 20% - 30% of the dry weight of wood. A central problem confronting the paper industry is the production of cellulose fibres from wood, which involves removal of lignin encrusting the fibres. The forage industry is also trying to explore the possibilities of transgenosis and obtaining new tree varieties with reduced lignin content to conserve process energy and to maximize the production of cellulose fibres.

Since lignin has been identified as the major component limiting forage digestibility, genetic and biotechnological reduction or modification of lignin in pasture plant is desirable goal in the development of new forages (Watson 1990). Several possibilities to improve wood quality by transgenosis have been proposed (Baucher et al. 1998, Grima-Pattenati and Goffner 1999).

The first experiments with the modification of lignin contents were performed on tobacco (Dwinudi et al.1994). Halpin et al. (1994) transformed tobacco with antisense sequence of cinnamyl alcohol dehydrogenase and demonstrated the production of transgenic plants with novel type of lignin, more accesible to chemical extraction and with changed (brown-red) colour. Similar change of lignin in poplar has been obtained by overexpressing the *Arabidopsis* gene encoding ferulate-5-hydroxylase (Franke et al. 2000).

3.6 Genetic Manipulation for Synthesis of Biodegradable Plastics

Introduction of appropriate foreign bacterial transgenes can give the plants the ability to synthesize polyhydroxyalkanolates (Poirier et al. 1995, Van der Leij and Withold 1995), particuarily poly-hydroxybutyrate-co-valerate (PHB/v) (Hanley et al. 2000). This compound is already biotechnologicaly produced by bacteria and the product, with trade name Biopol, is used in shampoo bottles. Monsanto developed the technology needed for oilseed rape to produce "plant Biopol" (Slater et al. 1997). The biosynthetic pathway used by the bacterium *Ralstonia eutropha* for the simplest C4 PHB uses three steps: The first enzymatic step involves β-ketothiolase encoded by the gene *phbA*, which condensates two acetyl-CoA molecules to form acetoacetyl/CaA. The second enzyme step uses the NADPH-dependent acetoacetyl/CoA reductase encoded for by *phbB* gene to produce 3-hydroxybutyryl/CoA, which is converted to polymer by a third protein, the PHA synthase encoded by the *phbC* gene.

Transgenic *Arabidopsis thaliana* plants synthesizing polyhydroxybutyrate (PHB) have been obtained by Nawrath et al. (1994). The PHB was accumulated in chloroplasts and its amount reached 14% of plant leaf dry weight, but PHB has certain disadvantages as industrial product in comparison with PHB/v, which required the additional presence of polyhydroxyvalerate in the polymer. To achieve this, *Brassica napus* plants were transformed with partially deregulated threonine deaminase gene from mutated strain of *E. coli* (Hanley et al. 2000). Instead of *phbA*, the gene coding for the second form of β-ketolase *BktB*, has been introduced into the plants. Altogether, four genes (*IlvA466*, *BktB*, *phbB* and *phbC*) have been introduced into the rape genome with seed specific promotors (Slater et al. 1998).

3.7 Genetic Manipulation of Spider Silk Protein Synthesis

Spider dragline is proteinaceous fiber which posses extremely high firmness. It is more firm than steel fiber or steel sheet of similar size. In order to be able to produce spider silk protein in sufficient quantity, the apropriate gene of the spider *Nephlia claviceps* has been cloned, sequenced and resynthesized to fulfil optimum similarity to plant genes. This new chimeric gene has been introduced into tobacco and potato genome (Scheller et al. 2001). Spider silk proteins accumulated in the endoplasmic reticulum of tobacco leaves and potato tuber cells and its fraction reached at least 2% of total soluble proteins. The recombinant protein exhibited heat stability, which was used for its purification.

4. CONCLUSIONS

Transgenosis is able to change basic physical properties of plants, although most of its possibilities in still have not been used in plant breeding. Changes of plant growth and development including plant architecture can be induced by original individual *A. tumefaciens* and *A. rhizogenes* morphoregulatory T-DNA genes. Transgenosis is able to change the composition of storage substances of plant seeds and tubers and of plant lignins. Gene manipulation can also carry into the plant genome the abilities of synthesis of new storage compounds like new types of polyfructans, starch, oils, biodegradable polymer polyhydroxybutyrate-valerate or proteinaceous spider fiber of remarkable physical properties. Plant varieties with modified physical properties released into the market up to now include tomatoes with prolonged shelf-life, carnations with blue flowers and carnations with prolonged vase-life. Several other types of transgenic plants have been induced.

REFERENCES

Akiyoshi, D.E., Morris, R.O., Hinz, R.O., Mischke, B.S., Kosuge, T., Garfinkel, D., Gordon, M.P., Nester, E.W., 1983. Cytokinin/auxin balance in crown gall tumors is regulated by specific loci in the T-DNA. *Proc. Natl. Acad. Sci. USA* **80**: 407-411.

Arrilage, I., Gil-Mascarell, R., Gilbert, C., Sales, E., Montesionos, C.,
Serrano, R., Moreno, V., 1998. Expression of the yeast HAL2 gene in tomatoe increases the *in vitro* salt tolerance of transgenic progenies. Plant Sci. **136**: 219-226.

Bachem, C.V.B., Speckman, G.J., Linde, P.C.G., Verghebben, F.T.M., Hunt, M.D., Steffens, J.C., Yabeau, M., 1997. Antisense expression of polyphenol oxidase genes inhibit enzymatic browning in potato tubers. *Biotechnology* **12**: 1101-1105.

Barro, F., Rooke, L., Békés, F., Gras, P., Tatham, A.S., Fido, R., Lazzeri, P.A., Shewry, P.R., Barcelo, P., 1997. Transformation of wheat with HMW subunit genes resulting in improved function properties. *Nature Biotechnol.* **15**: 1295-1299.

Barry, G.F., Rogers, S.G., Fralery, R.T., Brand, L., 1984. Identification of cloned cytokinin biosythetic gene. *Proc. Nat. Acad. Sci. USA* **81**: 4776-4780.

Batschauer, A., 1998. Photoreceptors of higher plants. *Planta* **206**: 479-492.

Baucher, M., Monties, B., Van Montagu, M., Boerjan, W., 1998. Biosynthesis and genetic engineering of lignin. *Crc. Rev. Plant Sci.* **17**: 125-197.

Bird, C.R., Ray, J.A., Fletcher, J.D., Boniwell, J.M., Bird, A.S., Teuliers, C., Blain, I., Bramley, P.M., Schuch, W., 1991. Using antisense RNA to study gene function: inhibition of carotenoid biosynthesis in transgenic tomatoes. *Nature Biotech.* **9**: 635-639, 1991.

Blechl, A.E., Anderson, O.D., 1996. Expression of a high-molecular-glutein gene in transgenic wheat. *Nature Biotechnol.* **14**: 875-879.

Bohnert, H.J., Nelson, D.E., Jensen, R.G., 1995. Adaptation to environmental stresses. *Plant Cell* **7**: 1099-1111.

Branlard, G., Autran, J.C., Monnevaux, P. 1989. High molecular weight glutein subunits of durum wheat (*T. durum*). *Theor. Appl. Genet.* **78**: 353-358.

Cardarelli, M., Mariotti, D., Pomponi, M., Spano, L., Capone, I., Costantino, P., 1987. *Agrobacterium rhizogenes* T-DNA genes capable of inducing hairy root phenotype. *Mol. Gen. Genet.* **209**: 475-480.

Chalfie, M., Yu, Y., Euskirchen, G., Ward, W.W., Prasher, D.C., 1994. Green fluorescent protein as a vital marker for gene expression. *Science* **263**: 663-664.

Chilton, M.D., Drummond, H.J., Merlo, D.J., Sciaky, D., Gordon, M.P., Nester, E.W., 1997. Stable incorporation of plasmid DNA into higher plant cells: the molecular basis of crown gall tumorigenesis. *Cell* **11**: 263-271.

Chrisley, M.C., 1997. Transgenic crop plants using *Agrobacterium rhizogenes*-mediated transformation. In: Deran, L.P.M. (Ed.): Hairy Roots: Culture and Application. Harwood Academic Publishers, Amsterdam, pp. 99-111.

Clack, T., Mathews, S., Sharrock, R.A., 1994. The phytochrome apoprotein fmily in *Arabidopsis* is encoded by five genes: the sequences and expression of PHYD and PHYE. *Plant Mol. Biol.* **25**: 413-427.

Damiani, F., Arcioni, S., 1991. Transformation of *Medicago arborea* L. with an *Agrobacterium rhizogenes* binary vector carrying the hygromycin resistance gene. *Plant Cell Rep.* **10**: 300-303.

Dehio, C., Grossmann, K., Schell, J., Schműlling, T., 1993. Phenotype and hormonal status of transgenic tobacco plants expressing the *rolA* gene of *Agrobacterium rhizogenes* T-DNA. *Plant Mol. Biol.* **23**: 1199-1210.

Dwinudi, U.N., Campbell, W.H., Zu, J., Datla, R.S.S., Bugos, R.C., Chiang, V.L., Podila, G.K., 1994. Modification of the lignin biosynthesis in transgenic *Nicotiana* through expression of antisense O-methyltransferase gene from *Populus*. *Plant Mol. Biol.* **26**: 61-71, 1994.

Ebskamp, M.J.M., Van der Meer, I.M., Spronk, B.A., Weisbeek, P.J., Smeekens, S.C.M., 1994. Accumulation of fructose polymers in transgenic tobacco. *Biotechnology* **12**: 272-275.

Edwards, A., Fulton, D.C., Hylton, C.M., Jobling, S.A., Gidlez, M., Rőssner, V., Martin, C., Smith, A.M., 1999. Specificity of starch synthase isoforms from potato. *Eur. J. Biochem.* **266**: 724-736.

Estruch, J.J., Chriqui, D., Grossmann, K., Schell, J., Spena, A., 1991a. The plant oncogene *rolC* is responsible for the release of cytokinins from glucoside conjugates. *EMBO J.* **10**: 2889-2895.

Estruch, J.J., Schell, J., Spena, A., 1991b. The protein encoded by the *rolB* plant oncogene hydrolyses indole glucosides. *EMBO J.* **10**: 3125-3128.

Fergason, V., 1994. High amylose and waxy corns. In: Hallauer, A.R. (Ed.): Speciality Corns. CRC Press, Pp. 55-77.

Flipse, E., Huisman, J.G., de Vries, B.J., Bergervoet, J.E.M., Jacobsen, E., Visser, R.G.F., 1994: Expression of a wild-type GBSS gene introduced into an amylose-free potato mutant by *Agrobacterium tumefaciens* and the inheritance of the insert at the microscopical level. *Theor. Appl. Genet.* **88**: 369-375.

Fover, C.H., Descourviers, P., Kunert, K.J., 1994. Protection against oxygen radicals: an important defence mechanism studies in transgenic plants. *Plant Cell Env.* **17**: 507-523.

Franke, R., McMichael, C., Meyer, K., Shirley, A.M., Cosumano, J.C., Chapple, C., 2000. Modified lignin in tobacco and poplar plants over-expressing the *Arabidopsis* gene ferulate 5-hydroxylase. *Plant J.* 22: 223-234.

Fray, R.G., Fraser, P.D., Valero, D., Hedden, P., Bramley, P.M., Galbright, D.W., Lambert, G.M., Grebenok, R.J., Sheen, J., 1995. Methods in Cell Biology. Acad. Press, San Diego.

Gan, S., Amasino, R.M. 1995. Inhibition of leaf senescence by the autoregulated production of cytokinin. *Science* **270**: 1986-1988.

Geigenberger, P., Hajirezan, M., Geiger, M., Dettig, V., Sonnewald, V., Stilt, M., 1988. Overexpression of pyrophosphorylase leads to increased sucrose degradation and starch synthesis, increased activities of enzymes for sucrose-starch conversion and increased levels of nucleotides in potato tubers. *Planta* **205**: 428-437.

Good, X., Kellog, W., Langhoff, D., Matsumura, W., Bestwick, R.K., 1994. Reduced ethylene synthesis by transgenic tomatoes expressing S-adenosylmethionine hydrolase. *Plant Mol. Biol.* **26**: 781-790, 1994.

Goto, F., Yashihara, T., Saiki, H., 2000. Iron accumulation and enhanced growth in transgenic lettuce plants expressing the iron-binding protein ferritin. *Theor. Appl. Genet.* **100**: 658-664.

Greene, T.W., 1998. Generation of up-regulated allosteric variants of potato ADP-glucose pyrophosphorylase by reversion genetics. *Proc. Natl. Acad. Sci. USA* **95**: 19322-10327.

Greyburn, W.S., Collins, G.B., Hildebrand, D.G.F., 1992. Fatty acid alternation by a ω9-desaturase in transgenic tobacco tissue. *Biotechnology* **10**: 675-679.

Grierson, S., Schuch, W., 1993. Control of ripening. *Phil. Trans. Roy. Soc. London* **342**: 241-250.

Grierson, D., 1995. Constitutive expression of a fruit phytoene synthase gene in transgenic tomatoes causes dwarfism by directing metabolites from the gibberellin pathway. *Plant J.* **8**: 693-701.

Grima-Pettenati, J., Goffner, D., 1999. Lignin genetic engineering revised. *Plant Sci.* **145**: 51-65.

Gupta, A.S., Heinen, J.L., Holaday, A.S., Burke, J.J., Allen, R.D., 1993. Incresed resistance to oxidative stress in transgenic plants that overexpress chloroplastic Cu/Zn supoeroxide dismutase. *Proc. Natl. Acad. Sci. USA* **90**: 1629-1633.

Halpin, C., Knight, G.A., Campbell, M.M., Boudert, A.M., Boon, J.J., Chabbert, B., Tolier, M.-T., Schuch, W., 1994. Manipulation of lignin quality by downregulation of cinnamyl alcohol dehydrogenase. *Plant J.* **6**: 339-350.

Hamad, T., Kodama, H., Nishimura, M., Iba, K., 1996. Modification of fatty acid composition by over- and antisense-expression of a microsomal ω-fatty acid desaturase gene in transgenic tobacco. *Transgenic Res.* **5**: 115-121.

Hamilton, A.J., Lycett, G.W., Grierson, D. 1990. Antisense gene that inhibits synthesis of the hormone ethylene in transgenic plants. *Nature* **346**: 284-287, 1990.

Handa, T., 1992. Genetic transformation of *Antirrhinum majus* L. and inheritance of altered phenotype induced by Ri T-DNA. *Plant Sci.* **81**: 199-207.

Hanley, Z., Slabas, T., Elborough, M., 2000. The use of plant biotechnology for the production of biodegradable plastics. *Trends in Plant Sci.* **5**: 45-49.

Haseloff, J., Siemering, K.R., Prasher, D.C., Hodge, S., 1997. Removal of a cryptic intron and subcellular localization of the green fluorescent protein are required to mark transgenic *Arabidopsis* plants brightly. *Proc. Natl. Acad. Sci. USA* **94**: 2122-2127.

He, G.Y., Rooke, I., Steele, F., Békés, F., Gras, P., Tatham, A.S., Fido, R., Barcelo, P. M., Shewry, P.R., Lazzeti, P.A., 1999. Transformation of pasta wheat (*Triticum durum* var. *durum*) with high-molecular-weight glutein subunit genes and modification of dought functionality. *Molec. Breeding* **5**: 377-386.

Heim, R., Prasher, D., Tsien, R., 1994. Wavelength mutations and posttranslational autooxidation of green fluorescent protein. *Proc. Nat. Acad. Sci. USA* **91**: 125012-12504.

Heim, R., Cubitt, A., Tsien, R., 1995. Improved green fluorescence. *Nature* **373**: 663-664.

Heim, R., Tsien, R.Y., 1996: Engineering green fluorescent protein for improved brightness, longer wavelengths and fluorescence resonance energy transfer. *Curr. Biol.* **6**: 178-182.

Hewelt, A., Prinsen, E., Schell, J., Van Onckelen, G., Schműlling, T., 1994. Promotor-tagging with a promotorless *ipt* gene leads to cytokinin-induced phenotypic viability in transgenic tobacco plants: implications and gene dosage effects. *Plant J.* **6**: 879-891, 1994.

Jang, I.-C., Nahm, B.H., Kim, J.-K., 1999. Subcellular targeting of green fluorescent protein to plastids in transgenic rice plants provides a high-level expression system. *Molec. Breeding* **5**: 453-561, 1999.

Jaglo-Ottosen, K.R., 1998. *Arabidopsis CBF1* overexpression induces *COR* genes and enhances freezing tolerance. *Science* **280**: 104-106.

Jefferson, R.A., Kavanagh, T.A., Bevan, M.W., 1987. GUS fusion: β-glucuronidase as a sensitive and versatile gene fusion marker in higher plants. *EMBO J.* **6**: 3901-3907.

Jobling, S.A., 2001. A minor form of starch branching enzyme in potato (*Solanum tuberosum*) tubers has a major effect on starch structure: Cloning and characterization of multiple forms of SBE A. *Plant J.* **18**: 163-171.

Kenward, K.D., Brandle, J., McPherson, J., Davies, P.L., 1999. Type II fish antifreeze protein accumulation in transgenic tobacco does not confer frost resistance. *Transgenic Res.* **8**: 105-117.

Klee, H.J., Horsch, R.B., Hinchee, M.A., Hein, M.B., Hoffman, N.L., 1987. The effect of overproduction of two *Agrobacterium tumefaciens* T-DNA gene auxin biosynthesis gene products in transgenic *Petunia* plants. *Genes Dev.* **1**: 86-96.

Knutzon, D.S., Thompson, G.A., Radke, S.E., Johnson, W.B., Knauf, V.C., Kridl, J.C., 1992. Modification of *Brassica* seed oil by antisense expression of a steraoyl-acyl-carrier protein desaturase gene. *Proc. Natl. Acad. Sci. USA* **89**: 2624-2628.

Kőhler, R.H., Zipfer, W.W., Hanson, M.R., 1997a. The green fluorescent protein as a marker to visualize plant mitochondria *in vivo*. *Plant J.* **11**: 613-621.

Kőhler, R.H., Cao, J., Zipfer, W.R., Hanson, M.R., 1997b. Exchange of protein molecules through connection between higher plant plastids. *Science* **276**: 2039-2042.

Kubo, A., Harada, K., Sonnewald, O., Willmitzer, L., 1999. The starch debranching enzymes isoamylase and pullulanase are both involved in amylopetin biosynthesis in rice endospem. *Plant Physiol.* **121**: 399-409.

Lilius, G., Holmberg, N., Bullowe, L., 1996. Enhaned NaCl stress tolerance in transgenic tobacco expressing bacterial choline dehydrogenase. *Biotechnology* **14**: 177-179.

Martin-Tanguy, J, Sun, L.Y., Burton, D., Vernoy, R., Rossin, N., Tepfer, D., 1996. Attenuation of the phenotype caused by the root-inducing, left-handed, transferred DNA and its *rolA* gene. *Plant Physiol.* **11**: 259-267.

Maurel, C., Barbierbrigoo, H., Spena, A., Tempé, J., Guern, J., 1991. Single *rol* genes from the *Agrobacterium rhizogenes* TL-DNA after some of the cellular responses to auxin in *Nicotiana tabacum*. *Plant Physiol.* **102**: 261-271.

McCormac, A.C., Cherry, J.R., Hershey, J.P., Vierstra, R.D., Smith, H. 1991. Photoresponses in transgenic tobacco plants carrying an oat phytochrome gene. *Planta* **185**: 162-170, 1991.

McKersie, B.D., Bowley, S.R., Harjanto, E., Leprince, O., 1996. Water-deficient tolerance and field performance of transgenic alfalfa overproducing superoxide dismutase. *Plant Physiol.* **111**: 1171-1177.

Műller-Rőber, B., Kossman, J., 1994. Approaches to influence starch quantity and starch quality in transgenic plants. *Plant Cell Env.* **17**: 601-613.

Műller-Rőber, B., Sonnewald, O., Willmitzer, L., 1992. Inhibition of the ADP-glucose pyrophosphorylase in transgenic potatoes leads to sugar-storing tubers and influences tuber formation and expression of tuber storage protein genes. *EMBO J.* **11**: 1229-1238.

Nawrath, C., Poirier, V., Somerville, C., 1994. Targeting of the polyhydroxybutyrate biosynthetic pathway to the plastids of *Arabidopsis thaliana* in high levels of polymer accumulation. *Proc. Natl. Acad. Sci. USA* **91**: 1260-1274.

Nilsson, O., Moritz, T., Imbault, N., Sandberg, G., Olsson, O., 1993. Hormonal characterization of transgenic tobacco plants expressing the *rolC* gene of *Agrobacterium rhizogenes* TL-DNA. *Plant Physiol.* **102**: 363-371.

Oeller, P.W., Min-Wong, L., Taylor, R., Pike, D., Theologis, A., 1991. Reversible inhibition of tomato fruit senescence by antisense RNA. *Science* **254**: 437-438.

Ondřej, M., Eder, J., Hrouda, M., Macháčková, I., Vlasák, J., 1989. Free auxin level and inheritance of introduced markers in tobacco transformed by binary vector based on A4 Ri plasmid. Biol. Plant. 31: 286-291.

Ondřej, M., Macháčková, I., Čatský, J., Eder, J., Hrouda, M., Pospíšilová, J., Synková, H., 1990. Potato transformation by T-DNA cytokinin synthesis gene. Biol. Plant. 32: 401-406.

Oono, Y., Kanawa, K., Uchimiya, H., 1990. Early transformation of transgenic tobacco plants possessing the *rolC* gene of *Agrobacterium* rhizogenes Ri plasmid. *Jpn. J. Genet.* **65**: 7-16.

Ow, D.W., Wood, K.V., DeLuca, M., deWet, J.R., Helinski, D.R., Howell, S.H., 1986. Transient and stable expression of the firefly luciferase gene in plant cells and transgenic plants. *Science* **234**: 856-859.

Reddy, A.S., Thomas, T.L., 1996. Expresion of a cyanobacterial ω6-desaturase gene results in γ-linoleic acid production in transgenic plants. *Nature Biotechnol.* **14**: 639-642.

Rőber, M., Geider, K., Műller-Rőber, B., Willmitzer, L., 1996. Synthesis of fructans in tubers of transgenic starch-deficient potato plants does not result in an increased allocation of carbohydrates. *Planta* **199**: 528-536.

Robson, P.R.H., McCormac, A.C., Irvine, A.S., Smith, H., 1996. Genetic engineering of harvest index in tobacco through overexpression of a phytochrome gene. *Nature Biotech.* **14**: 995-998.

Romano, C.P., Cooper, M.L., Klee, H.J., 1983. Uncouping auxin and ethylene effects in transgenic tobacco and *Arabidopsis thaliana* plants. *Plant Cell* **5**: 181-189.

Romano, C.P., Hein, M.B., Klee, H.J.: Inactivation of auxin in tobacco transformed with the indoleacetic acid-lysine synthetase gene of *Pseudomonas savastanoi*. *Genes and Dev.* **5**: 438-446, 1991.

Sakamoto, A., Murata, A., Murata, N., 1998. Metabolic engineering of rice leading to biosynthesis of glycinbetaine and tolerance to salt and cold. *Plant Mol. Biol.* **38**: 1011-1019.

Scheller, J., Gührs, K.-H., Grosse, F., Conrad, U., 2001. Production of spider silk protein in tobacco and potato. *Nature Biotech.* **19**: 573-577.

Schmülling, T., Schell, J., Spena, A., 1988. Single genes from *Agrobacterium rhizogenes* influence plant development. *EMBO J.* **7**: 2621-2629.

Schmülling, T., Beinsberger, S., DeGreef, J., Schell, J., Van Onckelen, H., Spena, A., 1989. Construction of a heat-inducible chimeric gene to increase the cytokinin content in transgenic plant tissue. *FEBS Lett.* **249**: 4012-406.

Schröder, G., Waffenschmidt, S., Weiler, E.W., Schröder, J., 1984. The region of Ti plasmid codes for an enzyme synthesizing indole-3-acetic acid. *Eur. J. Biochem.* **138**: 387-391.

Sharrock, R.A., Quail, P.H., 1990. Novel phytochrome sequences in *Arabidopsis thaliana*: structure, evolution and differential expression of a plant regulatory photoreceptor family. *Genes Dev.* **3**: 1745-1757.

Sidorov, V.A., Kasten, D., Pang, S.-Z., Hajdukiewicz, P.TJ., Staub, J.M., Nehra, N.S. (1999). Stable chloroplast transformation in potato: use of green fluorescent protein as a plastid marker. *Plant J.* **19**: 209-216.

Sitbon, F., Sundberg, B., Olsson, O., Sundberg, G., 1991. Free and conjugated indolacetic acid (IAA) contents in transgenic tobacco plants expressing the *iaaM* and *iaaH* biosynthetic genes from *Agrobacterium tumefaciens*. *Plant Physiol.* **95**: 480-485.

Sitbon, F., Hennion, S., Sundberg, B., Little, C.H.A., Olsson, O., Sandberg, G. 1992a. Transgenic tobacco plants coexpressing *Agrobacterium tumefaciens iaaM* and *iaaH* genes display altered growth and indoleacetic acid metabolism. *Plant Physiol.* **99**: 1062-1069.

Sitbon, F., Little, C.H.A., Olsson, O., Sandberg, G., 1992b. Correlation between expression of T-DNA IAA biosynthetic genes from developmentaly regulated promotors and the distribution of IAA in different organs of transgenic tobacco. *Physiol. Plant.* **85**: 679-688.

Slater, S., Houmiel, K.L., Traut, M., Mirsky, T.A., Padgatte, S.R., Grusz, K.J.I., 1998. Multiple ß-ketothiolases mediate poly(ß-hydroxyalkanolate) copolymer synthesis in *Ralstonia eutropha*. *J. Bacteriol.* **180**: 1979-1989.

Slater, S., Mirsky, J., Hormiel, K.L., Hao, M., Reiser, S.S., Taylor, N.B., Tran, M., Smith, H., Whitelam, G.C., 1997. The shade avoidance syndrome: multiple responses mediated by multiple phytochromes. *Plant Cell Env.* **20**: 840-844.

Slater, C.J., Kavakli, H.I., Okita, T.W., 2000. Engineering starch for increased quality and quantity. *Trends in Plant Sci.* **5**: 291-298.

Slightom, J., Durand-Tardif, M., Jouanin, L., Tepfer, D., 1986. Nucleotide sequence analysis of TL-DNA of *Agrobacterium rhizogenes* agropine type plasmid. *J. Biol. Physiol.* **261**: 108-121.

Smart,C.M., Scofield, S.R., Bevan, M.W., Dyer, T.A., 1991. Delayed leaf senescence in tobacco plants transformed with *tmr*, a gene for cytokinin production of *Agrobacterium*. *Cell* **3**: 647-656.

Smigocki, A.C., 1995. Expression of a wound-inducible cytokinin biosynthesis gene in transgenic tobacco: correlation of root expression with induction of cytokinin effects. *Plant Sci.* **109**: 153-163.

Smigocki, A.C., Owens, I.D., 1988. Cytokinin gene fused with a strong promotor enhances shoot organogenesis and zeatin levels in transformed plant cells. *Proc. Natl. Acad. Sci. USA* **85**: 5131-5135.

Smith, C.J.S., Watson, C.F., Ray, J., Bird, C.R., Morris, P.C., Schuch, W., Grierson, D., 1988. Antisense RNA inhibition of polygalacturonase gene expression in transgenic tomatoes. *Nature* **334**: 724-726.

Smith, C.J., Watson, C.F., Morris, P.C., Seymour, G.B., Gray, J.E., Arnold, C., Tucke, G.A., Schuch, W., Harding, S., Grierson, D., 199). Inheritance and the effect of ripening of antisense polygalacturonase genes in transgenic tomatoes. *Plant Mol. Biol.* **14**: 369-379.

Spena, A., Prinsen, E., Fladung, M., Schultye, S.C., Van Onckelen, H.: The indolacetic acid-lysine synthetase gene of *Pseudomonas syringae* subsp. *savastanoi* developmental alternations in transgenic tobacco and potato plants. *Mol. Gen. Genet.* **227**: 205-212, 1991.

Stark, D.M., Timmermann, K.P., Barry, G.P., Preiss, J., Kishore, G.M., 1992. Regulation of the amount of starch in plant tissues by ADP glucose pyrophosphorylase. *Science* **258**: 287-292.

Strobel, G.A., Nachmias, A., Hess, D., 1988. Improvement of the growth and yield of olive trees by transformation with the Ri plasmid of *Agrobacterium rhizogenes*. *Canad. J. Bot.* **66**: 2581-2585.

Suginuma, C., Akihama, T., 1995. Transformation of gentian with *Agrobacterium rhizogenes*. *Acta Hort.* **392**: 153-160.

Sweetlove, L.J., Burreel, M.M., Reews, T.A., 1996. Starch metabolism in tubers of transgenic potato (*Solanum tuberosum*) with increased ADP glucose pyrophosphorylase. *Biochem. J.* **320**: 493-498.

Takebe, T., Nakamura, T., Nomura, T., Hayashi, Y., Ishitani, M., Muramoto Y., Tanaka, A., Takebe, T., 1993. Glycinbetaine and the genetic engineering of salinity tolerance in plants. In: Satoh, K., Murata, N. (Eds.): Stress response in photosynthetic organisms. Elsevier Sci. Publ. Amsterdam, pp. 115-151.

Tanaka, Y., Hibino, T., Hayashi, Y., Tanaka, A., Takebe, T., 2000. Enhanced tolerance to salt stress in transgenic rice that overexpress chloroplast glutamine synthetase. *Plant Mol. Biol.* **43**: 103-111.

Tanaka, N., Matsumoto, T., 1993. Regeneration of *Ajuga* hairy roots with high productivity of 20-hydroxyecdysone. *Plant Cell Rep.* **13**: 87-90.

Tarcyznski, M.C., Jensen, R.G., Bohnert, H.J., 1993. Stress protection of transgenic tobacco by production of the osmolyte mannitol. *Science* **259**: 508-510.

Thomas, J.C., Smigocki, A.C., Bohnert, H.J., 1995. Light induced expression of *ipt* from *Agrobacterium tumefaciens* results in cytokinin accumulation and osmotic stress symptoms in transgenic tobacco. *Plant Mol. Biol.* **27**: 225-235.

Thomashow, L.S., Reeves, S., Thomashow, M.F., 1984. Crown gall oncogenesis: evidence that a T-DNA gene from the *Agrobacterium* Ti plasmid pTiA6 encodes an enzyme that catalyses synthesis of indole-3-acetic acid. *Proc. Natl. Acad. Sci. USA* **81**: 5071-5075.

Thomashow, M.F., Hugly, S., Buchholz, L.S., 1986. Molecular basis for the auxin-independent phenotype of crown gall tissues. *Science* **231**: 616-618.

Tuominen, H., Sitbon, F., Jacobsson, C., Sanberg, G., Olsson, G., Sandberg, B., 1995. Altered growth and wood characteristics in transgenic hybrid aspen expressing *Agrobacterium tumefaciens* T-DNA indolacetic acid-biosynthesis genes. *Plant Physiol.* **109**: 1179-1189.

Valentin, H.E., Rodriquez, D.J., Stone, D., Padgate, S.R., Kishore, G., Gruis, K.J., 1999. Metabolic engineering of *Arabidopsis* and *Brassica* for poly-3-hydroxybutyrate-co-3hydroxyvalerate copolymer production. *Nature Biotechnol.* **17**:1011-1016.

Van der Leij, F.R., Witholt, B., 1995. Strategies for the sustainable production of new degradable polyesters in plants: a review. *Canad. J. Microbiol.* **41** (Suppl.): 222-238.

Van Onckelen, H., Prinsen, E., Inzé, D., Rudelsheim, P., Van Lijsebettens, M., Follin, J., Greef, J., 1986. *Agrobacterium* T-DNA gene 1 codes for tryptophan-2-monooxygenase activity in tobacco crown gall cells. *FEBS Lett.* **198**: 357-360.

Vartanian, N., Berkaloff, A., 1989. Drought adaptability of *Agrobacterium rhizogenes* - induced roots in oilseed rape (*Brassica napus* var. *oleifera*). *Plant Cell Env.* **12**: 197-204.

Verhaven, K., van der Kop, D.A.M., Jacobsen, E., Feenstra, W.J., 1992. Complementation of the amylose-free starch mutant of potato (*Solanum tuberosum*) by the gene encoding granule-bound starch synthase. *Theor. Appl. Genet.* **82**: 289-295.

Visser, R.G.F., Hergerberg, M., van der Leij, F.R., Jacobsen, E., Witholt, B., Feenstra, W.J., 1989. Molecular cloning and partial characterization of the gene for granule-bond starch synthase from a wild type and an amylose-free potato (*Solanum tuberosum* L.) *Plant Sci.* **64**: 185-192.

Vlasák, J., Ondřej, M.: Construction and use of *Agrobacterium tumefaciens* binary vectors with *A. tumefaciens* C58 T-DNA genes. Folia Microbiol. 37: 227-340, 1992.

Watson, J.M., 1990. Genetic engineering of low-lignin pasture plants. In: Akin, D.E., Ljundgal, L.G., Wilson, J.R., Harris, P.J. (Ed.), 1990. Microbial and Plant Opportunities to Improve Lignocellulose Utilization by Ruminants. Pp. 215-227, Elsevier, New York.

Yan, L., Wang, G., Jiu, J., Peng, X., Xie, Y., Dai, J., Guo, S., Zhang, F., 1990. Transfer of *E. coli gutD* gene into maize and regeneration of salt-tolerant transgenic plants. *Sci.in China* **42**: 90-95.

Zhang, H.G., Blumwald, E., 2001. Transgenic salt tolerant tomato plants accumulate salt in the foliage but not in fruits. *Nature Biotechnol. 19*: 765-768.

Zhang, J., Van Toai, T., Preiszner, J., 2000. Development of flooding-tolerant *Arabidopsis thaliana* by autoregulated cytokinin production. *Molec. Breeding* **6**: 135-144.

Novel Agricultural Materials for Food and Feed. The Transgenic Crops: From Cereals to Potato

JÓZEF FORNAL, WIOLETTA BŁASZCZAK
and STANISŁAW GRUNDAS[1]
Institute of Animal Reproduction and Food Research,
Polish Academy of Sciences, Olsztyn, Poland
[1]Institute of Agrophysics, Polish Academy of Sciences, Lublin, Poland

In agricultural production, seeds, tubers, and fruits are usually the most economically important part of the plant as well as central and essential source of food and feed. A prime objective of plant physiology and genetic research is therefore to maximize the harvest yield. Increases in the yield of crop species resulted mainly from changes in assimilation by reducing the growth of non-harvestable organs (Dejardin et al., 1997), in particular starch and protein - the main components of the seeds and tubers upon which nutritional and industrial application seem to be dependent. Only since the 1940`s have yields of plants breeding significantly increased, through the introduction of hybrid cereal seed, chemical inputs, mechanization, and improved breeding methods (Shulman, 1999). However the quality of harvested grain and/or tuber and their main storage component, i.e. starch and protein has not been greatly improved using only the above-mentioned methods. Bearing in mind a very complicated mechanism of biology as well as physiology of plant organs, the influence of the agricultural factors only on the yield and quality of plant components seems to be not enough. The techniques for alternation of the biosynthesis of plant components or the ones which are based on the regulatory process of the storage products must be applied for modification of plant and its properties and for better filling of human needs (Schulman, 1999).

Table 1. Percentage of field trials according to modified property (Dunwell, 1999)

Category	Year			
	1990	1993	1996	1999
1. Herbicide tolerance	22	33	24	24
2. Insect resistance	21	22	23	9
3. Virus diseases	29	16	16	-
4. Bacterial diseases	5	5	1	-
5. Agronomic properties	3	2	3	6
6. Product quality	8	18	17	52
7. Nematode resistance	1	-	-	-
8. Fungal resistance	4	4	8	6
9. Others	6	2	5	3

Over the last time, genetically modified plants have moved from being a mysterious laboratory product to widely available commercial one. The potential of biotechnology is very evident to the commercial world and a great range of experimental material is being tested both in the laboratory and under field condition (Dunwell, 1998). Transgenic, or genetically modified (GM) crops are now very important commercial products of agricultural landscape mainly of North America. However, in certain parts of Europe there is still great reluctance to accept their introduction to the market. By 1996, 23 crops had received approval for commercial production in USA, 12 in Canada and 7 in Japan. In 1996 more than 2 million ha of genetically improved crops (maize, soybeans, cotton and potatoes) were planted only in North America (Franks, 1999). It is obvious that the commercial trends are towards vertical integration with a rapid movement of interest from the first generation of single gene traits, such as herbicide tolerance and insect resistance (Table 1), towards the second generation where the emphasis has moved to multiple gene inserts and product quality (Table 2) (Dunwell, 1999; Dale, 1999). The number of GM crops is all the time increasing, with soybean, maize and cotton still being on the top of the list. The area of cultivation of GM crops in 1997 and 1998 is put at 11 million and 27.8 million ha respectively, in countries including America, Canada, Argentina and Australia (Franks, 1999). Franks (1999) reported also that the rapid expansion in area is again illustrated by the increased acreage planted with Liberty Link oil seed rape (genetically improved for protection against glufosinate) from 16.000 ha in 1995 to 500.000 ha in 1997. Other genetically modified products such as: rice, wheat, rape, tomato, cucumber, melon, potatoes carnation, cichory, linseed, papaja, squash and sugar beet varieties also have been commercially released into large scale (www.agbios.com). Marked increase of global profit for the main GM crops was 119 mln for maize, 109 mln USD for soy and 81 mln for cotton (Kinsey, 1999). The global value of GM market is counted for 3 billions USD now, and it is estimated to reach 20 and 75 billions USD in the years 2010 and 2020, respectively (Stec, 2000).

From the agrophysical point of view, genetic modification may pertain to such properties decisive for the harvest process as stem hardness and elasticity, and mechanical resistance of pods (rapeseed, legumes) (Table 2). Greater shapeliness of cereal and legume grains is observed due to the lack of (or weaker) weeds and pest competitiveness. Shapeliness of GM grains could influence and make necessary revaluation of existing up to now numerical values of some physical properties of agricultural materials, as well as create the necessity for their new description, and in certain cases some modification of working parameters in unitary operations in technological process (cleaning, grading, drying).

Table 2. Some of the crop character being modified by genetic modification (Dale, 1999).

Category*	Crop characteristic
1	Herbicide tolerance to enable certain crop varieties to tolerance specific herbicides and in many cases reduce the number of the herbicide applications to achieve effective weed control (103 field trials/ decade)
2	Pest resistance (75)
3,4,8,9	Resistance to viral, bacterial and fungal diseases (93)
5	Plant architecture and flowering, including plant height, flowering time and flower colour, Increased tolerance to environmental stresses including cold, heat, water and saline soils, Reduction in seed losses trough shedding at harvest time, Modification in fruit and tuber ripening and storage (tomato, potato). Prolonged shelf-life and reduced dependence on the use of anti sprouting compound applied to stored tubers, Increase in the ability of certain plants to remove toxic metals from soils (bioremediation), such as an mining water sites (14)
6	Enhancement of vitamins, minerals and anticancer substances, Elimination of allergens from certain crops (rice, rapeseed, wheat, potatoes), Oil, starch and protein modification to provide sustainable supplies of raw materials for biodegradable plastics, detergents, lubricants, paper making, in addition to improvements in baking and brewing qualities (95)
10	Production of pharmaceutical substances (e.g. anticoagulant compounds, edible vaccines (16)

*The numbers description as in Table 1.

Taking into consideration the advance in the studies on transgenesis as well as its wide application (see above) in large scale breeding and considerable share in human diets, a review of expectation and possibilities of transgenesis should start with products of plant origin. It can be also difficult to predict how quickly new products will become available and be accepted by consumers especially that food is being considered not only as a source of basic nutrition but rather as a product with specific medicinal properties, the so-called 'functional food' (Dunwell, 1998).

The best and most practical way to establish the food safety is to verify whether a novel food and /or food components are substantially equivalent to the one analogous conventional food product (Madden, 1995). There are two basic criteria of similarity proposed. The first criterion includes the phenotype evaluation such as morphology, growth, resistance to diseases. The second one is related to chemical composition of the most important components (Ludwicki, 1998). Account should be also taken of the physical and physico-chemical properties of novel products compared to unmodified crop. Especially in the case of the GM crops it is necessary to estimate digital data which could be helpful for unitary operation within processing, e.g. during harvesting, transportation and storage.

Doubts appear if, despite designed modifications (e.g. increase of tolerance to herbicide, resistance to bacterial diseases or viruses, etc.) alternation in metabolic pathway of the major plant products can affect physical or physico-chemical properties of novel material. It is well known that some elements of grain morphology are also of great importance to physical and physico-chemical properties of a single grain and bulk grains or their behaviour during processing (transportation, filling or emptying) (Fornal, 1998). Widely applied, traditional separation or cleaning process are based on the geometrical features and on the surface properties (Figs. 1a & 1b), which in turn depend on grain moisture, species and variety (Molenda et al., 1995). It is obvious that the alternation in grain morphology (modification of cell walls or tissue composition, synthesis of non-starch polysaccharides), or main storage material and other constituents of grain endosperm, or seed cotyledons could lead to the changes in behaviour of novel crops during processing. The surface properties are also related to susceptibility of grain to water binding capacity (Fig. 1c), what results from cell wall composition. The modification of cell wall and cell content (Fig. 1d) could lead to an increase in water stress before harvesting and mechanical forces during handling what could result in damages of outer or even of inner part of a kernel (Fig. 1e). Fornal et al. (1993) found that the compactness of morphological components is also decisive for water sorption within a grain and legume seeds. Thus the internal structure together with processing parameters and initial moisture content of seeds seem to be important factors influencing the quality of seeds (Table 3).

Table 3. Percentage of faba bean seeds destroyed during drying

	Drying temperature [°C]			
	25	35	45	55
Damaged seed [%]	12.0	43.2	84.0	93.0

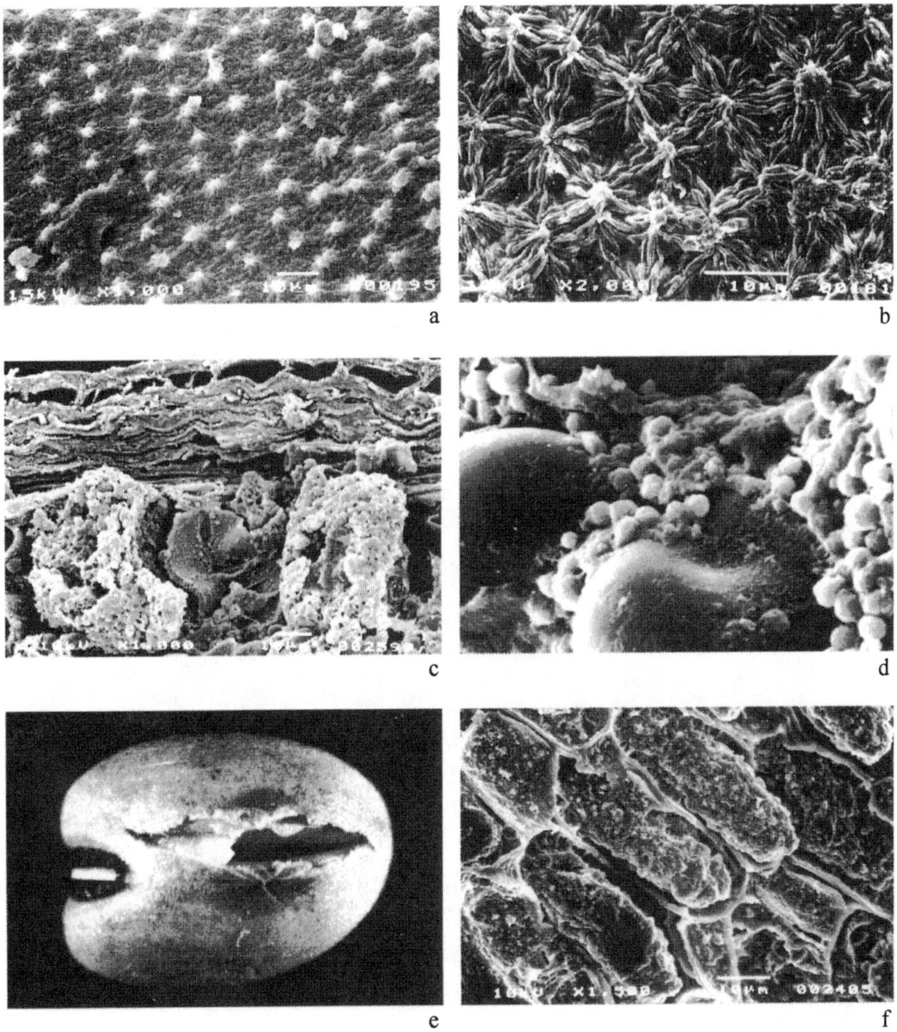

Figure 1. Structural elements of seed important for physical properties and changes occurring during storage that may be influenced by genetic modifications. (a) & (b) Surface morphology of *Lathyrus sativus*. (c) Seed coat with alleurone and suballeurone layers of wheat *Triticum aestivum*. (d) Protein bodies and starch granules in pea *Pisum sativum*. (e) An example of internal and external damages *in Vicia Faba* seed. (f) Rapeseed *Brassica napus* cotyledon structure organization.

Moreover, the weak mechanical properties and/or poor processability of many crops could result from improper post harvest storage and/or drying condition. However, the results presented by Fornal et al., (1994) and Sadowska et al., (1996) prove that behaviour of seeds during processing seemed to be connected mostly with the seed quality (Fig. 1f & Table 4).

Table 4. Rapeseed processing characteristic

Content	Sample no.				
[% d.w.]	1	2	3	4	5
Press-cake					
Water	7.8	7.8	6.7	7.2	8.3
Fat	19.3	18.3	23.3	21.3	27.9
Meal					
Water	12.9	14.7	9.2	10.0	10.8
Fat	4.4	3.2	18.8	18.6	27.9
Percolation	good	good	very bad	very bad	none

The evaluation of novel crops should also include technological processes that the novel product might undergo, including thermal processes (e.g. cooking), as well as the intake of novel crop. The GM grains with designed endosperm character (translucency or opacity) and starch content could also increase effectiveness of many processes and/or the end product quality. In the case of legumes, the thickness of husk or seed hull could significantly affect energy consumption during the dehulling process.

1. GENETICALLY MODIFIED CEREALS: NECESSITY OR A CHANCE?

Plant breeding has contributed to the increase in agricultural productivity. Due to introduction of the plant breeding based on molecular techniques into traditional breeding, the following strategies are made possible:

- to reduce the time necessary to develop new crop varieties from 12-15 to 6-7 years,
- to cut the varietal development cost,
- to advance contribution of plant breeding to the crop production.

Most of the GM crops developed up to now have influenced characteristic that are controlled by single genes (Tab. 1). Quantitative characteristic, yield, photosynthesis efficiency etc. are controlled by many genes and the ability to manipulate these characteristics would probably be a major step forward in the plant breeding technology [Franks, 1999].

It is possible to affect physico-chemical properties of cereals products. Cereal transformation by transgenic technologies is considerably more difficult and has limited application in those plants due to the fact that many physico-chemical properties of cereals are rather of bio-physical than bio-chemical nature. These properties are influenced by complex interaction with other components and the changes connected with environmental conditions (Shulman, 1999; Fornal et al., 1999; Igrejas et al., 2001). This type of modification is quite difficult since functional properties are more difficult to define in molecular aspect compared to other modifications.

The transgenic technologies make it possible to design some of the physico-chemical properties of grains as well as mechanical ones such as: hardness or softness of wheat grain (Figs. 2a & 2b). Grain hardness might be expected to be influenced by a range of biochemical factors and thus several genes. It is well known also that the grain hardness is directly related to a major gene located on chromosome 5D (MacRitchie et al., 1990). These authors informed also that in the seventies Simmonds et al., (1973) found some kind of relation between hardness of wheat and the presence of an 'adhesive' protein around the starch granules. On the basis of that correlation in the eighties Grenweel & Schofield (1986) demonstrated the presence of a protein of 15 kDa on the surface of starch granules. That low-molecular protein had been washed from soft wheat but not from hard-grained varieties, however. It is very probable that distribution of wheat grain softness protein – friabilin is different in small (< 5μm) and large (>5μm) starch granules (Figs. 2c & 2d). They suggested that the wheat grain-softness protein ('friabilin') affected the function of releasing the starch granules from the protein matrix of a soft wheat. Nowadays the genetic basis for two major texture classes (hard and soft) is well established as a single major gene the allelic expression of which is associated with mutations in the proteins puroindoline a and b (Bettge et al., 2000; Wilde et al., 1993). It is possible also to improve the quality of wheat for baking what seems to be a crucial advancement of biotechnology in designing food of new generation. Igrejas et al., (2001) showed that very low amounts of puroindolines (PIN-a), from 0.05 to 0.1%, have important effect on breadmaking. These compounds produce bread types with a fine crumb structure (Figs. 2e and 2f) resulting from the fine gas bubbles that are formed by puroindoline foams in the aqueous phase of dough. The low amounts of puroindolines are also an important factor for the changes in dough tenacity and extensibility (Figs. 3a & 3b). Miflin et al. (1999) reported that the changeability of gluten elasticity is connected with high-molecular fraction of glutenine, especially with the fractions number (3- 4%), share in total protein (6-12%) and their properties. Thus, the designing of cereals quality through biotechnology by means of encoding genes makes it possible to use for example wheat with gluten containing 20% of high molecular weight fraction as a baking improver. It makes the flour very strong and proper for improving of the properties of weak flours. Shewry et al. (1998) reported a qualitative effect of high-molecular subunits protein on the dough strength. The functional importance of the high molecular subunits and the demonstration of a relationship between gene expression and functional properties has made them a clear target for further genetic engineering studies. Among the major crop plants, both cereals and legumes are important sources of protein (10 – 15 %) for humans and their livestock. The cereal prolamins (the major storage proteins) have genetically reduced of lysine and threonine. Also lysine content is low of about 1.98 %, compared with a WHO recommended level of 5.5 % (Shewry et al., 1998). Since maize protein is considered to be of relatively poor quality due to the low lysine and tryptophane contents, researchers looked for new ways of improving its quality. Some suggestions for improving of the protein quality in the maize kernel involved:

a) decreasing the proportion of zein,
b) increasing the percentage of tryptophane,
c) increasing the germ within the kernel, or
d) increasing the percentage of lysine (Rand et al. in press).

The authors felt discouraged also in altering of the tryptophane and lysine contents, due to the difficulty of developing suitable analytical procedures. On the basis of the changes occurring in the amino acid composition, it was found that the quality of the novel High Protein maize was less nutritionally adequate than that of the low protein maize. Improving the protein quality of the maize kernel may be associated with potential negative effects, such as reduced yield. During studies on the behaviour of starches extracted from *Riso* mutant barley lines Tester et al., (1993), found that apart from low grain yields and weight the all mutants showed evidence of impaired starch synthesis. Selection for yield and agronomic traits simultaneous to selection for improved protein quality is essential. The examples of GM maize are experimental varieties of High Oil Maize (HOM). The HOM consists of an enlarged germ with higher nutritional content of fat, energy, protein and essential amino acids (Rand et al., in press). Shewry et al. (1998) reported that also in the case of barley some genetic techniques, focused on the exploitation of high lysine genes, were used to increase the lysine content. These attempts failed due to an inability to separate the high lysine character from deleterious effect on yield. The high lysine phenotype in barley lines is associated with a decrease in a proportion of prolamins and an increase in the proportion of more lysine-rich proteins. The contents of amino acids, such as methionine and tryptophan could be increased by expression of proteins reach in these amino acid, e.g. sunflower 2S albumin (SFA 8) and wheat puroindolines.

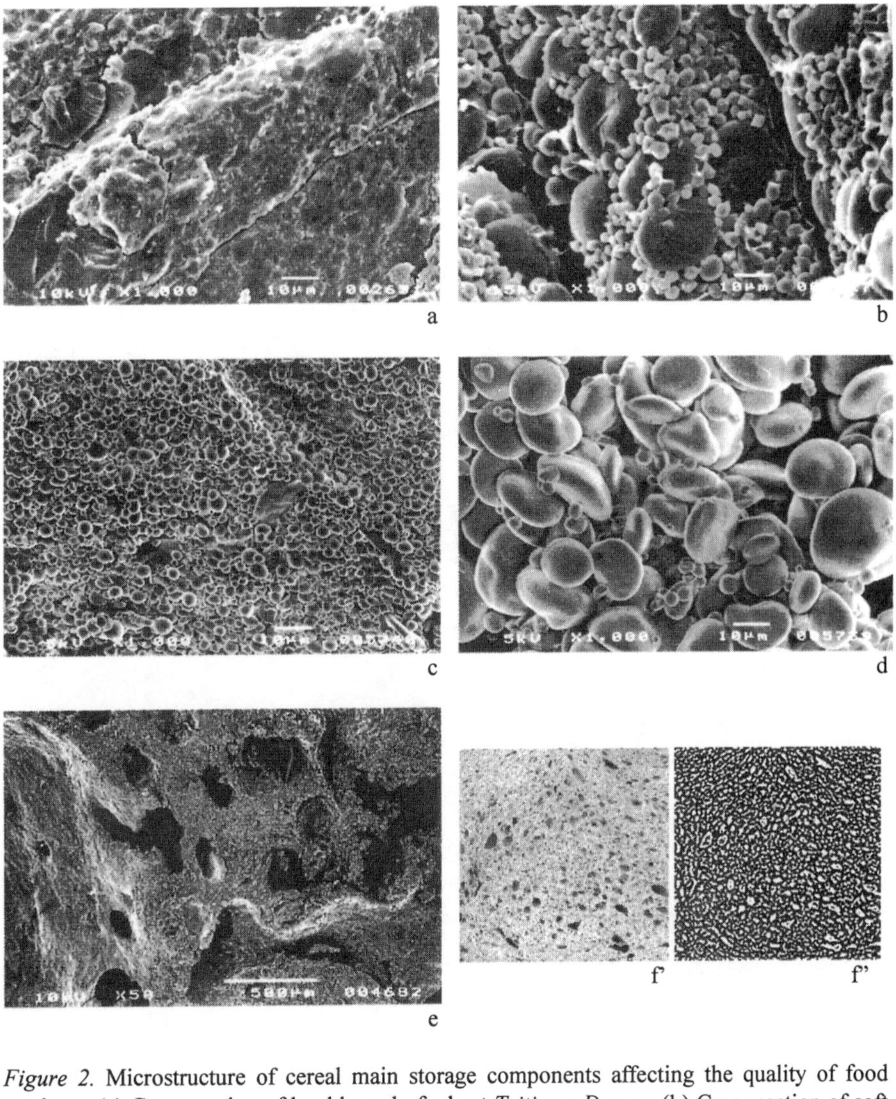

Figure 2. Microstructure of cereal main storage components affecting the quality of food products. (a) Cross section of hard kernel of wheat *Triticum Durum*. (b) Cross section of soft kernel of wheat *Triticum Aestivum*. (c) Small (type B) starch granules. (d) Large (type A) starch granules. (e) The distribution of gas cells in bread crumb. Real (f') and transformed (f'') images of bread crumb related to the Dallmann standards.

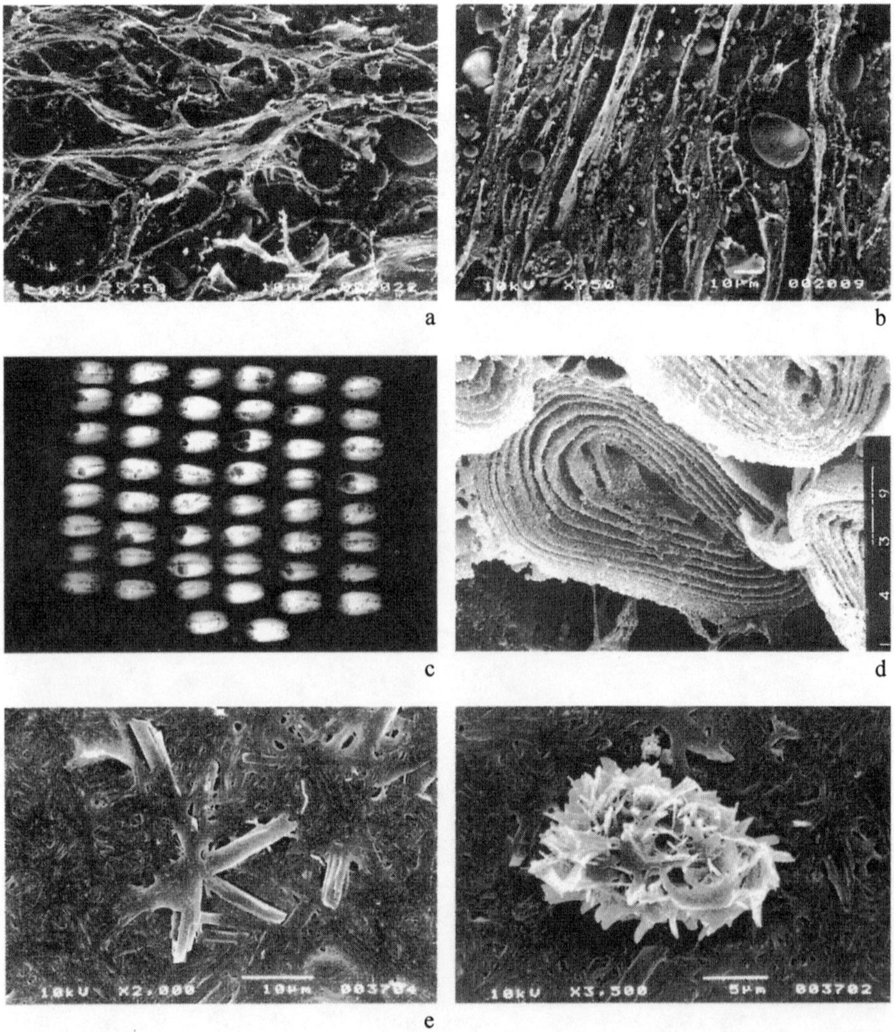

Figure 3. Microstructure of raw material and food products after selected processing. (a) Wheat dough characterised by weak structure of gluten matrix. (b) Wheat dough characterised by strong gluten structure. (c) Roentgenogram of wheat kernels infected by Granary weevil (*Sitophilus granarius*). (d) The view of starch granule after α-amylase treatment. (e) Fat crystals (needle - like) in margarine (f) Fat crystals (β-spherulites) in margarine.

Attempts are also made to eliminate gluten from wheat grains - the factor limiting consumption of cereals products by people suffering from coeliac diseases, dermatitis herpetioformis, or gluten enteropathy (Shewry et al., 1998). Statistics say that the number of people affected by gluten entheropathy in Europe varies between 1 per 4700 inhabitants to 1 per 300 ones, and in the USA 1 per 250 inhabitants. The most widespread adverse response to ingested wheat is coeliac disease, intolerance to specific sequences present in gluten proteins (Marsh, 1992). This is probably related to short peptide sequences which are rich in glutamine and proline. Marsh et al. (1995) and Kasarda (1998) proposed three different 'toxic' sequences of amino acids such as: GlnGlnGlnProPhePro; ProGlnGlnPro; ProSerGlnGlnPro, which could be responsible for coeliac disease and are widely distributed in wheat gluten proteins and related prolamins from barley (hordeins), rye (secaline) and oat (avenine). A gluten-free diet used as a prophylactic agent does not guarantee avoiding of the disease and among possible causes of that fact contamination during milling and prescription mistakes occurring during production are mentioned. Allergies to cereal seed proteins also occur, notably bakers asthma, a respiratory allergy to wheat flour, and dietary allergies to wheat and also to the rice proteins. In the case of bakers asthma, the suffering comes from production in the human organism specific immunoglobulin E (IgE) to salt soluble proteins, including members of the cereal family of α-amylase/trypsin inhibitors and gluten proteins (Sanchez-Monge et al., 1992). That disease is an immunological response, characterised by acute or chronic hypersensitivity, to allergenic factor. The factor responsible for the above-mentioned diseases has been already identified as low-molecular proteins of 14-16 kDa. The major allergenic proteins in rice are water soluble fractions of 14.5 kDa, other being α-amylase inhibitor from wheat endosperm and agglutinin of wheat germs (Franck-Oberaspach et al., 1997). It is hardly possible to eliminate the activity of the allergenic cereal components because of the presence of multiple proteins and epitopes (Shewry et al., 1998). On the other hand, the above-mentioned allergenic factors protect plants against insects (Fig. 3c) since they naturally act in kernels as insecticides or enzymes inhibitors (Franck-Oberaspach et al., 1997).

Biotechnology can be also used for improving of the enzymatic system of cereals, e.g. production of transgenic varieties with an increased activity of special kind of enzymes important from the point of view of baking and other branches of plant processing. The following enzymes are in the centre of interest of biotechnologists as the most significant ones influencing the quality of milling, baking and end-products:

- α-amylase – playing an important role in bread staling and responsible for the erosion of starch granules due to hydrolysing of two of the starch components (Fig. 3d),
- α-amylase and pentosanes – acting as bromates substitute in proper proportion,
- α-amylase/ lipase/ xylanase – crucial for the dough conditioning (strengthening),
- proteases (papaine, chymosine) – modifying gluten, shortening mixing time and substituting pirosulfate,
- glutaminase – allowing elimination of bromates (Shukla 1999).

Research is underway to develop more efficient post-harvest processing techniques to reduce the number of steps leading to the desired products (Dunwell, 1998). Another approach in the production of modified starch has become the focus of research, in which genetic engineering is used to produce novel starches in plants as well as plants with new technological properties. Therefore, there are possibilities of obtaining starch of determined and desired structure, properties and quantity:

- GM wheat starch revealing the properties of maize starch,
- GM starch revealing the properties of starch phosphates,
- GM starch revealing the properties of RS starch (amylase resistant one) (Miflin et al., 1999; Salamini, 1999),
- waxy corn and wheat with partially or totally reduced amount of amylose in starch (used in the production of frozen and microwaved foods, dressings as well as to prolong the shelf-life of baked goods) [Graybosch, 1998],
- GM maize with high amount of amylose in starch 50-70% (amylose extender),
- waxy sugary –1 and sugary-2 maize with partially reduced amount of amylose in starch (stable during freeze and freeze-thawing, mainly used as a gelling agent) (Schulman 1999).

The waxy mutation destroys expression of 60-kDa granular bound starch synthase I (GBSS I), thus resulting in elimination of amylose.

Nowadays, starches are modified to improve not only their utility in either chemical and enzymatic meaning. The alternation of starch quality by means of genetic engineering could be also important for food technology. In brewing the relative resistance of small (B-type) granules of barley to saccharification may complicate the malt production. Plant transformation technology and the increasing availability of appropriate genes make novel starch or biosynthesis of novel carbohydrates (e.g. fructans in shx mutant of barley, cyclodextrins in potato) in the plants possible.

Oats used for human food should have low kernel oil what will reduce the fat content of the end-product maintaining at the same time the nutraceutical, cholesterol-lowering effect of oat bran and oatmeal. The modifications of oat kernel composition could be significant as a means of developing novel food, feed and industrial products from oats that may increase the demand for oat and oat products. The main problem limiting the yield of the oat crop and restricting its production in many areas is of agronomic character including the lack of durable resistance to fungal and viral diseases (Somers, 1995; Clifford, 1995). Especially in case of viral diseases a high progress has been reported for resistance to barley yellow dwarf virus (BYDV) affecting crop yields. Based on promising results of Somers (1998) regarding study on the viral resistance in transgenic oat, it follows that further research is warranted on viral coat protein mechanism for BYDV resistance. Additionally, advantage is taken of genetic engineering to add or improve existing traits in oat by introducing genes from any organism into oat (e.g. the wheat HMW glutenin promoter (Dy10)). The opposite action to bringing new genes into oat from any organism or increasing expression of existing oat genes is to suppress oat gene expression using an antisense approach. These technique is very attractive because the antisense transgene is predicted to confer any specific kernel trait (Somers, 1995).

There have been known also some works on the expression of herbicide resistance in oat [Gressel, 1992].

Vitamin deficiency is rare in the developed world but there are still countries where people suffer from the lack of vitamins as well as other micronutrients. Dunwell (1999) reported that over 124 million children world-wide are deficient in vitamin A (carotenoids) and that improvement in dietary intake of this vitamin could prevent 1.3-2.5 million deaths among infants and small children. The expresion of *psy* gene in rice leads to the production of phytoen – a precursor of lycopene which in turn resulted in convertion to carotenoids, and thus to the accumulation of provitamin A in endosperm. An improved understanding of the biosynthetic pathway for vitamin C has also led to claims for its overexpression in plants e.g. by expressing the gene encoding L-galactono-α-lactone dehydrogenase enzyme (Burkhard et al., 1997; Bauw et al., 1998). Another approach is exemplified by introducing into rice a gene encoding the soybean ferritin as an iron-storage protein. The performed experiments showed that expression under control of a rice seed-storage protein promoter resulted in increasing the levels of iron in the GM seeds endosperm threefold compared to controls (Goto et al., 1999).

These above-presented examples of cereals-derived GM products are likely to be available on the market in the year 2020. The authors concentrated on showing the possibilities of genetic engineering in improving agronomic performance of crops as well as designing of nutritional value of novel cereal products.

2. "DESIGNED" LEGUME AND OIL SEEDS

Application of genetic engineering creates the possibility of obtaining 'designed plant' such as: soybean with changed level of linoleic acid and/or increased content of oleic one. Production of soybean with higher level of phytates makes it possible to better uptake of phosphorous from feed. Among of legume seeds only soybean already has been commercially released. However some works have been made related to increase of nutritional quality of other legumes such as lupin or lentil, also. Molving et al., (1997) reported that seed-specific expression of a sunflower seed albumin (SSA) into transgenic narrow-leafed lupin (*L. angustifolius*) resulted in increased content of methionine in seeds with simultaneous decreased content of cysteine. The expression of the yeast invertase gene in transgenic cotyledons of *V.narbonensis* caused reduction of sucrose, starch and protein content and increases of hexose level (Weber et al., 1998).

Many legumes contain relatively high concentration of lectins, which could be responsible for some diseases such as severe nausea, vomiting, diarrhoea. Especially in the case of African countries the level of cyanide-derivative compounds can lead to death and also to some neurological diseases resulting from the consumption of not-properly cooked legume seeds. Of course, thanks to genetic engineering it is possible to produce the novel plants with reduced level of some antinutritional carbohydrates, including the content of D- galactose containing oligosaccharides of sucrose (by expression of gene encoding enzymes that degrade α-galactosides), or to produce crops containing less sulphate and more total amino acid sulphur compared to non-transgenic ones (Kuchuk et al., 2001).

Genetic modification of canola seed allowed to increase lauric acid in the fatty acids content and made the seeds its valuable source. It was possible due to employing an enzyme preventing the production of long-chain fatty acids through cutting off the acids (which gained the chain longitude of C 12) from the complex of synthesising enzymes. It is well known that polyunsaturated fatty acids, present in small amount in some plants, are more valuable and crucial for humans. Therefore, one of the aims of biotechnology is to produce plants with an increased content of such components for wider consumption. The composition of plant oils as well as technology applied is of special importance for the quality of obtained products like spreads or margarine (Fornal & Błaszczak 2001a) (Figs. 3e & 3f). The second objective is the production of new highly valuable pharmaceutical oils and encosanoides precursors. It can be difficult, however, because of the necessity of obtaining transgenic plants which would produce proper enzymes – desaturases (Mifilin 1999; Butler et al., 1999). Biotechnology makes it possible to produce plants with greater proportion of saturated fats (suitable for margarine production), and ones which yield a high-temperature frying oil that contains a low proportion of saturated fat (Madden 1995).

3. "DESIGNED" NOVEL FRUITS AND VEGETABLES

In order to understand the ripening process of fruits and vegetables the antisense technology was used. Among of all fruits and vegetables tomatoes are characterised by relatively small set of genes, and due to that fact they are very often selected for work involving gene manipulation. The monitoring of the maturation of fruits and vegetables already on the cellular level might allow farmers in tropical countries to benefit from the demand for exotic fruits and independent on season temperature. It would be also possible to produce high-quality fruit products (juices and purees) because of less required processing and fewer additives resulting from slower decay of fruits during transport and storage.

The antisense rout has been also applied in order to slow down the softening process of some fruits and vegetables (Madden 1995). It is well known that an enzyme called polygalacturonase (PG) is the main one which forms fruit texture and is responsible for ensuring impression of hardness and freshness (Figs. 4a & 4b). A slightly higher PG activity result in progressive hydrolysing of pectins and breaking cell walls, what in turn, result in softening process and made the fruit mushy. Through the expression of antisense PG gene, in plants, it is possible to produce crops with the lowest level of polygaracturonase activity such as Flav Savr ^{Tm} tomato (produced in USA). To slow down the ripening of fruits, expression of gene inhibiting ethylene production during natural maturation was applied. That kind of gene manipulation might also be of advance to farmers who could pick the unripened fruits and vegetables and place them in ethylene atmosphere till distribution. Madden (1995) reported that the result of common work of French, Spanish, British and Greek researches on antisense technology in fruits is a modified melon called "Euromelon".

Genetically modified tomatoes as well as lettuce were produced with the intention of improving the palatability by expression of two genes for protein sweeteners. The protein sweeteners such as thaumatin and monellin naturally occur in tropical plants and thaumatin is 3000 times sweeter than sugar (Dunwell 1999; Maden 1995).

Modern biotechnology in plant breeding also included more sophisticated genetic improvements such as:

- avoiding of some technological processing – coffein - free coffee – which is normally removed by solvent extraction (Dunwell 1999);
- obtaining of apples and strawberries with special functional properties (health benefit)- limiting decay caused by Streptococcus mutants (by expression of the gene controlling the growth of bacteria),
- expression of alpha-lactoalbumin and lactoferrin proteins in transgenic tobacco (Dunwell, 1999),
- production of GM plants (maize, lettuce or carrots) to deliver vaccines (anti-viral vaccines), plant-derived insulin, glycoproteins as well as antigens/ antibodies.

4. POTATO TUBERS – AN EXCELLENT MATERIAL FOR GENETICAL ENGINEERING?

Potatoes are on the top of the most important food crops. The world production is about 287 800 kt per year (Europe: 81 140 kt -, Poland: 25 949 kt - 1998) (Bartens, 2001). Currently they are considered to be an important food for the ingestion of selected vitamins and minerals or other bioactive compounds (hemagglutinins, polyphenols, glycoalkaloids) (Table 5, 6).

Table 5. Proximate value for potato tuber constituents according to Rogan et al., (2000) (g/ 100g d.w.)

	Literature range
Total protein	7.1-14.62
Fat	0.1-0.8
Ash	2.2-9.5
Crude fibre	0.2-3.5
Total carbohydrates	84.5 (av.)
Calories	350 (av.)

Table 6. Nutritional parameters of potato tubers (Rogan et al., 2000)

	Literature range
Total solids (% f.w.)	16.8-26.8
Dextrose (% f.w.)	0.03-0.52
Sucrose % (f.w.)	0.05-0.88
Vitamin C (mg/ 100 g d.w.)	10.3-22.0
Soluble protein (% d.w.)	3.3-7.3
Total glycoalkaloids (mg/ 100 g f.w.)	2.5-16.1

Potatoes are also an excellent source of carbohydrates (starch and free sugars), free amino acids and high-quality protein. The consumption of potatoes is varying from country to country but they occupy a high position in human diet (140-400 g and more per capita per day).

The potato tubers could contain also significant amount of the naturally occurring toxicants - alkaloids. The representatives of the glycoalkaloids widely distributed in potato tubers (up to 95% of total amount) are mainly α-chaconine and α-solanine (Engel et al., 1996). With the use of genetic engineering techniques (see below), it was possible to significantly decrease (40%-60%) the amount of these undesirable substances.

The strategy of potato breeding is to introduce the genes responsible for resistance to pest into proven potato cultivars by genetic engineering. An advantage of this approach over traditional breeding is that a single characteristic may be added to a proven cultivar without changing other characteristics of the cultivar (Flis & Zimnoch-Guzowska, 2000).

The potato crop is susceptible to over 25 virus and viroid diseases, from which the most universally and economically important seem to be viruses of potato roll virus (PRLV), potato virus Y (PVY), potato virus X (PVX) (Hashimoto et al., 1999, Thomas et al., 1999, Madden, 1995), potato virus S (PVS), and potato virus A (PVA) (Singh, 1999). Thus, one of the main objectives of creating transgenic plants is attempt to engineer proper metabolic pathways for the improving of resistance toward above mentioned pathogens (Herbers & Sonnewald, 1999).

Deleterious effect of Potato Leafroll Virus (PLRV) pertains to viral perpetuation in tubers called net necrosis (Thomas et al., 1997, Mosley et al., 2000). Almost all commercial potato varieties could be infected with PLRV with world-wide yield losses estimated at 10%, but in some regions the yield losses can reach as much as 50%. During growing season this virus can develop net necrosis resulting in damage of phloem cells (Figs. 4c & 4d).

The necrotic part of tuber is characterised by network of coarse brown strands in the vascular ring of the flesh (Rogan et al., 2000). The attacked potato tubers are unacceptable for consumption as well as processing especially into chips and fried or canned potatoes (Pasche et al., 1997).

It is well known that expression of the virus coat protein (CP) gene in potato resulted in an increased resistance to infection with potato viruses such as: Y (PVY) or X one (PVX). The PVY virus is also considered one of the most damaging potato viruses because in some circumstances it causes economically significant yield reduction by as much as 80% (Bemster & Books, 1987). Smirnov et al., (1997) reported that the expression of antiviral protein – rootstock in potato displayed increased resistance against PVX and tobacco mosaic virus. It was found, however, that CP gene expression in potato is not enough to protect the plant against potato leafroll virus (PLRV). Most of the potato crops are infected with PLRV disease in the way of plant to plant spread accounts. For this reason much efforts should be direct to achieve the resistance to plant-to-plant spread of PLRV. Thomas et al., (1997) reported the possibility of modifying gene construct connected with transformation and plant expression vectors containing the PLRV and CP. These authors analysed transgenic potato lines containing a gene expression cassette with two copies of a PLRV/ CP gene in which the nucleotide sequence was modified to improve expression of the gene and the obtained results proved useful resistance to PLRV.

The commercial interest dictates that the majority of transgenic plants entering field trials should be resistant to herbicides, pathogens and also to insect pest. The most damaging insect pest of potato is Colorado potato beetle (CPB). If not controlled, it can lower potato yield even by 85% (Rogan et al., 2000). Control of CPB is almost totally focused on the application of proper insecticides depended on pesticidal characteristic and limitation. Many pesticides used for protection against aphids and other insect pest of potato are non-selective, e.g. killing beneficial occurring organisms that prey on aphids and CPB. The resistance of plans against insects was created by the expression of *Bacillus thuringiensis (Bt)* δ-endotoxins. The Bt toxins were the first insecticidal proteins discovered and they have been subject to extensive improving for expression in plants (McBride et al., 1995).

Nowadays, all transgenic clones which are subjected to transgenosis in order to improve resistance to the viruses, nematodes etc., insect pest as well as technological value could be directly evaluated in the field condition in respect to several agronomic and morphological traits and compared with measurements for non-transgenic analogous plants.

The evaluation and also further selection of novel crops is based on the consistence and substantial equivalence with traditionally obtained ones. Flis & Zimnoch-Guzowska (2000) found that the novel transgenic potato clones with increased resistance to some pests (e.g. PVY) showed a great variability (Table 7). These differentiation in phenotypes could be an important and also a necessary step in breeding strategy of transgenic plants in order to an improvement of the cultivar.

Table 7. Nutritional parameters of potato tubers (Flis & Zimnoch-Guzowska., 2000)

	Trait	Control	Transgenic R1 (n=15)	% dev.	Transgenic R2 (n=18)	% dev.
	Tuber yield [kg/hill]	3.7	4.3	53.3	3.7	22.2
	Tuber weigth [g]	62	69	40	60	11
	Starch content [%]	13.8	14.6	80.0	14.4	33.3
1-5 deformed-typical	Plant appearance	4.8	4.9	0.0	4.7	5.6
1-5 week-vigorous	Plant vigour	3.9	4.3	46.7	3.8	0.0
1-7 shorted-long	Tuber shape	4.0	4.0	6.7	4.1	11.1
1-9 irreg.-regular	Regularity of tuber shape	6.5	6.4	26.7	6.7	38.9
1-9 deep-shallow	Eye depth	7.0	6.9	26.7	7.0	11.1

In order to produce an improved potato clones the most important requirements should be met:

- the introduced transgene should have an acceptable expression of the improved property,
- maintaining of the key characteristic of the original cultivar by transgene plant.

Ganga & Corke [1999] reported that in many developing countries there are factors limiting the potatoes technological application. The potatoes differed in genotypes and for this reason they varied widely also in their physico-chemical properties and technological use as well. The varieties of potatoes grown in Asia are mostly table or cooking types, while these harvested in Europe or in USA are industrial/commercial. The biotechnology creates possibility of obtaining a new type of potato varieties that could be harvested irrespective of independent the place of growing and be suitable for home and industrial processing (cooking, frying, canned potatoes, frits and starch production).

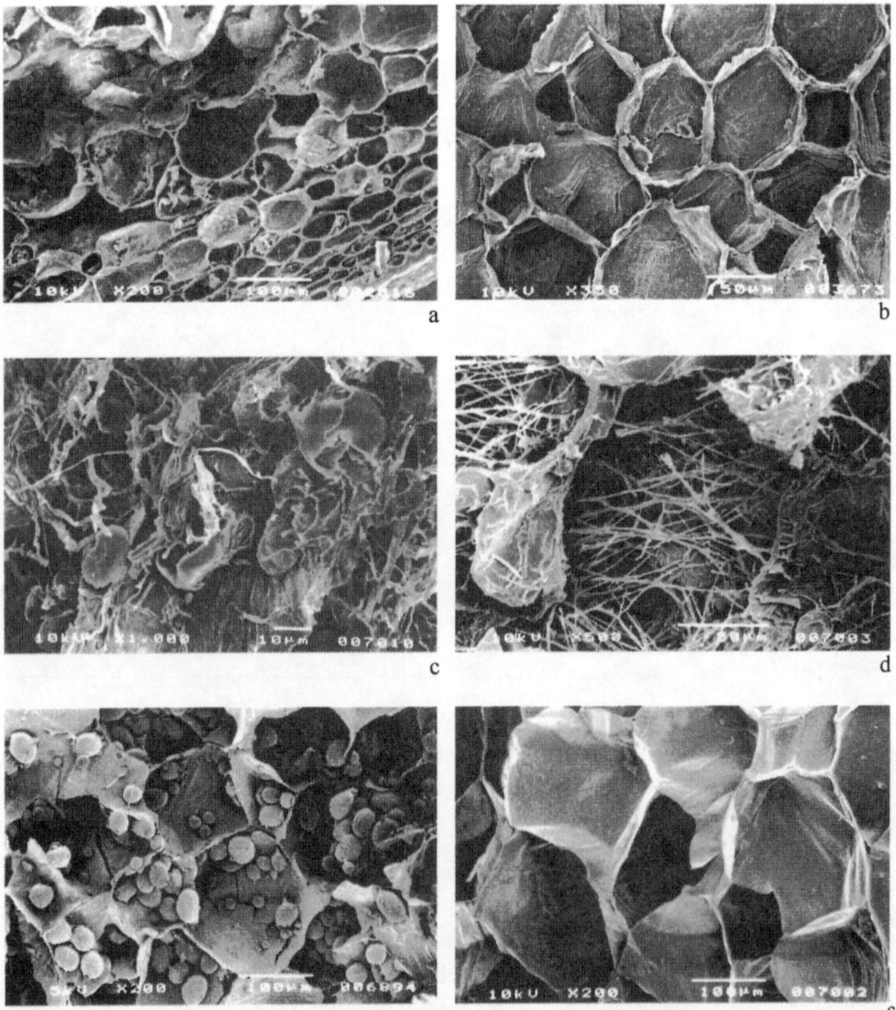

e f

Figure 4. Microstructure of selected fruits and vegetables. (a) Cross section through the apple tissues (peal and parenchyma). (b) The parenchyma tissue of an apple (enlarged). (c) The storage parenchyma tissue of potato tuber infected with potato virus. (d) The details of necrotic part in storage parenchyma tissue of potato tubers. (e) Storage parenchyma tissue of potato tubers reach in starch granules. (f) The cells of inner pith region of potato tuber.

However, there are still some doubts concerning safety of modified potatoes and other novel (GM) crops. Therefore the safety evaluation is of special importance in order to confirm that GM products are safe in terms of nutrition. The substantial equivalence apart from agronomic traits (morphology, growth, flowering, resistance to pests) includes also the chemical evaluation of the novel products, e.g. determination of the most important nutrients, antinutrients, toxicant and allergen compounds. The mentioned criteria of evaluation have to be comparable with those for unmodified crops. The compositional analysis of tubers from insect and virus resistant potato showed that these tubers are substantially equivalent to the tubers produced by conventional potato varieties.

Taking into account variability in structural organization of potato tubers (Figs. 4e & 4f), concentration of naturally occurring glycoalkaloids is different in outer and inner parts of tuber as well as is depends on its size, potato variety and also external conditions during vegetation (Engel et al., 1996, Edwards & Cobb, 1999). Some authors claim that the highest concentration of glycoalkaloids is located within the first millimeters from the outside of potato tubers and decreases towards to the inner tissues. Also, the concentration of glycoalkaloids was higher in small tubers what resulted from the surface to mass ratio. The techniques of genetic engineering such as expression of yeast derived invertase or antisense RNA expression were applied in order to reduce the amount of the mentioned compounds in potatoes. Only the first technique exerted the effect on the alkaloids content. However, application of that kind of technique led to an increased yield and an increased tuber size resulted from metabolic interference. The ongoing research aims also at obtaining the dilution effect of the alkaloids in potato due to direct increase of the size of tubers (Engel et al., 1996).

It is distinct that cell packing and cell to cell adhesion or cell content define mechanical properties (e.g. stiffness and density) of potatoes (Scanlon et al., 1998). Thus, the attention of scientists is also paid to the effect of genetic modification on physico-chemical properties and processability of potato tubers (Sadowska et al., 2002 unpublished data) (Table 8).

Table 8. Mechanical properties of potato tubers: fracture stress - F [kPa] and strain - D [%] after thermal treatment.

Sample	Raw		Cooked		Microvawed	
	F	D	F	D	F	D
GM1	1133.2	54.4	86.4	30.9	79.9	40.2
GM2	1337.4	54.3	93.4	25.9	86.1	38.6
Control (commercial)	1172.9	45.2	68.3	18.8	67.0	33.7
Control ("in vitro")	1116.0	43.0	72.9	20.9	73.2	44.7

Coefficient of variance: F (10, 25, 25%), D (7, 15, 50%)

The modified clones were obtained from the same variety, and the modification was performed in order to increase resistance to potato virus PVY^N only. As it was shown in Table 8 the methods of modification only slightly influenced mechanical properties of GM1 potato (sense orient) and GM2 (antisense orient) although the modification did not partain to physico-chemical properties of tuber. However the comparison of the mechanical properties of control samples with these of the both modified samples indicated differences in resistance to compression of the tubers, in some cases statistically significant. When analysing single tubers within variety, it was also found that differences within variety were more distinct than these between modified and not modified clones. Generally, one can say that this kind of modification with gene expression responsible for higher resistance to potato virus Y in sense orientation (GM1) and antysense orientation (GM2) might affect mechanical properties of the obtained clones.

Except for this "classical" modification leading to obtaining of the potato with higher resistance to insect and viruses, the trials of genetic modification were undertaken, and the expectation are directed on:

- resistance to drought (Dunwell 1999),
- modifying the action of starch synthase and starch branching enzymes – leading to changes in starch: mainly amount and its structure (amylopectin) (Jobling et al., 1999),
- designing the level of the phosphate groups in starch in order to modify starch properties (Blennow et al., 1999),
- suppressing the cold-induced sweetening of potatoes by reduction of UDPglucose pyrophosphorylase activity. That alteration of potato tubers, connected with sugar accumulation during cold storage, contributed to discoloration of chips or slices during frying and resulted in decrease in technological quality of tubers (Schulman, 1999).

Taking into consideration a variety of attributes achieved using GM potato, also the most sophisticated ones should be mentioned:

- expression of human milk protein β- casein in transgenic potato plants (Chong et al., 1997) or
- expression of cholera toxin B subunit oligomers (Arakawa et al., 1997).

Improving of the nutritional quality of potato protein by genetic engineering is also a goal to be achieved in modified potato. It is known that during tuber storage, the changes in the potato components resulted in decrease of the content of some compounds. Lowering of the level of such amino acids as: leucine, lysine and threonine, seriously decreases the nutritional value of potato (Dunwell, 1999). This is especially important in the context of an insufficient level of the above-mentioned amino acids for the nutrition of adults as well as children under 5 years (Hashimoto et al., 1999). In their work Hashimoto et al. (1999) showed also the possibility of reducing sulfur amino acid deficiency of the potato by introducing a gene of native soybean glycinin what also improves the functional properties of potato protein (gelling and emulsifying). These authors also found that the GM potatoes with modified glycinin genes seemed to be as safe for food as non-GM potatoes.

5. BENEFITS OR RISKS TO THE ENVIRONMENT?

The environmental context of biotechnology also seems to be controversial. From one side it is a greater risk of distortion of plant–soil system/ equilibrium, widespreading of genetic uniformity, and increased incidence of plant diseases. These could lead to the reduction of biodiversity. On the other hand biotechnology is a chance for bioremediation – the decontamination of naturally polluted land as well as consequence of industrial and other activity. It is possible to produce plants which would grew vigorously in soil containing toxic level of ionic mercury, cadmium or hydrocarbon pollutants (Dunwell, 1999). Introduction of the GM crops into the environment reduced the field applying of agrochemicals by 4mln kg in 1997- 1998, and resulted in the reduction of soil erosion by 90%. Related examples of GM plants may be beneficial for the reduction of the need for mechanical soil treatment and decrease in volunteer problems in agricultural rotation system.

Except of beneficial effects also the probability of harmful changes resulting from introduction of transgenic plants to the environment and its development should be stressed. The potential risks which could appear may be related to:

- greater reliance on herbicide for weed control,
- increase in herbicide use and its negative effect on native species.

6. CONCLUSION

Presented in this paper the examples of plant genetic improvements seem to be very interesting and promising regarding their present and/or future application in agriculture, technology and nutrition. However taking into account the human nature and resulting from it a fear from the GMO being introduced into food marked, it is necessary to give an answer to the following questions:

- should not we rather take the advantage of crops bred traditionally
- should we improve existing technologies in order to eliminate their weakest points
- should we produce a novel food realised the potential risk that is still possible?

Nowadays the meaning of biotechnology in creating of new food properties is very significant. On the other hand it is well know that only one thing we know about biotechnology for sure is the fact that we know not enough. Thus, if we decide to choose the way of biotechnology development towards GMO production the risk regarding safety of novel products should all the time be taken into consideration Bearing in mind one of the oldest but still valid the toxicological rule of Paracelcus (1493-1541) running as follows:

"All is the poison, nothing is free from poison, only the doses make that the things are not poisoning",

all the doubts concerning the use of GM crops as food and feed should be solved on the basis of detailed laboratory evaluation, safety control and proper labelling of novel food products.

REFERENCES

Arakawa, T., Chong, D.K.X., Merrit, L.J., and Langridge, H.R., 1997, Expression of cholera B subunit oligomers in transgenic potato plants. *Transgenic Research* **6**: 403-413.

Bartens., 2000, *Sugar and starch.* The 3rd ed. (Bartens Sp. z o. o., Słubice, Poland, edt.).

Bauw, G.J.C, Davey, M.W., Ostergaard, J., and van Montagu M.C.E., 1998, *Production of ascorbic acid in plants.* Patent Application WO 98/50558.

Bemster, A., and De Boks, J., 1987, Survey of properties and symptoms. In *Viruses of Potatoes and Seed Potato Production.* The 2nd ed. Wagenigen, The Nederlands,.

Bettge, A.D., and Morris, C.F., 2000, Relationship among grain hardness, pentosan fraction, and end-use quality of wheat. *Cereal Chemistry* **77**: 241-247.

Blennow, A., Bay-Smidt, A.M., and Moller B.L., 1999, Starch phosphate in relation to starch molecular structure and physico-chemical properties.In *The First International Workshop Devoted to Polysaccharide Design and Understanding of Starch Biosynthesis. Genetic Tailoring of Novel Starch Polymers,* September 16-20, Carry-le-rout, France,.

Burkhard, P.K., Byer P., Wunn, J., Kloti, A., Armstrong, G.A., Schledz, M., Linting, J., and Potrykus, I., 1997, Transgenic rice *Oryza sativa* endosperm expressing daffodil *Narcissus pseudonarcissus* phytoene, a key intermediate of provitamin A biosynthesis. *The Plant Journal* **11**:1071-1078.

Butler, D., Reichhard, T., Abbot, A., Dickson, D., and Saeguse A., 1999, Long-term effect of GM crops servers up food for though. *Nature* **398**:651-653.

Chong, D.K.X., Roberts, W., Arakawa, T., Illes ,K., Bagi, G., Slattery, C.W., and Langride, W.H.R., 1997, Expression of the human milk protein β-casein in transgenic potato plants. *Transgenic Research* **6**:289-296.

Clifford, B.C. 1995, Diseases pest and disorders of oats. In *The oat crop: production and utilization*. (R,W. Welch edt). New York Chapman and Hill, pp. 252-278.

Dale, P.J., 1999, Public reactions and scientific responses to transgenic crops. Commentary. *Current Opinion in Biotechnology* **10**:203-208.

Dejardin, C., Rochat, C., Wuilleme, S., and Boutin, J.P.,1997. Contribution of sucrose synthase , ADP-glucose pyrophosphorylase and starch synthase to starch synthesis in developing pea seed. *Plant, Cell and Environment* **20**:1421-1430.

Dunwell, J.M., 1998, Novel products from genetically modified crop plants: methods and future prospects. *International Journal of Food Science and Technology*: **33**:205-213.

Dunwell, J.M., 1999, Transgenic crops: The next generation, or an example of 2020 vision. *Annals of Botany*: **84**: 267-277.

Edwards, E.J., and Cobb, A.H., 1999, The effect of prior storage on the potential of potato tubers (*Solanum tuberosum* L) to accumulate glycoalkaloids and chlorophylls during light exposure, including artificial neural network modelling. *Journal of the Sciences of Food and Agriculture* **79**: 1289-1297.

Engel, K.H., Blaas, K., Gabriel, B., and Beckman, M., 1996, Modern biotechnology in plant breeding: Analysis of glycoalkaloids in transgenic potatoes. In: *Biotechnology for Improved Food and Flavor*. Washington: American Chemical Society.

Flis, B., and Zimnoch-Guzowska, E., 2000, Field performance of transgenic clones obtained from potato cv. Irga. *Journal of Applied Genetics* **42**: 81-90.

Fornal, J., 1998; The changes of plant materials microstructure during processing. *Polish Journal of Food and Nutrition Sciences* (3s), **7**: 9-21.

Fornal, J., Błaszczak, W., and Quattrucci, E., 1999, Genetically modified cereals: necessity or a chance. In: *Materie prime transgeniche, sicurezza alimentare e controllo qualita nell'industria cerealicola. Atti del Simposio,* ed. R.E. Cubadda & E. Marconi, 7-8 October, Campobasso, 1999, Italy, pp.246.

Fornal, J., and Błaszczak, W., 2001, Food of a new generation. The transgenic cereals. *Przegląd Piekarski i Cukierniczy* **7**: 12-16 *(in Polish)*.

Fornal, J., and Błaszczak, W., 2001a, Microstructure and functional food properties. *Przemysł Spożywczy* **8**: 34-37 (*in Polish*).

Fornal, J., Sadowska, J., Jaroch, R., Kaczyńska, B., and Winnicki, T., 1994, Effect of drying of rapeseed on their mechanical properties and technological usability. *International Agrophysics* **8**: 215-224.

Fornal, J., Sadowska, J., and Kaczyńska, B., 1993, Damage of faba bean seeds during drying. *Drying technology* **11**: 1293-1309.

Franck-Oberaspach, S.L., and Keller, B., 1997, Consequences of classical and biotechnological resistance breeding for food toxicology and allergenicity. *Plant Breeding* **116**: 1-17.

Franks, J.R., 1999, The status and prospects for genetically modified crops in Europe. *Food Policy* **24**: 565-584.

Ganga, Z.N., and Corke, H., 1999, Physical properties of starch of Asian – adapted potato varieties. *Journal of the Sciences of Food and Agriculture* 79: 1642-1646.

Goto, F., Yoshihara ,T., Shigemoto, N., Toki, S., and Takaiwa, F., 1999, Iron fortification of rice seed by soybean ferritin. *Nature Biotechnology* **17**: 282-286.

Graybosch, R.A., 1998, Waxy wheats: Origin, properties, and prospect. *Trends in Food Science and Technology* **9**: 135-142.

Greenwell, P., and Schofield, J.D., 1986, A starch granule protein associated with endosperm softness in wheat. *Cereal Chemistry* **63**: 376-380.

Gressel, J., 1992, Indiscriminate use of selectable markers-sowing wild oats? *Trends in Biotechnology* **10**: 382.

Hashimoto, W., Momma, K., Katsube, T., Ohkawa, Y., Ishige, T., Kioto, M., Utsumi, S., and Murata, K., 1999, Safety assessment of genetically engineering potatoes with designed soybean glycinin: compositional analyses of the potato tubers and digestibility of the newly expressed protein in transgenic potatoes. *Journal of the Science of Food and Agriculture* **79**: 1607-1612.

Herbers, K., and Sonnewald, U., 1999, Production of new/ modified proteins in transgenic plants. *Biotechnology* **10:** 163-168.

Igrejas, G., Gaborit, T., Oury, F.X., Chiron, H., Marion, D., and Branlard, G., 2001, Genetic and environmental effect on puroindoline-a and puroindoline-b content and their relationship to technological properties in French bread wheats. *Journal of Cereal Sciences* **34**: 37-47.

Jobling, S.A., Schwall, G.P., Westcott, R.J., Sidebottom, C.M., Debet, M., Gidley, M.J., Jeffcoat, R., and Safford, R., 1999, The role of starch branching enzymes in amylopectin biosynthesis in potato. *Genetic tailoring of Novel Starch Polymers. The First International Workshop Devoted to Polysaccharide Design and Understanding of Starch Biosynthesis.*, 16-20 September 1999, Carry-le-rout, France.

Kasarda, D.D., 1998, Gluten and gliadin precipitating factors in coeliac disease. *Proceedings of 76th International Symposium on Coeliac Disease*. 5-7 September, 1998, Tampere, Finland,

Kinsey, J.D., 1999, Genetically modified food and fibre: A speedy penetration or a false start? *Cereal Foods World* **44**: 487-489.

Kuchuk, N., Griga, M., Kosturkova, G., and Ilieva-Stoilova, M., 2001, Biotechnology. In *Carbohydrates in Grain Legume Seeds Improving Nutritional Quality and Agronomic Characteristics*, (C.L. Hedley, edt.) John Innes Centre, Norwich, UK.

Ludwicki, J.K., 1998, Genetically modified organism (GMO) – Toxicological aspects. *Roczniki PZH* **49**: 253-263 (*in Polish*).

MacRitchie, F., du Cros, D.L., and Wrigley, C.W., 1990, Flour polypeptides related to wheat Quality. In: *Advances in Cereal Sciences and Technology*, (Y. Pomeranz, edt.) Vol. X., Chapter 2, AACC, Minnesota:.

Madden, D., 1995, Food production. Plant biotechnology. In: *Food Technology and Introduction*. ILSI Europe concise monograph series.

Marsh, M.N., 1992, *Coeliac Disease*. (M.N., Marsh, edt.) .. Blackwell Scientific Publication, Oxford, pp. 371.

Marsh, M.N., Morgan, A., Ensari, A., Wardle, T., Lobley, R., Mills, C., and Aricchio, S., 1995, In vivo activity of peptides 31-41, 44-55, 56-68 of gliadin in gluten sensitive entheropathy (GSE). *Gastroenterology (s)* **108**: A871.

McBride, K.E., Svab, Z., Schaaf, D.J., Hogan, P.S., Stalker, D.M., and Maliga, P., 1995, Ampluification of chimeric *Bacillus thuringensis* cryl A(C) gene in chloroplast leads to an extraordinary level of an insecticidal protein in tobacco. *Biotechnology* **13**: 362-365.

Miflin, B., Napier, J., and Shewry, P., 1999, Improving plan product quality. *Nature Biotechnology* **17**: BV13-BV14.

Molenda, M., Horabik, J., Grochowicz, M., and Szot, B., 1995, Friction of wheat grain. *Acta Agrophisica* **4**: 1-89 (*in Polish*).

Molving, L., Tabe, L.M., Eggum, B.O., Moore, A.E., Craig, S., Spencer, D. and Higgins, T.J.V., 1997, Enhanced methionine levels and increased nutritive value of seeds of transgenic lupins (*Lupinus angustifolius* L.) expressing a sunflower seed albumine gene. *Proceedings of the National Academy of Sciences USA* **94**: 8393-8398.

Mosely, A.R., James, S.R., Hane, D.C., Rykbost, K.A., Shock, C.C., Charlton, B.A., Pavek, J.J., Love, S.L., Corsini, D.L., and Thornton, R.E., 2000, Umatilla Russet: full season long Russet for processing and fresh market use. *American Journal of Potato Research* **77**: 83-87.

Pasche, J.S., Secor, G. S., and Gudmestad, N.C., 1997, *Potato leafroll virus and net necrosis*. Department of Plant Pathology North Dacota State University.

Rand, N., Noy, Y., and Stilborn, H. *Genetic improvement of maize*. Miloubar Feed Mill M.P. Ashrat Israel 25102.

Rogan, J.R., Bookout, J.T., Duncan, D.R., Fuchs, R.L., Lavrik, P.B., Love, S.L., Mueth, M., Olson, T., Owens, E., Raymont, P., and Zalewski, J., 2000, Compositional analysis of tubers from insect and virus resistant potato plants. *Journal of Agricultural and Food Chemistry* **48**: 5936-5945.

Sadowska, J., Błaszczak, W., Palacios, M.A., and Fornal, J., 2002, *Influence of cooking and microwave irradiation on microstructure and rheological properties of genetic modified potato tubers* - unpublished data. Department of Physical Properties of Food, Institute of Animal Reproduction and Food Research, Polish Academy of Sciences, Tuwima 10, 10-747 Olsztyn, Poland.

Sadowska, J., Fornal, J., Ostaszyk, A., and Szmatowicz, B., 1996, Drying condition and processability of dried rapeseed. *Journal of Sciences of Food and Agriculture*; **72**: 257-262.

Salamini, F., 1999, North-South innovation transfer. *Nature Biotechnology* **17**: BV11-BV12.

Sanchez-Monge, R., Gomez, L., Barber, D., Lopez-Otin, C., Armentia, A., and Salcedo, G., 1992, Wheat and barley allergens associated with baker's asthma. Glycosylated subunits of the amylase-inhibitor family have enhanced IgE-binding capacity. *Biochemical Journal*; **281**: 401-405.

Scanlon, S.G., Day, A.J., and Povey, M.J.W., 1998, Shear stiffness and density in potato parenchyma. *International Journal of Food Science and Technology* **33**: 461-464.

Schulman, A.H.,1999, Chemistry, biosynthesis, and engineering of starch and other carbohydrates. In *Molecular Biotechnology for Plant Food Production*. Paredes Lopez, O., Technomic Publishing Co., Inc. Lancaster Pa USA.

Shewry, P.R., Fido, R., Tatham, A.S., Rooke, L., Lazzeri, P., and Barcelo, P., 1998, Manipulation of the protein composition and quality of wheat and other cereals by genetic engineering. *Genetic Engineering in Cereals, Selected papers from the ICC-AACC Symposium, 16th ICC Conference*, (J.W., van der Kamp, and R.N., Chibbar, edts) 12 May, Vienna, Austria:

Shukla, T.P., 1999, Genetic modification and enzymes for cereal processing. *Cereal Food World* **44**: 28-29.

Simmonds, D.H., Barlow K.K., and Wrigley, C.W., 1973, The biochemical basis of grain hardness in wheat. *Cereal Chemistry* **50**: 553-562.

Singh, R.P., 1999, Development of the molecular methods for potato virus and viroid detection and prevention. *Genome* **42**: 592-604.

Smirnov, S., Shulaev, V., and Tumer , N.E., 1997, Expression of pokeweed antiviral protein in transgenic plants induces virus resistance in grafted wild type plants independent of salicylic acid accumulation and pathogenesis-related protein synthesis. *Plant Physiology*, **114**: 1113-1121.

Somers, D.A., 1998, Application of genetic engineering to oat improvement. In *Genetic engineering in cereals. Selected papers from the ICC-AACC Symposium, 16th ICC Conference.*, (J.W., van der Kamp, and R.N., Chibbar, edts) 12 May, Vienna, Austria:

Stec, W.J., 2000, The tendency development of the world biotechnology. *Biotechnologia* **1**: 15-30 (*in Polish*).

Tester, R.F., Morrison, W.R., and Shulman A.H., 1993, Swelling and geletinization of cereal starches. Riso mutants of Bomi and Carlsberg II barley cultivars. *Journal of Cereal Sciences* **17**: 1-9.

Thomas, P.E., Kaniewski, W.K., and Lawson, E.C., 1997, Reduced field spread of potato leafroll in potatoes transformed with the potato leafroll virus coat protein gene. *Plant Diseases*; **12**: 1447-1453.

Weber, H., Heim, U., Golombek, S., Borisjuk, L., and Wobus, U., 1998, Metabolic control of legume seed development altering carbohydrate metabolism in transgenic *vicia narbonensis*." In: *Opportunities for High Quality, Healthy and Added-Value Crops to Meet European Demands. Proceedings 3rd European Conference on Grain Legumes*, Valladolid, Spain..

Wilde, P.J., Clark, D.C., and Marion, D., 1993, Influence of competitive adsorption of a lysopalmitoylphosphatidylcholine on the functional properties of puroindoline, a lipid-binding protein isolated from wheat flour. *Journal of Agriculture and Food Chemistry* **41**: 1570-1576.

PART 3

FRUITS

Physical Properties of Fruit Firmness and Chemical Structure of Cell Walls during Fruit Softening

NAOKI SAKURAI
Department of Environmental Studies, Faculty of Integrated Arts and Sciences
Hiroshima University, Hiroshima, Japan

1. INTRODUCTION

Fruit softening is essentially a chemical and biological process that occurs during ripening or after harvest, but the precise and comprehensive mechanism by which fruit resumes physical softening is still unresolved. Before an attempt to describe key events involved in the fruit softening, two specific cellular regions of plant should be identified. In plant tissues, there is an *apoplast* and a *symplast* (Fig. 1). Symplast is the region within cell membrane, and apoplast is the space external to the cell membrane but within the plant tissue (Munch 1930, Sakurai 1998). The apoplast consists of intercellular space, vessel, tracheids and free air space within the plant. As predicted by the prefix "sym-" in symplast, most of plant cells are interconnected by plasmodesmata, cytoplasmic strands that traverse the cell wall. Small molecules such as water, amino acid, sugars and ions, can easily transport to cell to cell through plasmodesmata. Therefore, if one cell loses water, the water loss directly affect the water status of the neighboring cells.

The first candidate thought to contribute to losses of fruit firmness was the degradation of pectic polysaccharides in fruit cell walls. There has been considerable evidence to show that pectins are degraded into small fragments during fruit softening (Huber 1983, Brady 1986, John and Dey 1986, Fischer and Bennett 1991, Brummell and Harpster 2001), but molecular technique has revealed that (1) tomato in which the gene for pectin degrading enzyme (polygalacturonase, PG) was suppressed and only less than 1% of pectin was degraded still exhibited normal fruit softening (Smith et al. 1990), (2) there was a mutant of tomato with suppressed softening; even when PG gene was introduced into the mutant and enhanced pectin degradation, the fruit retained a hard texture (Giovannoni et al. 1989). These two observations suggested that the role of pectin degradation in loss was not directly related to fruit firmness (Chapple and Carpita 1998, Brummell and Harpster 2001). While these observation do not deny a role of pectin degradation in the regulation of fruit texture, a simple relationship does not exist. Fruit firmness can be physically measured as an elastic parameter, the force required for the intrusion of probe into the fruit tissue to a fixed depth. Since in the above studies, only the elasticity of transgenic fruit was evaluated, there is still a possible role of pectin degradation in the regulation of physical properties of fruit other than elasticity. One good example is a transgenic tomato that became commercially available in Davis, California in 1991. The tomato named Flavor Saver contained a construct for the anti-sense gene of PG with the goal of promotion of shelf life. The tomato did tended to soften even with the suppressed pectin degradation, but there were some changes in texture which affected fruit quality as compared with that of its parent variety (Brummell and Harpster 2001).

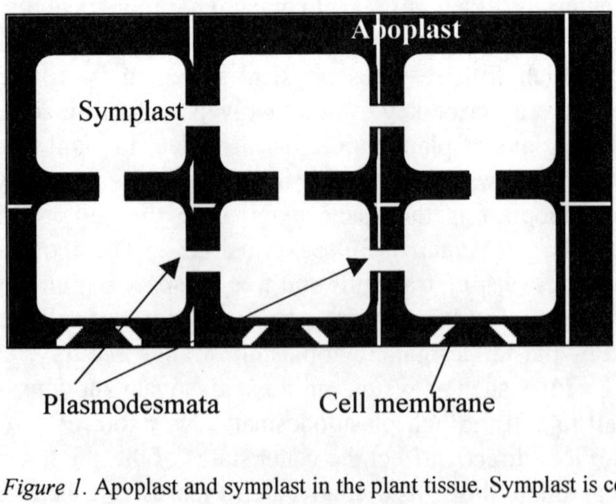

Figure 1. Apoplast and symplast in the plant tissue. Symplast is connected by plasmodesmata

After extensive study on the role of pectins, the trend has been to consider other wall components for structural roles in fruit firmness.

Xyloglucan degradation has been a major form. For pectin degradation, the most obvious enzyme is PG, but the enzyme that degrades xyloglucan remains unequivocal. Cellulose, another main component of cell walls, was also reported to be degraded during some fruit softening. To ascertain that a chemical change is essential or direct cause for fruit softening, one first must find the enzyme responsible for these changes and then clone the genes for the enzyme or proteins. Over-expression or suppression of this gene affords a clues for the cause-and-result relation in fruit softening.

Before concluding that the above biochemical factors are responsible for the loss of fruit firmness, one should appropriately evaluate the physical changes in fruits of transgenic plants. In studies of transgenic modification little attention has been paid, for example, whether or not the very gene causes to alteration of viscosity or elasticity or both. Establishment of a reliable and versatile measuring technique is essential for evaluation of firmness for future studies on ripening of transgenic fruits.

2. MEASUREMENT OF FRUIT FIRMNESS

2.1 Destructive Method

There are two basic approaches with techniques for fruit firmness measurement. One is destructive and the other non-destructive. Most of analytical studies for fruit quality still continue to use a Magness-Taylor puncture tester (Magness and Taylor 1925). In this case, force required for penetrating a cylindrical probe into fruit flesh to certain fixed depth is recorded. This method is simple, inexpensive, and practical, but the value depends on tester's experience and labor (Harker et al. 1996). A more sophisticated approach but using the same principle is an Instron type device. A horizontal stage on which a sliced fruit sample is subjected to a fixed probe driven into fruit. The force required to penetrate the fruit is recorded as a firmness index (Kojima et al. 1991, 1992). Flat probe is also used for compression of excised fruit (Arhens, and Huber 1990). Since the rate of travel is precisely controlled, the force data are more reliable and reproducible than those by a Magness-Taylor puncture tester.

A Universal Testing Machine (e.g. Instron) was adapted to fruit tissue (Kojima et al. 1991, 1992). When a conical probe is introduced into the excised fruit tissue to a pre-fixed depth, the probe is stopped and the decay of stress imposed by the probe was monitored. A stress-relaxation curve was simulated by a following simple equation (Yamamoto et al. 1970),

$$S/S_0 = R \ln \frac{t+T_m}{t+T_0}$$

where S is stress at time t, S_0, initial stress at time = 0, R, relaxation rate, T_m, maximum relaxation time,T_0, minimum relaxation time and c, the residual stress. T_0 and T_m may be regarded as the viscous indices, and the initial stress S_0 as the elastic index (Sakurai 1991). The method was used to classified several tomato mutants into groups with viscous and elastic characteristics (Sakurai and Nevins 1992). Brummell et al. (1999) determined the effect of trans-gene (*Cel2*, 1,4-β-glucanase) on tomato flesh texture using the stress-relaxation analysis and reported that this gene affected neither viscosity nor elasticity of transgenic tomato. Although this technique evaluates elasticity and viscosity separately, a destructive method requiring penetratin into fruit is not appropriate for fruit sorting line.

2.2 Non-destructive Method

Two methods for non-destructive measurement of fruit firmness have been explored. One is a measure of the response to vibration and the other is determine the response to impact (Chen and Sun 1991). Both methods detect fruit vibration signal. The impact method can be destructive for some fruits like peach, strawberry, tomato and pear, causing exterior and interior damage and fruit lose any commercial value. The method does have application for hard fruits like apple, avocado, melon, watermelon, and pumpkin. Vibration signal can be detected by an accelerometer, microphone, piezoelectric transducer or laser. The detection by an accelerometer and piezoelectric transducer require contact with the fruit. Detection by a microphone or laser represent remote sensing. For speed of sorting, remote sensing is superior.

Vibration tests are based on the resonant theory. The second resonant peak is usually used as an index for fruit firmness (Abbott et al. 1968). Finney (1970) placed an apple horizontally on a vibration generator to impose a sine wave signal and identified the second resonance. Cooke (1970) analyzed the resonant behavior of an intact fruit as an oscillating sphere using geophysics. Based on the assumption that the mode of vibration at the second resonance was a ${}_0T_2$ mode, he proposed an equation to estimate elasticity index E – proportional to modulus of elasticity of the fruit - as follows,

$$E = f_2^{\,2} m^{2/3}$$

where f_2 is the second resonant frequency and m is mass of fruit (Cooke 1972). Later Terasaki et al. (2001b) directly measured the vibration mode of apple on the vibrating stage by electronic speckle pattern interferometry and concluded that the mode at second resonance is ${}_0S_2$ mode. The final equation, however, is essentially the same.

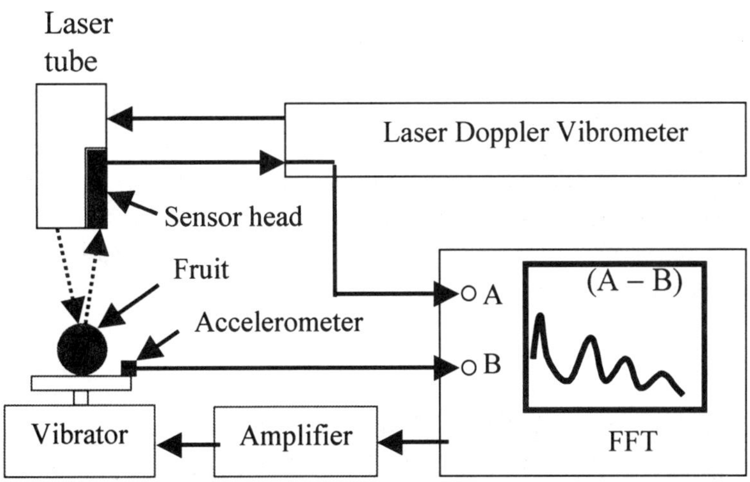

Figure 2. Schematic diagram of laser Doppler vibrometer

Vibration of fruit can be detected by several methods, but the remote sensing by laser beam is the most versatile (Muramatsu et al. 1999) (Fig. 2). They recorded the spectrum of resonant frequencies of apple by a laser Doppler detection and compared the results with the signal from an accelerometer and found that even the small weight of an accelerometer (more than 1 g) on a fruit interferes with the resonance data and compromise detection limits (Muramatsu et al. 1997). The relationship of the damping ratio to viscosity was proposed by Drake (1962), but it was not until the theoretical analysis of damping constant (c) by Terasaki et al. (2001a) that the expression could be formulated as follows,

$$c = 2\varsigma\pi fm \qquad (\varsigma = \omega / 2\omega_r)$$

where ω_r is a resonance angular frequency and $\Delta\omega$ is a half power value at 1/1.41 below the peak top. Ripe kiwifruit was described in terms of elasticity and viscosity when measured by a laser Doppler method (Terasaki et al. 2001c).

Yamamoto et al. (1980) first theoretically described impact method using a wooden pendulum as a hitting device. Yamamoto and Haginuma (1984) later proposed the equation to calculate shear modulus (μ) as follows,

$$\mu = Cf^{2/3} m^{2/3} \rho^{1/3}$$

where C is a constant related to a type of fruit, and ρ is the fruit density. Fruit was placed on a rigid surface and hit by a wooden pendulum. A microphone monitored the sound produced by the fruit after impact and this signal was analyzed using an FFT (fast Fourier transformation) algorithm to obtain the resonance frequency spectrum. This impact method was applied to apple (Shmulevich et al. 1996) and avocado (Galili et al. 1998) as a practical way for fruit sorting.

3. PHYSICAL PROPERTIES RELATED TO THE DIRECT LOSS OF TURGOR PRESSURE

Turgor pressure is defined as the difference between the water potentials of symplast and apoplast (Nobel 1983). Usually the water potential at the symplast is more negative than the apoplast and hence the difference is a negative value, leading to uptake of water from apoplast. This potential sustains a positive turgor pressure. Turgor pressure is opposed by the physical structure of the cell walls, creating cell wall pressure. Its magnitude is the same as the turgor pressure, but in an opposite direction. Any increase in the osmotic concentration of apoplast or decrease in the symplast reduces the water potential difference between apoplast and symplast, thus reduces the turgor pressure. The involvement of direct loss of turgor in fruit softening has not received special attention. Three possible mechanisms accompanying for a loss of fruit turgor pressure are described below. Any decrease in turgor pressure of fruit cells may lead to, at least partly, softening.

3.1 Decrease in Osmotic Concentration of Symplast

There are several ways to decrease osmotic concentrations of symplast. One is a decrease in the amount of solutes in the symplast. Synthesis of insoluble polymer, such as starch, from soluble monomers, glucose, and/or sucrose from glucose and fructose, reduces osmotic concentration of symplast. Second is consumption of solutes in symplast for example, by respiration. Generally, an increase in ambient temperature enhances the rate of fruit respiration. Third is a leakage of solutes from symplast to apoplast through a leaky cell membrane that has less compromised permeability. Mesocarp tissue of cucumber exhibits a greater rate of electrolyte leakage under low phosphorus fertilization than under high phosphorus regime (Knowles et al. 2001). Storage of tomato at low temperature also increased electrolyte leakage from the tissue (McDonald et al. 1999). This causes an increase in the apoplastic osmotic concentration and a simultaneous decrease in the symplastic concentration.

3.2 Increase in Apoplastic Osmotic Concentration

Degradation of insoluble or soluble polymer that generates smaller soluble fragments in apoplast increases the osmotic concentration of apoplast. Any increase in the osmotic concentration at the apoplast results in decrease in the difference between apoplast and symplast water potentials, leading to loss of turgor. Most pectins and hemicelluloses undergo the degradation during fruit softening. Such degradation of polysaccharides can increase the osmotic concentration of apoplast. But this effect is attenuated as the monomeric sugars are absorbed by the symplast. The direct measurement of turgor pressure of tomato by the pressure probe technique revealed that the turgor was much lower than that expected by the osmotic concentration of symplast, suggesting the presence of solute accumulation in the apoplast (Shackel et al. 1991). Leakage of solute from symplast through leaky cell membrane as described above also increases the osmotic concentration of apoplast.

3.3 Evaporation

Simple evaporation of water from fruit takes place from the cell walls, i.e., apoplast, through outer surface, although fruit skin has much less number of stomates than leaves. Sabarez and Price (1999) simulated the dehydration process of prune fruit by a diffusion model and claimed that the surface resistance reduced the water loss in the early stage of dehydration. The rate of water loss from mature cherry fruit was constant for several days at 1C (Knoche et al. 2000). Loss of water from the apoplast causes an increase in its osmotic concentration. This reduces the water potential difference between symplast and apoplast and the result is a decrease in turgor pressure. The difference between water potential of symplast and apoplast is balanced by the cell wall pressure, so that water is immediately moved from symplast to apoplast as the apoplastic concentration increases. This water movement directly decreases turgor pressure. As water loss takes place, the shrunken volume of symplast compensates for the adjusted cell wall pressure as the cell shrinks. Cell walls of shrunken cells have less wall pressure. To compensate for water loss, surface coatings have been used to maintain fruit quality after harvest and to reduce skin susceptibility to abrasion damage (Amarante et al. 2001).

4. CHEMICAL MECHANISM OF CELL WALL WEAKENING

The cell walls play specific roles in fruit development. After fertilization, the fruit increases cell number and develops a small immature fruit with a fixed number of small-sized cells. The mass of fruit, then, increases to a final size by expansion of individual cells. During the cell expansion in fruit development, cell walls are synthesized and at the same time cell walls must loosen to accommodate expansion. Until the fruit develops its final size, fruit remains hard and inedible, mainly because of rigidity of the cell wall.

After the fruit attains the final size, the fruit undergoes a softening process and the same cell walls are modified again, but without cell expansion. In fruit development two types of cell wall metabolism occur; one is wall loosening accompanying cell expansion, and the other is loosening in the absence of cell expansion. There is no plausible explanation as to why softening of cell walls during the later stages of fruit ripening does not cause increases in fruit size. Perhaps the rigid epidermal tissues may restrict the potential increase in cell size through compression, even if the walls beneath the epidermis are loosened. Fruit tissue, however, does not expand appreciably when it is peeled; hence there is little evidence for tissue compression caused by the epidermis.

4.1 Polysaccharides

Plant cell walls consist of three major polymers; polysaccharides, proteins, and lignin. Polysaccharides are further classified into three groups by their susceptibility to chelating reagent and concentrated alkaline solution. They are *pectins, hemicelluloses* and *cellulose.* Chelating reagents such as EDTA extracts acidic polysaccharides, pectins. The reason why EDTA extracts pectins from cell walls is because most of pectin is bound as a continuous complex by so-called calcium bridges to form a supra-molecular structure. Such supra-molecules are not readily extracted from cell walls by simple hot water treatment. Since pectins are released by chelating reagent, it suggests that pectins in the cell walls are not covalently bound to other cell wall polysaccharides. Typically hemicelluloses are extracted by concentrated alkali from the wall materials that had been treated with a chelating reagent. Alkaline breaks weak bonds in the wall to free hemicellulose. The residual cell wall material is cellulose. The effectiveness of concentrated alkaline for the hemicellulose extraction suggests that the hemicelluloses are bound to cellulose or other polymer by labile bonds. Contribution of these three groups of polysaccharides to the fruit softening will be discussed below.

4.1.1 Pectin

Pectin is synthesized as methylester form and secreted from the symplast to apoplast via Golgi apparatus (Buchanan et al. 2000). Therefore, the newly formed pectin molecules lack the ability to bind Ca^{++} ion (Fig. 3). There is no electrostatic repulsion of pectin molecules, because methyl group neutralizes the carboxyl group. Backbone of pectin molecules is composed of galacturonic acid (GalA) and rhamnose (Brett and Waldron 1996). One rhamnose is inserted between several contiguous GalA residues. Because of the angular configuration at the linkage which rhamnose is inserted, the molecule kinks. Therefore, the molecule tends to shrink in solution, but is highly soluble (>20mg/ml). Rhamnose may link to another neutral small side chain, arabinogalactan, at the third position.

Pectin metyhlesterase (PME) cleaves ester linkages between methyl and carboxyl group of GalA. The enzyme is consistently secreted to apoplast. It removes methyl group from the pectin in the cell walls to reveal the carboxyl group. Partially de-methylated pectin is more soluble than that fully methylated (>50 mg/ml). Free carboxyl groups tend to repel each other and this force extends the pectin molecule. Persistence length of the pectin molecules without methoxy group (34 nm) is longer than that of 0.6 methyl-esterification (30 nm) (Yalpani 1988). In addition, free carboxyl groups acquire the ability to bind Ca^{++}, if it is present in apoplast. Calcium chelates two adjacent pectic molecules to form gel. Moderately edited pectin with an appropriate ratio of free to methyl carboxyl group forms gel structure together with an adequate amount of Ca^{++} ion. The gel structure of pectin affords significant viscous and elastic properties to the cell walls. This gel forming ability of pectin has been considered responsible for a major role of pectin in fruit softening.

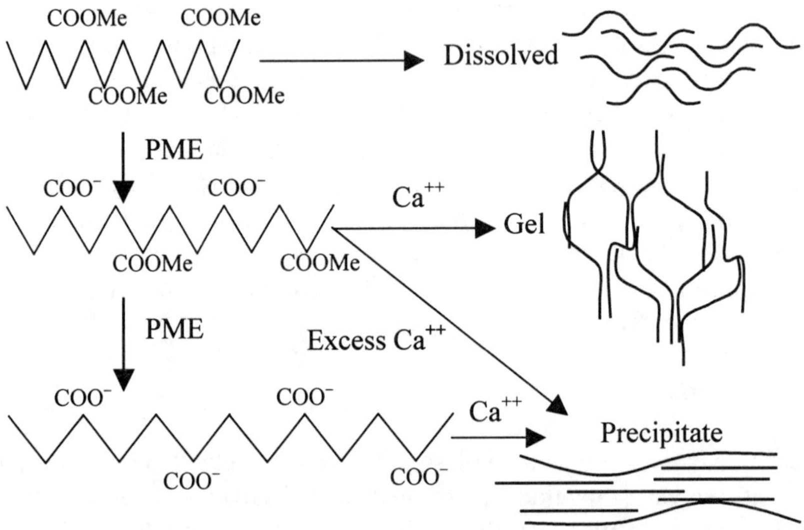

Figure 3. Changes in acidity of pectin by pectin methylesterase (PME), extension of pectin and its binding capacity to calcium ion

Excess Ca^{++} ion, however, precipitates pectin molecules even with an appropriate ratio of free to methyl carboxyl group in the pectin. Acidic conditions in apoplast shrinks the gel, because half of carboxyl groups are not free at pH 4.7 or under. Further trimming of methyl group by PME extends the molecules by encouraging intramolecular repellent force and at the same time binds more Ca^{++} ions. This leads to precipitation of pectic molecules (Yalpani 1988). Therefore, the physical status of pectin in the wall is affected by PME activity, amount of free apoplastic Ca^{++} ion and apoplastic pH. Most of studies on the fruit pectin during softening seem to ignore the amount of Ca^{++} in apoplast. In part this is because the amount of "free" Ca^{++} is difficult to estimate. In apricot fruit, GalA content was about 1 mole per kg of alcohol-insoluble materials, but the Ca^{++} ion was only 0.04-0.07 mole, while Mg^{++} ion was five times higher than Ca^{++} (Femenia et al. 1998). In tomato, the Mg^{++} ion concentration (8 to 12 mM) was also two to three times higher than Ca^{++} (4 to 5 mM) (Almeida and Huber 1999).

Depolimerization of pectin is caused by endo-polygalacturonase (PG). This enzyme cleaves pectic polymer along the backbone. When PG hydrolyzes internal linkage of the pectin gel, it results in a drastic decrease in viscous and elastic properties of cell walls. PME is regarded as a initiator of

PG action, because methylesterified pectin is resistant to PG action. (Pressey and Avants 1982, Seymour et al. 1987, Koch and Nevins 1989). As described in the Introduction, PG is not clearly involved in fruit softening, when the fruit firmness is measured by conventional means. However, since penetration, vibration or compression methods determined largely the elastic property of the fruit (Chen and Sun 1991), there is still a possibility that PG-mediated pectin alteration affects viscosity of fruit cell walls. In kiwifruit, the mean molecular weight of pectin increased in the early phase of ripening, and paralleled with the increase in a viscous parameter measured by laser Doppler method (Terasaki et al. 2001a).

As described in the Introduction, the molecular techniques suggest that pectin is not the sole determinant of fruit softening. Irradiation by gamma- or X-ray sources softened mature-green and pink fruit of tomato, but this softening was not associated with changes in PG, PME and galactosidase activity (Assi et al. 1997). The irradiation treatment did cause electrolyte leakage that must relates to a loss of turgor described in section 3.1.

4.1.2 Hemicellulose

As the name infers, hemicellulose is a name of mixture of heterogeneous polysaccharides, including xyloglucans, arabinoxylans, glucomannan, and arabinogalactans. These polysaccharides are not extracted by chelating reagents, but are released with concentrated alkaline solution. Therefore these polysaccharides are considered to be bound to cellulose microfibrils or each other via hydrogen bonds. Of these polysaccharides, xyloglucan has been extensively studied in relation to fruit softening (Table 1).

Average molecular weight of newly synthesized xyloglucan may be 500 kDa whose length (700 nm, in stretched form) is potentially enough to cross-link adjacent cellulose microfibrils that are typically 20-40 nm apart (McCann et al. 1990). If xyloglucan molecules cross-link several cellulose microfibrils comprising a highly crystalline structure with a high Young's modulus and are degraded by enzyme action, the elasticity of cell walls should be greatly affected (Rose and Bennett 1999). Xyloglucan degradation was well characterized in elongating stem tissue treated with a plant hormone (Nishitani and Masuda 1983, Wakabayashi et al. 1991). The prime enzymes responsible for the xyloglucan modification remains controversial; partly because an enzyme that modifies xyloglucan was not clearly identified before. Recently an enzyme responsible for the specific degradation of xyloglucan (Xyloglucan hydrolase, XGH) was identified however (Tabuchi et al. 2001).

Table 1. Species of fruit that exhibits degradation of xyloglucan during fruit softening and presence or absence of fucose (Fuc) residue

Fruit	Reference	Method	Fuc
Tomato	Tong & Gross (1988)	GPC	nd
	Sakurai & Nevins (1993)	GLC	N
Muskmelon	McCollum et al. (1989)	GPC	N
	Rose et al. (1998)	KI-I_2	Y
Kiwifruit	Radgewell et al. (1991)	GPC	Y
	Terasaki et al. (2001a)	GLC	Y
Avocado	O'Donughue and Huber (1992)	KI-I_2	nd
	Sakurai & Nevins (1997)	GLC	Y
Persimmon	Cutillas-Iturralde et al. (1994)	KI-I_2	Y
Pepino	O'Donoghue et al. (1997)	KI-I_2	nd
Grape	Yakushiji et al. (2000)	GLC	Y

GPC, gel-permeation chromatography; KI-I_2, specific KI-I_2 colorimetric determination coupled with GPC; GLC, gas-liquid chromatography analysis coupled with GPC. Fuc, fucose. nd, not determined. N, no; Y, yes.

Endo-xyloglucan transferase [EXGT (Nishitani and Tominaga 1992), or xyloglucan endo-transferase (XET) (Fry 1992)] is an enzyme that cleaves a xyloglucan molecule internally and re-connects one of the split ends of the molecule to another xyloglucan. Therefore, this activity is not hydrolytic activity that splits one xyloglucan molecule into two parts, but transglycosylase activity. A large gene family of xyloglucan endotransferases exists in higher plants.

The complete nucleotide sequence of the *Arabidopsis* genome revealed that 33 genes are identified as members of the EXGT gene family (Yokoyama and Nishitani 2001). Each gene independently responds to plant hormones and was expressed in different tissues, suggesting that each has a function specific to a limited stage of growth and development of *Arabidopsis*. The classical northern technique covered a large region of the mRNA including consensus regions and therefore would detect unspecifically types of EXGT gene family involving hydrolase type genes. Since the EXGT family has been classified into three subfamilies (Nishitani 1997), subfamily III includes EXGT with hydrolytic activity, while the other two families lack. It was agreed that this enzyme was to be renamed as xyloglucan endotransglucosylase/hydrolase (XT/H) at the 9th International Cell Wall Meeting held in Toulouse, France (2001).

The conventional EXGT exhibited no activity in the hydrolysis of xyloglucan, but a further study found an EXGT family that could cleave xyloglucan, with a final molecular weight that exceeds more than 60 kDa. (Tabuchi et al. 1997, Schroder et al. 1997). Therefore, this family does not belong to the conventional endo-type of hydrolase. A novel group of xyloglucan-degrading enzymes was recently found in azuki bean (Tabuchi et al. 2001). This new enzyme has no transferase activity, but it is responsible for the xyloglucan degradation often found on the hormone-induced elongation of plant stem segments. Also, this enzyme can be responsible for the xyloglucan degradation during fruit softening as shown in Table 1.

The hypothesis that xyloglucan degradation is responsible for the fruit softening is based on the assumption that xyloglucan molecules interconnect more than two cellulose microfibrils (Hayashi 1989). The bonding between cellulose and xyloglucan is generally regarded as hydrogen bonding, but there is no unequivocal evidence that xyloglucan directly attaches to the cellulose microfibrils by hydrogen bonding. High concentration of urea used to break most of the hydrogen bonds is not capable of releasing much xyloglucan from the plant cell walls (Hayashi 1989). The requirement of concentrated alkali to extract xyloglucan from plant cell walls infers that the bonding power between xyloglucan and cellulose is not solely hydrogen bonding, but more complex. The xyloglucan could be enmeshed in cellulose microfibril matrix (Rose and Bennett 1999). If so, one should consider a coordinated weaving mechanism of cellulose and xyloglucan at the cell membrane where cellulose is believed to be synthesized. Xyloglucan is synthesized in Golgi vesicles and secreted to the apoplast, while cellulose is synthesized at cell membrane by a supra-molecular complex of proteins termed a terminal complex and the product is extruded from membrane into apoplast. A consideration of a weaving mechanism requires an interaction between terminal complex for cellulose synthesis at the cell membrane and secreted vesicles containing xyloglucan molecules. Since the mechanism for cellulose synthesis remained largely unknown, it is not possible to demonstrate the 3-D structure of xyloglucan and cellulose microfibril interaction.

One characteristic of xyloglucan is the high degree of xylose residues attached to 1,4-β-glucan backbone (Fig. 4). The ratio of xylose (Xyl) to glucose residue is generally 3 to 4 in xyloglucans (Hayashi 1989). Typically unsubstituted 1,4-β-glucans exhibit strong intra-molecular hydrogen bonding generating rigid and linear molecular arrangement, but the substitution of Xyl residue on the glucan backbone of xyloglucan precludes the intra-molecular hydrogen bonding. If the polysaccharide backbone is freely flexible, the polymer forms a random coil. Xyl substitution also suppresses the formation of random coils in xyloglucan, because it interferes with the free rotation of two adjacent glucose (Glc) residues at 1,4-β-linkage. Therefore, it seems that shape of xyloglucan is neither random coil nor stretched polymer, but rather a flexible rod. Pea xyloglucan mainly consists of XXLG and XXFG, while storage xyloglucan of nasturtium seed mainly consists of XLLG. Nasturtium xyloglucan does not exist in cell walls nor does plays any role in cell wall softening. The binding capacity of pea xyloglucan to cellulose was two times more than that of the nasturtium. The tri-saccharide side chain (including F) was proposed to be responsible for the binding capacity to cellulose (Levy et al. 1997).

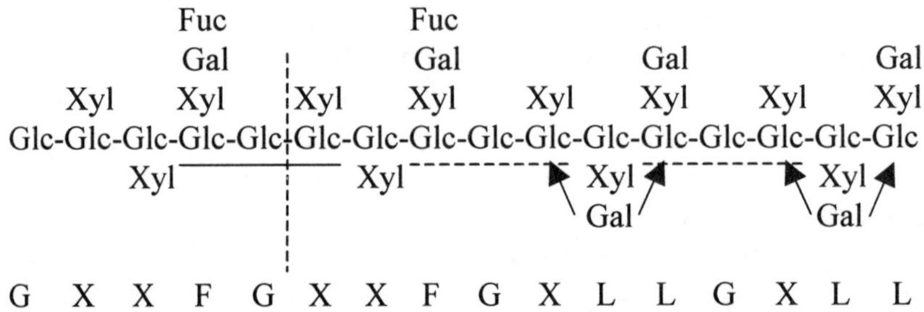

Figure 4. Structure of typical xylogluan. The bottom letters are abbreviations of each unit of mono- di- tri- and tetra-saccharide. G, glucose; X, Xyl-Glc; F, Fuc-Gal-Xyl-Glc; L, Gal-Xyl-Glc (Fry et al. 1993). Vertical dashed line shows the site for ET/H action. Horizontal solid line shows the flat and smooth site of three consecutive Glc residues that may contribute hydrogen bonding to cellulose. Dashed horizontal line shows the three consecutive Glc residues that may be affect by the next Gal residue (shown by arrows).

In Fig. 4, one notes that the three consecutive glucose residues (FGX) show a relatively smooth surface along backbone. In fact, XT/H enzyme reacts the linkage between G-X, suggesting that the enzyme can access the site from this smooth side without hindrance by Xyl residues. If xyloglucan interacts with cellulose by hydrogen bonding, this smooth face of xyloglucan should also contribute to the bonding. In the case of nasturtium seed xyloglucan, terminal Fuc of F site is missing and additional Gal is attached to one of X site, resulting in XLLG unit. Since the terminal Gal residue can rotate freely, the Gal residue of left L site (X<u>L</u>LG) may overlap the Glc residue of the next X site as shown by an arrow (Fig. 4) and reduce the accessibility of three consecutive Glc residues to cellulose molecule in the formation of a hydrogen bond. In fruit, the nasturtium-type xyloglucan is lacking. Tomato xyloglucan lacks Fuc residue, but retains X<u>X</u>LG unit. The second site is not L but X. Therefore the tomato xyloglucan with a smooth surface of three consecutive Glc may be degraded during softening.

Another hypothetical explanation for the xyloglucan-cellulose interaction is glucomannan. Glucomannan is composed of Glc and mannose (Man) at the ratio of 1 to 2~3 and seems relatively small sized polymer, ca. 10 kDa (Sakurai et al. 1987), but the number of consecutive Glc or Man is still unknown. The polymer is linear, exhibits significant crystalline structure and interacts with cellulose molecules (Chanzy and Vuong 1985). Since the

Glc and Man residues in glucomannan are linked by 1,4-β-linkage, the backbone is smooth and linear and potentially interacts with cellulose microfibrils *and* xyloglucan. Therefore it is probable that cellulose is first bound to glucomannan as the first stage in the crystalline complex and then xyloglucan binds to glucomannan. In other words, xyloglucan may bind to cellulose over a layer of glucomannan. Man was often reported to exist in cellulose fraction (residual fraction after concentrated alkaline extraction) in tomato (Tong and Gross 1989) and melon (Rose et al. 1998). The presence of Man residue in cellulose fraction after alkaline extraction infers that the glucomannan binds to cellulose more tightly than xyloglucan.

The contribution of xyloglucan to physical properties of the cell walls may be primarily elasticity, since it is proposed to interconnect cellulose microfibrils. However, degradation of xyloglucan affects viscosity of cell walls, probably because pectic polymer can move more loosely in the broken xyloglucan-cellulose network. In tomato, xyloglucan degradation parallels the decrease in elasticity and viscosity of tomato (Kojima et al. 1992). In kiwifruit, xyloglucan degradation parallels the decrease in elasticity but there is not a strong correlation with a decrease in elasticity (Terasaki et al. 2001a). Xyloglucan was degraded in the early stages of softening with a decrease in the average molecular weight then remained constant for several days, but the elasticity continued to decrease, suggesting that xyloglucan degradation alone does not completely account for changes in fruit elasticity during softening. Furthermore viscosity actually increased while xyloglucan degraded. In this case, the molecular weight of pectin was correlated with increase in viscosity. Therefore both pectin and xyloglucan degradation are involved in the expression of the physical properties of fruit. Furthermore when persimmons were stored at room temperature, the fruit underwent softening but did not exhibit any xyloglucan degradation (Tsuchida and Sakurai, unpublished data). Enhancement or suppression of XT/H (or more precisely xyloglucan endohydrolase) by gene manipulation should afford the clues about the xyloglucan role in fruit softening process.

4.1.3 Cellulose

There are some fruits in which cellulose content actually decreases during softening, such as tomato (Huber 1985, Sakurai and Nevins 1993), grape (Yakushiji et al. 2001). The characterization of cellulose polymer during fruit softening is difficult, mainly because of the insolubility of cellulose. In fruit softening experiment, the molecular weight of cellulose is not usually determined. There is only one report describing the predicted changes in cellulose molecular weight during fruit softening (O'Donoghue et al. 1994). They demonstrated that molecular distribution of whole avocado cell walls in the solvent (LiCl/dimethylacetamide). In this solvent cellulose molecule can be dissociated and changed during the ripening, suggesting that linear (probably cellulose) polymer degraded. Avocado has long been known as the fruit with a very high "cellulase" activity and the extracted activity increased as the fruit ripen (O'Donoghue et al. 1994). The enzyme can hydrolyze an artificially modified water-soluble cellulose (carboxymethyl cellulose) in endo-wise manner, but cannot hydrolyze native crystalline cellulose. The enzyme was named Cx-cellulase. Furthermore Hatfield and Nevins (1986) claimed that Cx-cellulase activity was not responsible for the degradation of cellulose in avocado cell walls. Therefore the cellulase in avocado fruit was thought to degrade amorphous (non-crystalline) region of cellulose (O'Donoghue et al. 1994) or xyloglucan (Hatfield and Nevins 1986), but the xyloglucan-degrading activity was very low (O'Donoghue and Huber 1992).

These discrepancies may be resolved, if one could show that "cellulase" hydrolyzes glucomannan molecules. Because there is no evidence to show the extent of consecutive Glc residues in glucomannan backbone, the activity by cellulase is unclear. But there is a possibility that sequence of several consecutive Glc residues exist in glucomannan bound to cellulose and could be digested by the cellulase found in avocado fruit. In fact, O'Donoghue et al. (1994) observed a slight increase (42.7 to 46.0%) in crystallinity of cellulose after cellulase treatment and reported a significant increase in the crystallinity as fruit firmness decreased, suggesting that degradation of non-crystalline glucomannan by the cellulase enhanced the extent of crystallinity of cellulose. Even if the cellulase hydrolyzes Glc-Glc linkage in glucomannan, there might be no released product from glucomannan, such as Glc or Man (Hatfield and Nevins 1986), because the bulk of the glucomannan remains bound to cellulose.

4.2 Lignin

There is little known about the role of lignin during fruit softening. In peach, Alba et al. (2000) reported that the lignin content measured by colorimetric method increased in the end of first stage of the fruit development, but disappeared before the second stage of development when the peach softened. Assis et al. (2001) measured lignin content of cherimoya fruit by gravimetrical methods, and found that under the normal softening condition, the lignin content declined as the fruit firmness decreased, while cherimoya fruits that were treated with high concentration of CO_2 exhibited the similar softening but without a drastic decline of lignin, suggesting that the lignin content was not associated with the fruit softening. These findings seem to negate the possible involvement of lignin synthesis or degradation in fruit softening process, and suggest that lignin may not be degraded during fruit softening. Lignin is difficult to digest even upon fungal attack and considered to be a primary factor in resistance to microbial degradation of cellulose (Blanchette 1994)

Ferulic acid (FA) is one of the precursor of lignin, but FA itself is not incorporated into lignin itself. It is known to be bound to arabinose (Ara) residue in a slightly acidic polysaccharide, glucuronoarabinoxylan (GAX). When two FA residues attached to Ara residues of two GAXs, and two FAs are linked each other by peroxidase action, two GAX molecules interconnected affect the physical properties of cell walls, such as crispness after cooking (Waldron et al. 1997). The resultant di-ferulic acid (diFA) seems not to be precisely regulated in the apoplast (Kamisaka et al. 1990). The ratio of FA to diFA remains relatively constant during the growth, suggesting that the cross-bridge between two GAX molecules by peroxidase simply depends on the sequence and abundance of FA associated with GAX. In pineapple, cell wall-degrading enzymes released the FA-bound Ara residue, suggesting that the fruit cell walls also accumulated FA in the cell walls (Smith and Harris 2001).

4.3 Structural Protein

Extensins were the first characterized protein specifically localized in plant cell walls (Showalter 1993). The name was coined as the protein responsible for cell extension, but the accumulation of extensins commenced after the growth ceased. Therefore the involvement of extensins in regulation of plant cell extension was untenable. The extensins are unusually rich in hydroxyproline (Hyp) as an amino acid component.

Later another two groups of cell wall proteins were reported; one is proline (Pro)-rich protein, and the other is glycine (Gly)-rich protein. Since Hyp and Pro have an internal ring structure that afford a rigidity to the molecule, the protein rich in Hyp or Pro has a relatively rigid structure, like collagen known for its high content of Gly, Pro, and Hyp. Therefore, Hyp- or Pro-rich proteins may participate in the rigidity or inflexibility of cell walls. If extensin or Pro-rich proteins were degraded during the fruit softening, the elasticity and/or viscosity of fruit cell walls would be reduced. Information of changes in the level of extensins or Pro-rich protein is very limited for fruit softening. Infection of pathogen elicits the production and immobilization of extensins and Pro-rich protein that enhance the strength of cell walls to inhibit further infection of pathogen (Sheng and Showalter 1994). Role of Gly-rich protein in growth and development is largely unknown.

4.4 Cell Separation

Even if the rigidity or stiffness of cell walls remains unchanged, fruit firmness could possibly be reduced by the loss of intercellular adhesion of fruit cells. The mechanism of cell adhesion is not un-equivocally clarified, but arabinogalactan side chains and xylose residues associated with pectin are regarded as the key adhesive substance (Kikuchi et al. 1996). When such arabinogalactan or xylose residues were excised by enzymes such as arabinosidase or galactosidase, both of which has been found in many fruits (Brummell and Harpster 2001), cells within a fruit could move, leading to softening. It has been claimed that middle lamella between two adjacent cells is rich in pectin, but the treatment with polygalacturonase that digests the granular materials in the middle lamella still retained a meshwork structure (Tamura and Senda 1992), suggesting that pectin is not as abundant in middle lamella but is concentrated in the cell corners (Carpita and McCann 2000).

5. FUNCTIONAL PROTEIN FACTORS FOR LOSS OF FRUIT FIRMNESS

There are three classes of functional proteins in the cell walls associated with fruit softening. One is the hydrolase that splits linkages of polysaccharides that might be involved in the cell wall softening of fruit. Second is expansin, not shown to be an enzyme but may act to destroy hydrogen bonds between cellulose and some essential polymer (Cosgrove 2000). The third is another protein without enzyme activity, yieldin (Okamoto-Nakazato et al. 2001).

5.1 Hydrolases

Polysaccharide hydrolases are classified into two groups, endo- and exo-type. Exo-type hydrolases remove one or more sugar units from the non-reducing terminal end of polymer by splitting the glycosyl linkage with the addition of one water molecule (X–O–P ––> X–OH HO–P, where X is the terminal sugar unit, and P is the remaining polymer). Therefore, the reaction product is often monomeric sugar. When the polymer is comprised of only one type of sugar, such as Ara, the polysaccharide is called araban and is completely degraded into many monomeric Ara by an exo-type arabanase. Endo-type hydrolase splits the araban at internal sites. Initially endo-type hydrolases do not produce monomeric sugar as a reaction product.

The viscosity of polysaccharide solution is not only due to its concentration but also to its average molecular size. Viscosity is produced by the friction or entanglement of polysaccharides (Doi and Edwards 1978). In a concentrated solution of polysaccharides with a high molecular size, endo-type hydrolases rapidlly reduce viscosity than exo-type one (Fig. 5).

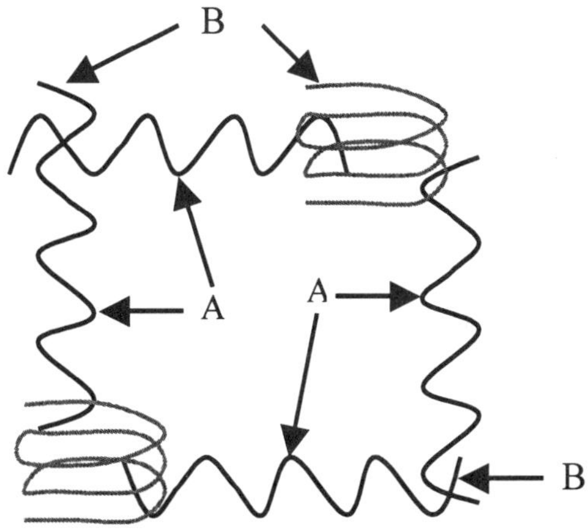

Figure 5. Schematic presentation of effect of endo-type hydrolase on the change in viscosity of concentrated solution of high molecular weight of polysaccharides. A, reaction site for endo-type hydrolase. B, reaction site for exo-type hydrolase. There are two types of polysaccharides in this figure.

If two polysaccharide molecules interact by entanglement, these molecules restrict the free movement of its counterpart. Two reactions by one exo-hydrolase, for example, would release two monomeric unit from each polysaccharide. The extent of this reaction does not essentially alter the physical property of the solution. When an endo-type hydrolase mediates two reactions one on each polysaccharide, this reaction essentially cancels the entanglement and allows the restricted molecules to move more freely. This is the mechanism by which the endo-type hydrolases reduce viscosity of a concentrated polysaccharides solution. There is another aspect of endo-hydrolase in the reaction. When endo-hydrolase splits the polysaccharide into two parts, it exposes one new non-reducing end accessible to an exo-hydrolase. In the presence of an exo-hydrolase, the endo-hydrolase accelerates the degradation much more quickly than in the absence of exo-hydrolase. Another important role of exo-hydrolase is a trimming function. Usually endo-type hydrolases require two or three unsubstituted sugar units within the substrate polysaccharide. If the polymer is highly branched, endo type enzyme access is restricted (Fig. 6). Therefore trimming of side chain is an important prerequisite for the action of endo-type enzyme. Trimming of branched chains may also reduce the friction to the neighboring molecules.

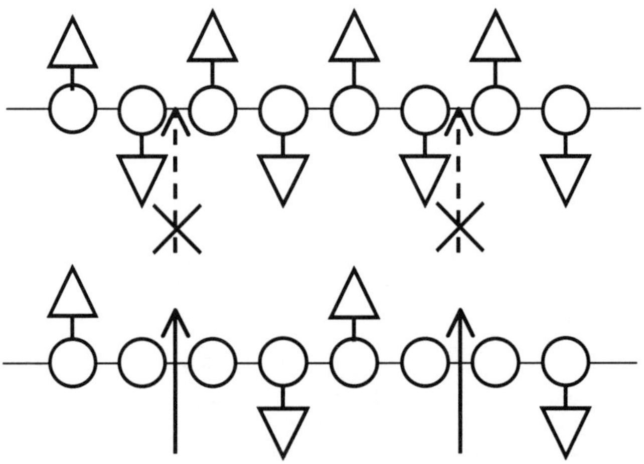

Figure 6. Schematic representation of accessibility of endo-type hydrolase to a highly branched polysaccharide and sparsely branched polysaccharide. O , backbone sugar residue. Δ, sugar residue of side chain. Trimming of side chain allows endo-hydrolase to access more easily to the backbone chain.

5.1.1 Pectinmethylesterase (PME)

As mentioned in section 4.1.1, PME enhances the negative charge of pectin, leading to stretching the molecules by intramolecular repulsion force. This extension of molecule may increase the viscosity, because the extended form of molecule more easily interacts with another neighboring molecules. This possibility has not been taken into account the role of PME in fruit softening.

5.1.2 Arabinofuranosidase

Arabinofuranosidase represents an exo-type hydrolase that acts on araban or non-reducing end of Ara side chain. Furanosidase means that the enzyme acts on the substrate consisting of arabinofuranoside. Pentoses, like Ara, and Xyl, may be configured as pyranose or furanose in solution. Thus far there is no report of endo-type arabanase in plant. Arabinofuranosidase was found in carrot cell cultures (Konno et al. 1987, 1994), radish seeds (Hata et al. 1992), fruits of pear (Tateishi et al. 1996) and goldenberry (Trinchero et al. 1999). The enzyme activity increased from pre-ripe to ripe stage 4 fold in pear (Tateishi et al. 2001a) and 50% in goldenberry. Treatment of tomato fruit with a plant hormone at high concentration to elicit ethylene formation caused an increase in arabinofuranosidase activity (Sozzi et al. 2000), suggesting that the wounding response is involved in the increase.

5.1.3 Galactosidase

β-Galactosidase is an exo-hydrolase that cleaves galactan or non-reducing end of Gal unit. Galactan found in plant cell walls has a β-configuration (Brett and Waldron 1996). Endo- β–galactanase enzyme has not been found in plant tissue, but does exist in fungi (Sorensen et al. 2000). Over expression of fungal endo- β–galactanase in transgenic potato reduced the Gal content of the pectin (Sorensen et al. 2000), but there was no explanation of physical properties in the transgenic potato tuber. There are numerous reports to show the increase in β–galactosidase activity during fruit ripening (see Tateishi et al. 2001b). There are several isoforms of β–galactosidase and related genes in tomato (Smith and Gross 2000), but only one of them was involved in the degradation of pectin during fruit ripening of tomato (Carington and Pressy 1996), papaya (Ali et al. 1998), 'D'Anjou' pear (Perdue et al. 1998) and Japanese pear (Kitagawa et al. 1995). mRNA level of β–galactosidase was monitored by Northern analysis in two species of grape during ripening, and the level increased before and after veraison, when the grape suddenly soften, in one species, but not changed in the other species (Nunan et al. 2001). These results suggest that β–galactosidase indirectly modifies the texture quality of fruit during softening.

5.1.4 Polygalacturonase (PG)

Exo- and endo-type PG (polygalacturonase) hydrolase are capable of digesting polygalacturonan. Genome analysis of *Arabidopsis thaliana*, suggest the presence of over 50 genes related to PG, but in tomato only one PG gene was expressed in ripening fruit (Della Penna et al. 1986) and its expression was elicited by ethylene (Sitrit and Bennett 1998).

5.1.5 Endo-glucanase

Endo-1,4- β–glucanase (EG) digests 1,4- β–glucan, "cellulose". Fungi responsible for degrading plant tissues (mostly cell walls) have a complete cassette of cellulase digesting enzymes. EG is only one of them. It, therefore, assumes that the EG alone is not capable of digesting cellulose molecules completely in plant cell walls. The next question may be, then, what is the substrate for EG. Cellulose is crystalline and non-crystalline structure. The non-crystalline portion is the amorphous region. The cellulose-digesting system of fungus also includes a component to release the hydrogen bonding between parallel glucan chains. Single glucan chain partially separating from the other crystalline molecules becomes susceptible to EG. Another possible substrate is within the region where xyloglucan might be interwoven into the glucan chains. These non-crystalline region of cellulose molecules are candidates for EG degradation. In the past, it was assumed that EG could degrade xyloglucan, but now that seems unlikely, unless all or most branched sugars such as Xyl, and Gal or Fuc are removed.

Several EGs have been found in higher plants, but the temporal and spatial expression within tissues is a critical factor to ascertain if EG is involved in cellulose degradation. Membrane-bound type EG has been identified (Brummell et al. 1997b) and regarded to function in cellulose synthesis (Sato et al. 2001). Moreover most EGs in higher plants do not have cellulose-binding site within the enzyme (Brummell and Harpster 2001).

Two biochemically different EGs were found in strawberry fruits (Trainotti et al. 1999). The levels of two mRNA encoding two divergent EGs were determined by northern analysis. One was expressed after large green stage and the other after white stage, suggesting these EGs are involved in the fruit expansion but not in the fruit softening. There are 5 reported genes of EG in tomato fruit, *LeCel1, 2, 4, 5* and *7* (Brummell and Harpster 2001), and these genes were expressed independently. *LeCel1* and *2* seem involved in the fruit expansion, and in the abscission zone (Lashbrook et al. 1998). Antisense suppression of tomato *Cel2* accumulation did not affect stress-relaxation parameters of fruit (Brummell et al. 1999), indicating that *Cel2* might not be involved in the fruit softening. *LeCel4* was expressed in a rapidly expanding tissues such as leaf, hypocotyls and pistil (Brummell et al. 1997a). *LeCel5* was expressed at the light red stage but also at abscission zone (Kalaitzis et al. 1999). mRNA level of *LeCel7* increased to mature green stage but declined during the softening stage after pink stage (Catala et al. 2000). *LeCel1* was expressed in the non softening mutant, *rin* subjected to propylene (an ethylene analog) treatment, suggesting that *LeCel1* was not involved in the softening process of tomato (Gonzalez-Bosch et al. 1996). *LeCel2* was somewhat expressed in another mutant, *Nr* (never ripe), after harvest. These data suggest that EG genes are not directly associated with softening of tomato fruit, but is associated with the other facets of fruit development.

5.1.6 Xyloglucan Hydrolase (XGH)

As mentioned in the section 3.2.2, xyloglucan endotranglucosylase/hydrolase (XT/H) might be a candidate for xyloglucan degradation during fruit softening, but recently more plausible candidate was found as xyloglucan hydrolase (XGH) (Tabuchi et al. 2001). So far there is no report about the XGH activity or gene in fruit ripening process.

5.2 Expansin

Expansin is the wall-modulating protein that apparently has no enzyme activity (Cosgrove 2000). Expansin was extracted from cucumber cell walls by salt treatment, and the extracted protein(s) were capable of loosening the cell wall under *in vitro* condition (McQueen-Mason et al. 1992). The function of expansin was implicated in the separation of a non-cellulosic glycan molecule from cellulose microfibrils by affecting hydrogen bonds (MaQueen-Mason and Cosgrove 1994). A number of expansin-related genes, at least 25, exist in *Arabidopsis thaliana* (Cosgrove 2000). Studies on expansin focus on expression of expansin gene and not on function of each expansin protein. There are seven reported expansin genes in tomato, *LeExp1, 2, 3, 4, 5, 6,* and *7* (Brummell and Harpster 2001). *LeExp1* was expressed in softening fruit of tomato after mature green stage, but also expressed in a mutant tomato that does not soften (Rose et al. 1997), suggesting the complex association of *LeExp1* with the softening. Other member such as *LeExp2, 3, 4, 5* and *6* were expressed before mature green stage (Brummell et al. 1999, Catala et al. 2000), suggesting that these genes control cell expansion during the fruit development. Physical measurement of transgenic tomato in which *LeExp1* is over-expressed or suppressed by an antisense technique is anticipated.

5.3 Yieldin

Yieldin is a cell wall protein that reduces the yield threshold of cell walls (Okamoto-Nakazato et al. 2000a, b). Yield threshold is defined as follows (Lockhart 1965),

$$\text{Relative growth rate} = \phi\,(P\text{-}Y)$$

where ϕ is extensibility of the cell wall, P is turgor pressure, and Y is the yield threshold of the cell wall. When Y declines, (P-Y), the effective turgor is more positive to extend the cell more. The sequence of yieldin is different from expansin and the protein is localized in the cell walls of actively growing cells (Okamoto-Nakazato et al. 2001). If this type of protein is synthesized and secreted to fruit cell walls during ripening, the protein might regulate the physical properties of the fruit firmness. There is so far no report about the expression of the yieldin in fruit tissue during softening.

6. FUTURE PERCEPTIONS

In nearly 30 years of research the role of pectin degradation in fruit softening process is continuous to be assessed. We recognize there are many candidates for the fruit softening including various hydrolases of cell wall polysaccharides, functional proteins and other functioning elements in the cell walls, but the complete scheme of fruit softening process seems to be highly complex and well organized. It follows that changes in fruit texture during softening is divergent. In fruit, changes in one component of cell wall polysaccharides alters some aspect of fruit texture, while that in another component causes a different change. It, therefore, is very important to have a measuring system to evaluate physical properties of fruit texture. It is imperative that objective evaluation of fruit elasticity, viscosity and other physical properties be considered. Elasticity and viscosity alone may not account for all the properties accounting for fruit texture. Crispness and crunchiness are also very important character of fruit texture., both of them being indirectly related to elasticity or viscosity (Vickers 1991). It is no doubt that ethylene is an essential key hormone to commence the fruit softening process, but there are many genes responding to the presence of ethylene. Some of them are expressed in parallel with the change in fruit firmness, but such an association does not guarantee a causal relationship between the physical change and expressin of a gene. Loss-of-function and gain-of-function using manipulation of genes is a promising technique to ascertain the role of individual gene in a specific change of fruit texture.

REFERENCES

Abbott, J. A., Bachmann, G. S., Childers, N. F., Fitzgerald, J. V. and Matusik, J. F., 1968. Sonic techniques for measuring texture of fruits and vegetables. *Food Technol.* **22**(5): 635-646

Alba, M.C., de Forchetti, S.M. and Tigier, H.A., 2000. Phenoloxidase of peach (*Prunus persica*) endocarp: Its relationship with peroxidases and lignification. *Physiol. Plant.* **109**: 382-387.

Ali, Z.M., Ng, S., Othman, R., Goh, L. and Lazan, H., 1998. Isolation , characterization and significance of papaya galactosidases to cell wall modification and fruit softening during ripening. *Physiol. Plant.* **104**: 105-115.

Almeida, D.P.F. and Huber, D.J., 1999. Apoplastic pH and inorganic ion levels in tomato fruit: a potential means for regulation of cell wall metabolism. *Physiol. Plant.* **105**: 506-512.

Amarante, C., Banks, N.H., and Ganesh, S., 2001. Relationship between character of skin cover of coated pears and permeance to water vapor and gasses. *Postharvest Biol. Tech.* **21**: 291-301.

Arhens, M.J. and Huber, J.D., 1990. Physiology and firmness determination of ripening tomato fruit. *Phsyiol. Plant.* **78**: 8-14.

Armstrong, P. R., Stone, M. L. and Brusewitz, G. H., 1997. Peach firmness determination using two different nondestructive vibrational sensing instruments. *Transactions of the ASAE* **40**(3): 699-703

Arpaia M. L., Labavitch, J. M., Greve, C. and Kadaer, A.A., 1987. Changes in the cell wall components of kiwifruit during storage in air or controlled atmosphere. *J. Amer. Soc. Hort. Sci.* **112**(3): 474-481.

Assi, N.E., Huber, D.J. and Brechit, J.K., 1997. Irradiation-induced changes in tomato fruit and pericarp firmness, electrolyte efflux, and cell wall enzyme activity as influenced by ripening stage. *J. Amer. Soc. Hort. Sci.* **122**: 100-106.

Assis, J.S., Maldonado, R., Munoz, T., Escribano, M.I., 2001. Effect of high carbon dioxide concentration on PAL activity and phenolic contents in ripening cherimoya fruit. *Postharvest Biol. Tech.* **23**: 033-39.

Blanchette, R.A. 1994. Lignin biodegradation in cell walls of woody plants. In *Host Wall Alteration s by Parasitic Fungi* (eds. O. Petrini and G.B. Ouellette), APS press, The American Phytopathological Society, St. Paul, Minnesota, p. 55-65.

Brady, C. J. 1986. Fruit ripening. *Annual Rev. Plant Physiol.* **38**: 155-178.

Brett C. and Waldron, K., 1996. Physiology and biochemistry of plant cell walls. Chapman & Hall, London, pp. 255.

Brett, C. and Waldron K., 1996. *Physiology and Biochemistry of Plant Cell Walls.* Chapman & Hall, London, pp. 255.

Brummell.., D.A., Lashbrook, C.C. and Bennett, A.B., 1994. Plant endo-β-glucanase: structure, properties and physiological function. In, *Enzymatic Conversion of Biomass for Fuels Production* (American Chemical Society Symposium Series 566, M.E. Himmel, J.O. Baker and R.P. Overend Eds.), American Chemical Society, pp. 100-129.

Brummell, D.A., Bird. C.R., Schuch, W., and Bennett, A.B., 1997a. An endo-1,4-β-glucanase expressed at high levels in rapidly expanding tissues. *Plant Mol. Biol.* **33**: 87-95.

Brummell, D.A. and Harpster, M.H., 2001. Cell wall metabolism in fruit softening and quality and its manipulation in transgenic plants. *Plant Mol. Biol. 47*: 311-340.

Brummell, D.A., Catala, C., Lashbrook, C.C. and Bennett, A.B., 1997b. A membrane-anchored E-type endo-1,4-β-glucanase is localized on Golgi and plasma membranes of higher plants. *Proc. Natl. Acad. Sci.* **94**: 4794-4799.

Brummell, D.A., Hall, B.D. and Bennett, A.B., 1999. Antisense suppression of tomato endo-1,4-β-glucanase Cel2 mRNA accumulation increases the force required to break fruit abscission zones but does not affect fruit softening. *Plant Mol. Biol.* **40**: 615-622.

Buchanan, B.B., Gruissem, W. and Jones, R.L., 2000. *Biochemistry and molecular biology of plants.* American Society of Plant Physiologists, Rockville, Maryland, pp. 1367.

Carpita, N. and McCann, M., 2000. The cell wall. In *Biochemistry and molecular biology of plants* (B.B. Buchanan, W. Gruissem and R. J. Jones eds.), American Soc. Plant Physiologists, Rockville, Maryland, p. 52-108.

Carrington, C.M. and Pressy, R., 1996. β-galactosidase II activity in relation to changes in cell wall galactosyl composition during tomato ripening. *J. Amer. Soc. Hort. Sci.* **121**: 132-136.

Catala, C., Rose, J.K.C. and Bennett, A.B., 2000. Auxin-regulated genes encoding cell wall-modifying proteins are expressed during early tomato fruit growth. *Plant Physiol.* **122**: 527-534.

Chanzy, H. and Vuong, R., 1985. Ultrastructure and morphology of crystalline polysaccharides. In *Polysaccharides* (E.D.T. Atkins ed.), VCH Verlagsgesellschaft, Weinheim, Germany, p. 41-71.

Chapple, C. and Carpita, N., 1998. Plant cel! walls as targets for biotechnology. *Current Opinion Plant Biol.* **1**: 179-185.

Chen, P. and Sun, Z., 1991. A review of non-destructive methods for quality evaluation and sorting of agricultural products. *J. Agri. Engng. Res.* **49**: 85-98.

Cooke, J. R., 1970. A theoretical analysis of the resonance of intact apples. *ASAE Paper* **No. 70-345,** ASAE St. Joseph, Mich. 49085

Cooke, J. R., 1972. An interpretation of the resonant behavior of intact fruit and vegetables. *Transactions of the ASAE* **15**(6):1075-1080

Cosgrove, D.J., 2000. Loosening of plant cell walls by expansins. *Nature* **407**: 321-326.

Cutillas-Iturralde, A., Zarra, I., Fry, S. C. and Lorences, E. P., 1994. Implication of persimmon fruit hemicellulose metabolism in the softening process. Importance of xyloglucan endotransglycosylase. *Physiol. Plant.* **91**: 169-176.

Della-Penna, D., Alexander, D.C. and Bennett, A.B., 1986. Molecular cloning of tomato fruit polygalacturonase: analysis of polygalacturonase mRNA levels during ripening. *Proc. Natl. Acad. Sci. USA* **83**: 6420-6424.

Doi, M. and Edwards, S.F., 1978. Dynamics of concentrated polymer systems par I. Brownian motion in the equilibrium state. *J. Chem. Soc. Faraday Trans.* **II. 74**: 1789-1801.

Drake, B. 1962. Automatic recording of vibrational properties of foodstuffs. *J. Food Sci.,* **27**: 182-188.

Femenia, A., Sanchez, E.S., Simal, S. and Carmen, R., 1998. Developmental ripening-related effects on the cell wall of apricot (*Purunus armeniaca*) fruit. *J. Sci. Food Agric.* **77**: 487-493.

Finney, E. E. Jr., 1970. Mechanical resonance within red delicious apples and its relation to fruit texture. *Transaction of the ASAE* **13**(2): 177

Fischer, R. L. and Bennet, A. B., 1991. Role of cell wall hydrolases in fruit ripening. *Annu. Rev. Plant Physiol.* **42**: 675-703.

Fry, S. C., York, W. S., Albersheim, P., Darvill, A., Hayashi, T., Joseleau, J.-P., Kato, Y., Lorences, E.P., Maclachlan, G.A., MacNeil, M., Mort, A.J., Reid, J.S.G., Seitz, H.U., Selevendran, R.R., Voragen, A.G.J., and White, A.R., 1993. An unambiguous nomenclature for xyloglucan- derived oligosaccharides. *Physiol. Plant.* **89**: 1-3.

Fry, S.C., Smith, R.C., Renwick, K.F., Martin, D.J., Hodge, S.K., and Matthews, K.J., 1992. Xyloglucan endotransglycosylase, a new wall-loosening enzyme activity from plants. *Biochem.* **282**: 821-828.

Galili, N., Shmulevich, I. and Benichou, N., 1998. Acoustic testing of avocado for fruit ripeness evaluation. *Transactions of the ASAE* **41**: 399-407.

Giovannoni, J.J., Della Penna, D., Bennett, A. and Fischer, R. L., 1989. Expression of a chimeric polygalacturonase gene in transgenic rin (ripening inhibitor) tomato fruit results in polyuronide degradation but not fruit softening. Plant Cell 1: 53-63.

Gonzalez-Bosch, C., Brummell, D.A. Bennett, A.B., 1996. Differential expression of two endo-1,4-β-glucanase genes in a pericarp and locules of wild-type and mutant tomato fruit. *Plant Physiol.* **111**:1313-1319.

Harker, F.R., Maindonald, J.H., Jackson, P.J., 1996. Penetrometer measurement of apple and kiwifruit firmness: operator and instrument differences. *J. Am. Soc. Hort. Sci.* **121**:927-936.

Hata, K., Tanaka, M., Tsumuraya, Y. and Hashimoto, Y., 1992. β-L-arabinofuranosidase from radish (*Raphanus sativus* L.) seeds. *Plant Physiol.* **100**: 388-396.

Hatfield, R. and Nevins, D.J., 1986. Characterization of the hydrolytic activity of avocado cellulase. *Plant Cell Physiol.* **27**: 541-552.

Hayashi, T., 1989. Xyloglucans in the primary cell wall. *Annu. Rev. Plant Physiol.* **40**:139-168.

Huber, D. J. 1983. The role of cell wall hydrolases in fruit softening. *Hortic. Rev. 5*: 169-219.

Huber, D.J., 1985. A possible role for cellulose and Cx-cellulase activity in locular gel formation in tomato fruit. *HortScience* **20**: 442-443.

John, M. A. and Dey, P. M., 1986. Postharvest changes in fruit cell wall. *Adv. Food Res.* **30**: 139-193

Kalaitzis, P., Hong, S., Solomos, T. and Tucker, M.L., 1999. Molecular characterization of a tomato endo-β-glucanase gene expressed in mature pistils, abscission zone and fruit. *Plant Cell Physiol.* **40**: 905-908.

Kamisaka, S., Takeda, S., Takahashi, K. and Shibata, K., 1990. Diferulic and ferulic acid in the cell wall of Avena coleoptiles – their relationships to mechanical properties of the cell wall. *Physiol. Plant.* **78**: 1-7.

Kikuchi, A., Edashige, Y., Ishii, T., Fujii, T. and Satoh, S., 1996. Variations in the structure of neutral sugar chains in the pectic polysaccharides of morphologically different carrot calli and correlations with the size of cell cluster. *Planta* **198**: 634-639.

Kitagawa, Y., Kanayama, Y., and Yamaki, S., 1995. Isolation of β-galactosidase fraction from Japanese pear: Activity against native cell wall polysaccharides. *Physiol. Plant.* **93**: 545-550.

Knoche, M., Peschel, S., Hinz, M., and Bukovac, M.J., 2000. Studies on water transport through the sweet cherry fruit surface: characterizing conductance of the curricular membrane using pericarp segments. *Planta* **212**: 127-135.

Knowles, L., Trimble, M.R. and Knowles, N.R., 2001. Phosphorus status affects postharvest respiration, membrane permeability and lipid chemistry of European seedless cucumber fruit (*Cucumis sativus* L.). *Postharvest Biol. Tech.* **21**: 279-188.

Koch, J.L. and Nevins, D.J., 1989. Tomato fruit cell wall. I. Use of purified tomato polygalacturonase and pectinmethylesterase to identify developmental changes in pectins. *Plant Physiol.* **91**: 816-822.

Kojima, K., N. Sakurai, S. Kuraishi, R. Yamamoto and A. Inaba, A., 1992. Physical measurement of firmness of banana fruit pulp: determination of optimum conditions for measurement. *Postharvest. Biol. Tech.* **2**: 41-49

Kojima, K., Sakurai, N., Kuraishi, S., Yamamoto, R., and Nevins, D. J., 1991. Novel technique for measuring tissue firmness within tomato (*Lycopersicon esculentum* Mill.) fruit. *Plant Physiol.* **96**: 545-550

Konno, H., Tanaka, R. and Katoh, K., 1994. An extracellular β-L-arabinofuranosidase secreted from cell suspension cultures of carrot. *Physiol. Plant.* **91**: 454-460.

Konno, H., Yamasaki, Y., and Katoh, K., 1987. Purification of an β-L-arabinofuranosidase from carrot cell cultures and its involvement in arabinose-rich polymer degradation. *Physiol. Plant.* **69**: 405-412.

Lashbrook, C.C., Giovannoni, J.J., Hall, B.D., Fischer, R.L. and Bennett, A.B., 1998. Transgenic analysis of tomato endo-1,4-β-glucanase gene function. Role of *cel1* in floral abscission. *Plant J.* **13**: 303-310.

Levy, S, Maclachlan, G. and Staehlin, L.A., 1997. Xyloglucan sidechains modulate binding to cellulose during *in vitro* binding assays as predicted by conformational dynamics simulations. *Plant J.* **11**: 373-386.

Lockhart, J.A., 1965. An analysis of irreversible plant cell elongation. *J. Theor. Biol.* **8**: 264-275.

Magness , J. R. and Taylor, G. F., 1925. An improved type pressure tester for the determination of fruit maturity. *USDA Agric. Cir.* 350.

McCann, M.C., Wells, B. and Roberts, K., 1990. Direct visualization of cross-links in the primary cell wall. *J. Cell Sci.* **96**: 323-334.

McCollum, T.G., Huber, D.J., Cantliffle, D.J., 1989. Modification of polyuronides and hemicelluloses during muskmelon fruit softening. *Physiol. Plant.* **76**: 303-308.

McDonald, R.E., McCollum, T.G. and Galdwin, E.A., 1999. Temperature of water heat treatments influences tomato fruit quality flowing low-temperature storage. *Postharvest Biol. Tech.* **16**: 147-155.

McQueen-Mason, S.J. and Cosgrove, D.J., 1994. Disruption by hydrogen bonding between plant cell wall polymers by proteins that induce wall extension. *Proc. Natl. Acad. Sci. USA* **91**: 6574-6578.

McQueen-Mason, S.J., Durachko, D.M. and Cosgrove, D.J., 1992. Two endogenous proteins that induce cell wall extension in plants. *Plant Cell* **4**: 1425-1433.

Munch, E., 1930. *Die Stoffbewegungen in der Pflanze*. Gustav Fischer, Jena, pp. 234.

Muramatsu, N., Sakurai, N., Wada, N,, Yamamoto, R., Tanaka, K. Asakura, T., Ishikawa-Takano, Y., and Nevins, D. J., 1999. Remote sensing of fruit textural changes with a laser Doppler vibrometer. *J. Amer. Soc. Hort. Sci.* **125**: 120-127.

Muramatsu, N., Sakurai, N., Wada, N,, Yamamoto, R., Tanaka, K., Ishikawa-Takano, Y. and Nevins, D. J., 1997. Critical comparison of an accelerometer and a laser Doppler vibrometer for measuring fruit firmness. *HortTechnol,* **7**(4): 434-438.

Nishitani, K., 1997. The role of endoxyloglucan transferase in the organization of plant cell walls. *Int. Rev. Cytol.* **173**: 157-206.

Nishitani, K. and Masuda, Y., 1983. Auxin-induced changes in the cell wall xyloglucans: Effects of auxin on the two different subfractions of xyloglucans in the epicotyls cell wall of *Vigna angularis. Plant Cell Physiol.* **24**: 345-355.

Nishitani, K. and Tominaga, R., 1992. Endo-xyloglucan transferase, a novel class of glycosyltransferase that catalyzes transfer of a segment of xyloglucan molecule to another xyloglucan molecule. *J. Biol. Chem.* **267**: 21058-21064.

Nobel, P.S., 1983. *Biophysical plant physiology and ecology*. W.H. Freeman and Company, San Francisco, pp. 608.

Nunan, K.J., Davies, C., Robinson, S.P. and Fincher, G.B., 2001. Expression patterns of cell wall-modifying enzymes during grape berry development. *Planta* **214**: 257-264.

Okamoto-Nakazato, A., Nakamura, T. and Okamoto, H., 2000b. The isolation of wall-bound proteins regulating yield threshold tension in glycerinated hollow cylinders of cowpea hypocotyls, *Plant. Cell Environ.* **23**: 145-154.

Okamoto-Nakazato, A., Takahashi, K., Katoh-Semba, R. and Katou, K., 2001. Distribution of yieldin, a regulatory protein of the cell wall yield threshold, in etiolated cowpea seedlings. *Plant Cell Physiol.* **42**: 952-958.

Okamoto-Nakazato, A., Takahashi, K., Kido, N., Owaribe, K., and Katou, K., 2000a. Molecular cloning of yieldins regulating the yield threshold of cowpea cell walls; cDNA cloning and characterization of recombinant yieldin. *Plant Cell Environ.* **23**: 155-164.

O'Donoghue, E. M. and Huber, D. J., 1992. Modification of matrix polysaccharides during avocado (*Persea americana*) fruit ripening: an assessment of the role of Cx-cellulase. *Physiol. Plant.* **86**: 33-42.

O'Donoghue, E. M., D. J. Huber, J. D. Timpa, G. W. Erdos and J. K. Brecht 1994. Influence of avocado (*Persea americana*) Cx-cellulase on the structural features of avocado cellulose. *Planta.* **194**: 573-584.

O'Donoghue, E.M., Somerfield, S.D., de Vre, L.A. and Heyes, J.A., 1997. Developmental and ripening-related effects on the cell wall of pepino (*Solanum muricatum*) fruit. *J. Sci. Food Agric.* **73**: 455-463. *Plant Cell Environ.* **23**: 145-154.

Perdue, D., Neven, L. and Drake, S., 1998. Cell wall β-galactosidase in ripening 'D'Anjou' pears. *J. Food Quality* **21**: 341-353.

Pressey, R. and Abants, J.K., 1982. Solubilization of cell walls by tomato polygalacturonases: effects of pectinesterases. *J. Food Biochem.* **6**: 57-74.

Redgwell, R. J., Melton, V. and Brasch, D. J., 1991. Cell wall polysaccharides kiwifruit (*Actinidia deliciosa*): effect of ripening on the structural features of cell-wall materials. *Carbohydr. Res.* **209**: 191-202.

Rose, J. K. C., Hadfield, K. A., Labavitch, J. M. and Bennett, A. B., 1998. Temporal sequence of cell wall cell wall disassembly in rapidly ripening melon fruit. *Plant Physiol.* **117**: 345-361

Rose, J.K.C. and Bennett, A.B., 1999. Cooperative disassembly of the cellulose-xyloglucan network of plant cell walls: parallels between cell expansion and fruit ripening. *Trends Plant Sci.* **4**: 176-183.

Rose, J.K.C., Lee, H.H. and Bennett, A.B., 1997. Expression of a divergent expansin gene is fruit-specific and ripening related. *Proc. Natl. Acad. Sci. USA* **94**: 5955-5960.

Sabarez, H.T. and Price, W.E., 1999. A diffusion model for prune dehydration. *J. Food Enginr.* **42**: 167-172.

Sakurai, N., 1991. Cell wall functions in growth and development. *Bot. Mag. Tokyo* **104**: 235-251.

Sakurai, N., 1998. Dynamic function and regulation of apoplast in the plant body. *J. Plant Res.* **111**: 133-148.

Sakurai, N. and Nevins, D. J., 1992. Evaluation of stress-relaxation in fruit tissue. *Hort Technol.* **2**: 398-402

Sakurai, N. and Nevins, D. J., 1993. Changes in physical properties and cell wall polysaccharides of tomato (*Lycopersicon esculentum*) pericarp tissue. *Physiol Plant.* **89**: 681-686

Sakurai, N. and Nevins, D.J., 1997. Relationship between fruit softening and wall polysaccharides in Avocado (*Persea americana* Mill) mesocarp tissues. *Plant Cell Physiol.* **38**: 603-610.

Sakurai, N., Tanaka, S. and Kuraishi, S., 1987. Changes in wall polysaccharides of squash (*Cucurbita maxima* Duch.) hypocotyls under water stress condition II. Composition of pectic and hemicellulosic polysaccharides. *Plant Cell Physiol.* **28**: 1059-1070.

Sato, S., Kato, T., Kakegawa, K., Ishii, T., Liu, Y., Awano, T., Takabe, K., Nishiyama, Y., Kuga, S., Sato, S., Nakamura, Y., Tabata, S. and Shibata, D., 2001. Role of the putative membrane-bound endo-1,4-β-glucanase KORRIGAN in cell elongation and cellulose synthesis in *Arabidopsis thaliana*. *Plant Cell Physiol.* **42**: 251-263.

Schroder, R., Atkinson, R.G., Langenkamper, G. and Redgwell, R.J., 1997. Biochemical and molecular characterization of xyloglucan endotransglycosylase from ripe kiwi fruit. *Planta* **204**: 242-251.

Seymour, G.B., Lasslett, Y. and Tucker, G.A., 1987. Differential effects of pectolytic enzymes on tomato polyuronides in vivo and in vitro. *Phytochem.* **26**: 3137-3139.

Shackel, K.A., Greve, C., Labavitch, J.M. and Ahmadi, H., 1991. Cell turgor changes associated with ripening in tomato pericarp tissue. *Plant Physiol.* **97**: 814-816.

Sheng, J. and Showalter, A.M., 1994. Plant cell wall structural proteins: regulated expression and roles in fungal infection. In *Host wall alterations by parasitic fungi* (O. Pentrini and G.B. Ouellette edts.), The American Phytopathological Society, St. Paul, Minnesota, pp. 91-102.

Shmulevich, I., Galili, V. and Rosebfeld, D., 1996. Detection of fruit firmness by frequency analysis. *Transactions of the ASAE* **39**(3): 1047-1055

Showalter□A. M., 1993. Structure and function of plant cell wall proteins. *Plant Cell* **5**: 9-23.

Sitrit, D.L. and Bennett, A.B., 1998. Regulation of tomato fruit polygalacturonase mRNA accumulation by ethylene: A reexamination. *Plant Physiol.* **116**: 1145-1150.

Smith, B.G. and Harris, P.J., 2001. Ferulic acid is esterified to glucuronoarabinoxylans in pineapple cell walls. *Phytochem.* **56**: 513-519.

Smith, C.J., Watson, C. F., Morris, P. C., Bird, C. R., Seymour, G. B., Gray, J.E., Aenold, C., Tucke, G. A., Schuch, W., Harding, S. and Grierson, D., 1990. Inheritance and effect in ripening of antisense polygalacturonase genes in transgenic tomatoes. *Plant Mol. Biol.* **14**: 369-379.

Smith, D.L. and Gross, K.C., 2000. A family of at least seven β-galactosidase genes is expressed during tomato fruit development. *Plant Physiol.* **123**: 1173-1183.

Sorensen, S.O., Pauly, M., Bush, M., Skjot, M., McCann, M.C., Borkhardt, B. and Ulvskov, P., 2000. Pectin engineering: Modification of potato pectin by in vivo expression of an endo-1,4-β-D-galactanase. *Proc. Natl. Acad. Sci. USA* **97**: 7639-7644.

Sozzi, G.O., Trinchero, G.D. and Fraschina, A.A., 2000. Ethylene and glucosidase promotion in GA3- and IAA-treated tomato fruit (*Lycopersicon esculentum* Mill.). *J. Plant Growth Regul.* **19**: 359-368.

Tabuchi, A., Kamisaka, S. and Hoson, T., 1997. Purification of xyloglucan hydolase/endotransferase from cell walls of azuki bean epicotyls. *Plant Cell Physiol.* **38**: 653-658.

Tabuchi, A., Mori, H. Kamisaka, S. and Hoson, T., 2001. A new type of endo-xyloglucan transferase devoted to xyloglucan hydrolysis in the cell wall of azuki bean epicotyls. *Plant Cell Physiol.* **42**: 154-161.

Tamura, S. and Senda, T., 1992. Fine structure of the cell wall of carrot parenchyma revealed by quick-freeze, deep-etch electron microscopy. *J. Electron Microsc.* **41**: 91-98.

Tateishi, A., Inoue, H. and Yamaki, S., 2001a. Fluctuations in activities of three β-galactosidase isoforms from ripening avocado (*Persea americana*) fruit and their different activities against its cell wall polysaccharides as substrates. *J. Japan. Soc. Hort. Sci.* **70**: 586-592.

Tateishi, A., Kanayama, Y. and Yamaki, S., 1996. β-L-arabinofuranosidase from cell walls of Japanese pear fruits. *Phytochem.* **42**: 295-299.

Tateishi, A., Onoue, H., Sibata H., and Yamaki, S., 2001b. Molecular cloning of β-galactanase from Japanese pear (*Pyrus pyriflia*) and its gene expression with fruit ripening. *Plant Cell Physiol.* **42**: 492-498.

Terasaki, S., Sakurai, N., R. Yamamoto, R., Wada, N. and Nevins, D.J., 2001a. Changes in cell wall polysaccharides of kiwifruit and the visco-elastic properties detected by a laser Doppler method. *J. Japan. Soc. Hort. Sci.* **70**: 572-580.

Terasaki, S., Sakurai, N., Wada, N., Yamanishi, T., Yamamoto, R. and Nevins, D.J., 2001b. Analysis of the vibration mode of apple tissue using electronic speckle pattern interferometry. *Transactions of ASAE* **44**: 1697-1705.

Terasaki, S., Wada, N., Sakurai, N., Muramatsu, N., Yamamoto, R., and Nevins, D.J., 2001c. Nondestructive measurement of kiwifruit ripeness using a laser Doppler vibrometer. *Transactions of ASAE* **44**: 81-87.

Tong, C.B.S. and Gross, K.C., 1988. Glycosyl-linkage composition of tomato fruit cell wall hemicellulosic fractions during ripening. *Physiol. Plant.* **74**: 365-370.

Trianotti, L., Ferrarese, L., Vecchia, F.D., Rascio, N. and Casadoro, G., 1999. Two different endo-β-glucanases contribute to the softening of the strawberry fruits. *J. Plant Physiol.* **154**: 355-362.

Vickers, Z., 1991. Sound perception and food quality. *J. Food Sci.* **14**: 87-96.

Wakabayashi, K., Sakurai, N. and Kuraishi, S., 1991. Differential effect of auxin on molecular weight distributions xyloglucans in cell walls of outer and inner tissues from segments of dark grown squash (*Cucurbita maxima* Duch.) hypocotyls. *Plant Physiol.* **95**: 1070-1076.

Waldron, K.W., Smith, A.C., Parr, A.J., Ng, A. and Parker, M.L., 1997. New approaches to understanding and controlling cell separation in relation to fruit and vegetable texture. *Trends in Food Sci. Tech.* **8**: 213-221.

Yakushiji, H., Sakurai, N. and Morinaga, K., 2001. Changes in cell-wall polysaccharides from the mesocarp of grape berries during veraison. *Physiol. Plant.* **111**: 188-195.

Yalpani, M., 1988. *Polysaccharides.* Elsevier, Amsterdam, pp. 499.

Yamamoto, H. and Haginuma, S., 1984. Estimation of the dynamic Young's modulus of apple flesh from the natural frequency of an intact apple. *Rept. Natl. Food Res. Inst.* **44**: 30-35.

Yamamoto, H., Iwamoto, M. and Haginuma, S., 1980. Acoustic impulse response method for measuring natural frequency of intact fruits and preliminary applications to internal quality evaluation of apples and watermelons. *J. Texture Studies* **11**:117-136.

Yamamoto, R., Shinozaki, K. and Masuda, Y., 1970. Stress-relaxation properties of plant cell walls with special reference to auxin action. *Plant Cell Physiol.* **11**:947-956.

Yokoyama, R. and Nishitani, K., 2001. A comprchensive expression analysis of all members of a gene family encoding cell-wall enzymes allowed us to predict *cis*-regulatory regions involved in cell-wall construction in specific organs of Arabidopsis. *Plant Cell Physiol.* **42**: 1025-1033.

Firmness and Softening of Fruits and Vegetables

JOSSE DE BAERDEMAEKER, PÁL T. JANCSÓK,
and BERT E. VERLINDEN
Katholieke Universiteit Leuven, Department of Agro-Engineering and –Economics and Flanders Centre of Postharvest Technology, Leuven, Belgium

1. INTRODUCTION

The term texture is related to the feeling of food within the mouth and as such it includes a wide range of attributes that can be measured with instrumental methods or with sensory tests. Texture properties arise from structural elements and the way they respond to forces or deformations (in the mouth), eventually resulting in breakdown of the structure and the flow of the material.

The macroscopic texture properties of fruit are determined by its cellular and histological properties such as cell wall elasticity, cell turgor pressure, and pectin content. These properties will manifest themselves as a complex of sensations as there are firmness, crispness, juiciness, ripeness, toughness and many more. The mechanical properties of the fruit are an important basic element to texture and the interaction of a person with the fruit in the mouth causes rupture and failure of the fruit tissue. Often, texture measurements are based on a mechanical failure test of the fruit or of a fruit tissue sample. Mechanical characteristics are one of the important attributes of texture and these will be the main concern of this paper.

1.1 Instrumental Texture Measurement

A number of different techniques are available commercially or can be found in the literature to measure mechanical texture attributes (Harker *et al.*, 1997, Abbot *et al.*, 1997). They involve measurements on tissue samples, destructive measurements on whole fruit or (quasi-non)-destructive measurements on intact fruit.

1.1.1 Measurements on Tissue Samples

The aim of the measurements on tissue samples is to determine the mechanical properties of the flesh of the fruits and vegetables by eliminating the disturbing effects like the shape of the products, the effect of the skin, etc. The measurements are mostly carried out on a small, regularly shaped sample and they can be compression, tension, torsion or beam bending examinations. They can be performed until failure or in some cases loading and unloading is repeated over several cycles. The latter is to obtain a 'texture profile' or imitation of the chewing. In most cases each loading will yield different values for elastic modulus or failure stress. Also creep or relaxation experiments can be conducted to obtain visco-elastic properties of the samples and relate these with texture. The storage (or elastic) modulus and the loss (or viscous) modulus can also be obtained from dynamic excitation of fruit flesh samples. It should be observed that the loading rate may significantly influence elastic modulus measurements or failure criteria. For example in case of apple tissue samples the elastic modulus measured by slow deformation is much lower than the elastic modulus measured by fast deformations. These failure criteria can be maximum stress, maximum strain or strain energy. Figure 1 shows an example of an apparatus to carry out tensile tests. This test is especially suited to measure very soft samples that would disintegrate when clamped between jaws in a traditional tensile test. Furthermore the tensile test yields clean ruptured surfaces which cannot be obtained in a compression type test (Verlinden *et al.*, 1996). Such tests are used to measure the texture properties related to mealiness and blanching.

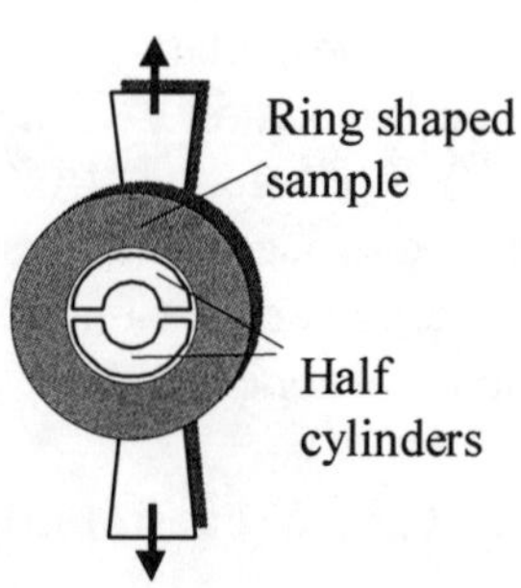

Figure 1. Apparatus to measure the tension of ring shaped food samples.

The advantage of these methods on samples using controlled loads is that they give objective view of the material itself, and that the test methods can easily be standardised. Disadvantages are the destructive way of the examinations which can produce a lot of wastes and does not allow on line examinations.

1.1.2 Destructive Methods on Whole Fruits

Destructive measurements can be carried out not only on small samples but also on whole fruits or vegetables. Such measurements are for example the puncture tests or penetrometer reading and they involve pushing a cylindrical probe with a convex tip into the fruit flesh. The force is measured during the penetration, and usually the maximum penetration depth is held constant. The fruit peel is removed beforehand to eliminate the effect of the peel breakage. Penetrometer tests (e.g., the Magness-Taylor) are performed using hand held devices. In this case the readings are slightly dependent on the experience of the measuring person. To overcome the subjective effect of the handheld device universal testing machines with a constant penetration speed between 1 to 4 mm/s are often used. During a puncture test the force usually increases steadily until a maximum, at which point rupture initiates. After that the force remains constant or decreases up to the maximum deformation (Fig. 2).

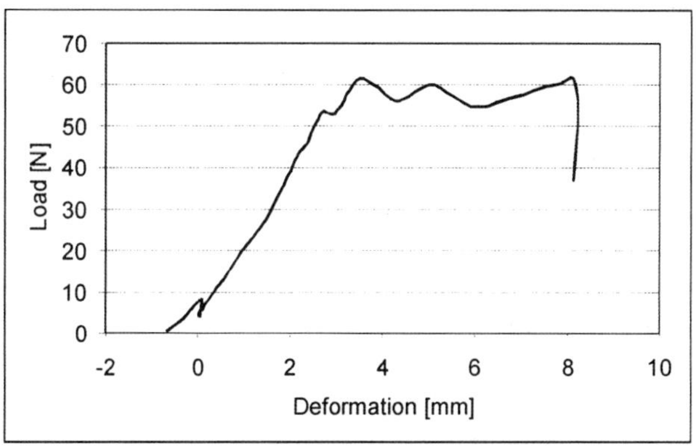

Figure 2. Load-Deformation curve of an apple during puncture. The puncture head was pushed to 8 mm into the apple.

Another kind of firmness measurement method was suggested by Studman and Yuwana (1992). In this test called twist test a thin rectangular blade is radially mounted on a rod and inserted in the fruit. The fruit and rod are twisted until the blade crushes the flesh. This twist test appears to be more related to shear rupture than the puncture test. Different firmness estimation methods may measure different physical properties so the different methods are hardly comparable with each other. It can happen, that according to one test a fruit is harder than the other however if measured with an alternative method the second fruit is harder. Those difficulties in the comparisons of the different methods are more emphasized when a new non-destructive method is related to the existing destructive measurements.

1.2 Non-destructive Firmness Measurements

Non-destructive tests have been developed in order to monitor fruit firmness changes during storage or to be able to grade fruit according to firmness. Some methods can be considered as based on the elastic contact theory. One is the whole fruit compression between two flat plates. The results of such measurements are influenced by fruit size or radius of curvature. This can be overcome by the use of hard spherical indenters with a radius that is much smaller than the fruit radius as in the Zwik test where a hard steal ball is pressed against the fruit surface. Usually the maximum indentation is small and kept constant for a type of apparatus. This allows the use of the required indentation force as a measure of the elastic modulus of the fruit flesh.

A special case of local indentation is the 'air-puff method' that uses compressed air from a nozzle to deform the fruit surface. The deformation is measured using laser based distance sensors (Hung *et al.*, 1999). In dynamic contact force measurements under well-defined loading conditions, the fruit firmness is derived from the peak force, the contact duration or a combination of these. Sometimes this contact force is analysed in the frequency domain where measurements on firm fruits contain higher frequency components than the one on soft fruits. All the mentioned methods measure mostly the fruit properties near the point of contact.

Another method is based on whole fruit vibrations. (Cooke and Rand 1973) Fruits are excited either by impact or by some electromagnetic excitation device. The resulting fruit vibrations are detected and the resonance frequencies determined. In practice the stiffness factor (SF) is used and calculated as

$$SF = f^2 m^{2/3} \qquad (1)$$

where f is the lowest spherical resonance frequency and m is the mass of the fruit. This factor is closely related to the elastic properties of fruits and assumes a spherical body. The method yields a kind of overall measure of fruit firmness rather than a local indication. The interesting thing is that the vibratory tests can be performed by different types of equipment. In Fig. 3 a comparison is made of the firmness results obtained by measuring the fruit vibrations, and the resonance frequencies using either contact sensors or non-contact microphones. This allows for the introduction of equipment independent measures of firmness.

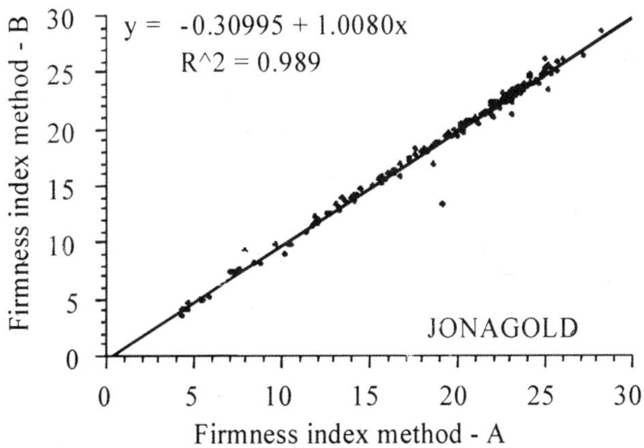

Figure 3. Comparison of the fruit firmness index on the basis of resonance frequency obtained with two different instruments, (apples, var. Jonagold)

It should be noted that this method works best for nearly spherical fruit. For other fruit shapes the method can still be used but the appropriate mode shapes and vibration modes have to be measured and analysed (Fig. 4). The different vibration modes give information about the different parts of the fruit (Dewulf *et al.*, 1999). For example in case of pears the texture properties of the conical top can be measured by the lowest bending mode. In this mode shape, the top part of the pear is deformed and bent. To gain information about the spherical bottom the oblate-prolate mode can be used. In this mode the bottom part of the pear becomes flatter in one direction and wider in the perpendicular direction. Using the compression mode when the whole body of the pear is vibrating in longitudinal direction, information can be gained about the whole fruit (Dewulf *et al.*, 1999).

In order to relate the resonant frequencies of a non-spherical fruit to the texture properties, the stiffness factor equation must be expanded and terms about the shape have to be inserted. Such expansion can be the following:

$$F = \frac{f^2 m^{2/3}}{(a_S \cdot S + b_S)^2} \tag{2}$$

where F is the stiffness related factor, f is the resonance frequency, S is a global shape descriptor (Length/Diameter ratio, Fourier shape descriptor, etc.) a_S and the b_S are the appropriate constants determined by experiments (Jancsók *et al.*, 2001). Based on finite element modelling of the vibration (modal analysis) of Conference pears, the influence of the global shape on the resonance frequency of pears was investigated. Different shape descriptors like Length/Diameter (L/D) ratio, Fourier shape descriptors were examined. The L/D ratio is an easy to measure, but rather rough shape descriptor. In order to be able to compare the resonant frequencies of fruit with different mass the resulting resonance frequencies were normalised. A strong linear correlation was found between the L/D and the normalised frequencies. The higher the L/D ratio, the lower the frequency becomes in the first bending and compression modes. The resonant frequency in the oblate-prolate mode was less influenced by the L/D ratio of the fruits.

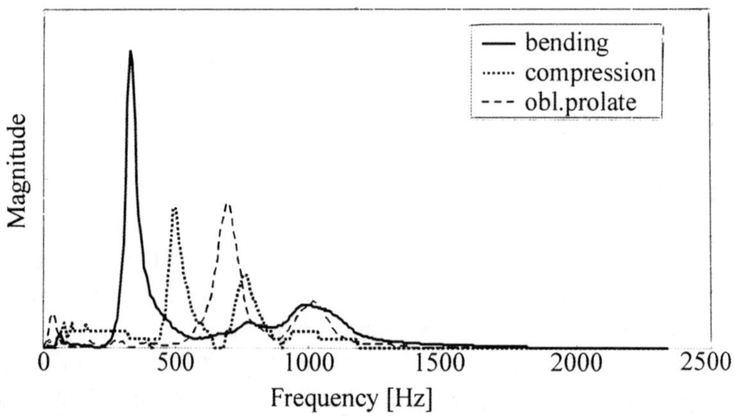

Figure 4. The resonant frequencies of a pear variety Conference.

The texture properties of the fruits and vegetables are also related to the optical properties of the flesh. The scattering of laser light falling onto the fruit changes as the fruit internal structure and the firmness change. For tomatoes and apples of different stages of maturity it was found that the size of the scattered light spot increased with storage time and also with decreasing firmness (Tu *et al.*, 2000a, 2000b). Similar effects are also observed during heat treatment of vegetables or as a result of turgor changes in intact fruits.

2. INSTRUMENTED SENSORY ANALYSIS OF TEXTURE

Since 1963 biting and chewing sounds have been recorded and studied to obtain more information on food texture. Various techniques have been used to record the sounds. Drake (1963) compared different pick-up devices: a microphone pressed against the cheek, the microphone held 2 inches in front of the open mouth and a hearing-aid earphone connected to the ear canal via short plastic tubing. Vickers and Bourne (1976) used a microphone held 1 inch in front of the mouth or held against the outer ear immediately above the opening of the ear canal. Dacremont *et al.* (1991) estimated bone conduction by recording chewing sounds with a microphone pressed against the subject's cheek, near the maxillar angle. Air conduction was measured with a microphone held in front of the ear canal opening at 8 cm. De Belie *et al.* (2000a) recorded bone conducted chewing sounds by placing a microphone, fixed in an earmuff, over the ear canal at the side where the apple pieces are to be chewed.

All techniques might give an empirical measure of differences between foodstuffs. Nevertheless some disadvantages are still connected with the different techniques and it is not always easy to interpret the sounds in terms of the actual vibrations of the foodstuffs and the jaw. The soft tissues in the mouth tend to absorb or damp especially the higher frequencies of the sound (Vickers 1991). Interference between chewing movements in general and specific food crushing sounds can occur. Kapur (1971) measured the resonance frequency of the jaw to be about 160 Hz. The bone conducted sound travelling through teeth and jaws to the ear is therefore amplified at this frequency. On the other hand, when air conducted sounds are recorded with a microphone held in front of the open mouth, this does not necessarily give a true picture of normal chewing.

Seymour and Hamann (1988) used mean sound pressure, pressure level and intensity in different frequency ranges, of sounds produced while the sample was fragmented by a shear/compression cell. Also combinations of mechanical and acoustic parameters have been determined (Vickers 1987, Seymour and Hamann 1988)

A study of Harker *et al.* (1997) suggested that chewing sounds could also differentiate between different textures in apples. The energy of the first bite appeared to be the best predictor of the sensory attributes crispness (defined as the amount and pitch of sound generated when the sample is first bitten with the front teeth) and crunchiness (the amount of sound generated when chewing with the back teeth). De Belie *et al.* (2000a) analysed chewing sounds to detect crispness differences in apples and found results that correspond to sensory analysis.

3. TEXTURE CHANGE DURING STORAGE

3.1 Modelling Texture Changes

Non-destructive measurements make it possible to monitor firmness changes of fruit after harvest when stored under different conditions. If a good model for firmness change can be developed then this can help to predict firmness changes or to sort out fruit according to expected shelf life.

The relationship of cell wall metabolism to the loss of fruit texture that occurs during shelf-life and storage has stimulated considerable research on changes in fruit cell wall composition. In many fruits tissue softening is related to the breakdown of the cell wall-pectin complex. Studies on the chemical composition of the cell wall-pectin complex in relation to the mechanical properties of the tissue have been concerned mainly with changes in pectic substances. The role of calcium forming stable pectin bridges and the linkage of pectins with cellulose and hemicellulose is weakly investigated. Considerably less is known regarding the role in texture degradation during shelf-life and storage of cell wall polymers other than polyuronides (Siddiqui *et al.*, 1996).

The breakdown of the cell wall-pectin complex has been generally attributed to the activity of the several enzymes (Siddiqui *et al.*, 1996). It is realised that the enzymes like peroxidase (Ingham *et al.*, 1998), polygalacturonase and pectin methylesterase (Massiot *et al.*, 1994) contribute to the degradation of texture of the fruit. These enzymes among others have been reported to be present and active in the cell walls but their roles are poorly understood (Klein *et al.*, 1995).

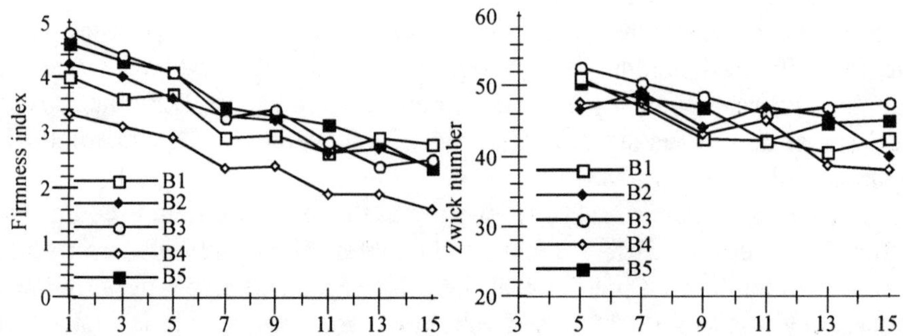

Figure 5. The firmness evolution of tomatoes measured by the acoustic resonance method (left) and by the small local indentation, or Zwick method (right) (B1-B5 are different individual tomatoes).

Work is ongoing to model all these processes on the basis of enzyme kinetics, but a number of simple models can be of help for post-harvest texture management.

In Fig. 5, the firmness evolution of tomatoes is given as measured by the (acoustic resonance method or by the small local indentation (or Zwick) method. Note the more consistent measurement of the evolution when using the acoustic method.

Research on avocado's and apples (Galili and De Baerdemaeker, 1996) has shown that fruits with an initially higher firmness, remain firmer after storage and during shelf life. This evolution can mathematically be described which gives the possibility to predict the further firmness evolution. For a given temperature, the firmness S changes as a function of time can be expressed as an exponential function. This assumption was also made by Thai *et al.* (1990).

$$S = S_0 \exp(-\alpha . t) \tag{3}$$

With S: firmness

S_0 : firmness at the initial measurement time 0

α : deterioration constant, temperature dependent

t : time (in days)

The model above assumes that all individual fruits travel the same hypothetical absolute path of maturity change although they may not be at the same stage simultaneously. Therefore, one can find for each tomato a time shift τ, which is a function of the individual firmness S_0 at the initial measurement. This time shift τ essentially shifts the function S along the time axis so that it coincides with the hypothetical absolute path.

Because ripening is enhanced by temperature, it can be expected that for the same season and the same segment, the firmness decrease is smaller at lower temperatures. However, care must be taken to avoid chilling injury.

The influence of temperature is often expressed in the form of an Arrhenius equation:

$$\alpha = \alpha^{ref} \exp\left(-\frac{E_a}{R_g}\left(\frac{1}{T} - \frac{1}{T_{ref}} \right) \right) \tag{4}$$

with α^{ref} : deterioration constant at a reference temperature

E_a : activation energy (J/mol)

R_g : universal gas constant (8.314J/mol K)

T, T_{ref} : absolute temperature and reference temperature (K)

For red picked tomatoes and with a reference temperature T_{ref} of 15°C, the values of the other constants are $\alpha^{ref} = 0.0320 \pm 0.0004 \times 10^6$ 1/day and $E_a = 82.36 \pm 33.24$ kJ/mol.

This activation energy can be considered as a normal value for temperature-dependent processes in fruits and vegetables, e.g., growth of floral stalks in chicory, or texture change in potatoes or celery (Verlinden, 1996).

In Table 1, deterioration (or softening) constants are given for apples shortly after harvest. The model allows for a quantitative comparison between the varieties. Similar data were obtained for softening during shelf life of normal or ULO stored fruits. In general, the ULO stored fruits had a higher firmness at the end of the storage, but during the shelf life they had a higher deterioration or softening rate.

Table 1. Deterioration constant α (average and 95% confidence interval) for different apple varieties stored at 20°C and 65 % RH

Variety	α (10^6 1/day)
Jonagold	0.0540 ± 0.0018
Boskoop	0.0950 ± 0.0090
Elstar	0.0690 ± 0.0140
Golden	0.0760 ± 0.0033
Braeburn	0.0730 ±0.0003

The remaining shelf life of the product, Δt_{shelf} (in days) can here be defined as the time until the product will have a firmness $S_{\lim}$ that is considered as no longer acceptable to clients or consumers.

$$\Delta t_{shelf} = -(1/\alpha)\ln\left(S_{\lim}/S_0\right) \tag{5}$$

Mealiness is an important texture feature of apple, which is characterised by texture deterioration, resulting in soft, dry and mealy fruits. Mealiness reduces the quality of the fruit, and, hence, its commercial value. Mealiness in apple develops during storage and depends on air composition, temperature and relative humidity and can be induced by applying high temperature (20°C) and high relative humidity (95%) (Barreiro *et al.*, 1998). Apples in refrigerated stores under normal air conditions develop much faster a mealy texture than apples under ultra low oxygen (ULO) conditions (De Smedt, 2000). Apart from the storage conditions, harvesting date and fruit size also influence the development of mealiness. Apples of larger size or of late harvest are more susceptible to mealiness (Barreiro *et al.*, 1998, De Smedt, 2000). Hatfield and Knee, (1988) and De Smedt *et al.* (1998)

concluded that the cells in mealy apples become more rounded, and, as a consequence, the amount of intercellular spaces is larger in mealy apples than in fresh apples. Mealiness in apple is related to the relative strength of the cell wall compared to the strength of the middle lamella (Harker and Hallett, 1992, De Smedt *et al.*, 1998). Sensory analysis has shown that mealiness can be characterised by a combination of a lack of crispiness, a lack of juiciness and a lack of hardness (Barreiro *et al.*, 1998).

3.2 Texture Change During Processing

The physical, chemical and biochemical processes that occur during thermal treatment of vegetative tissue lead to changes in mechanical properties and texture. Again, these mechanical changes can be described by summary models while it is nearly impossible to make detailed models of all the processes that lead to the changes.

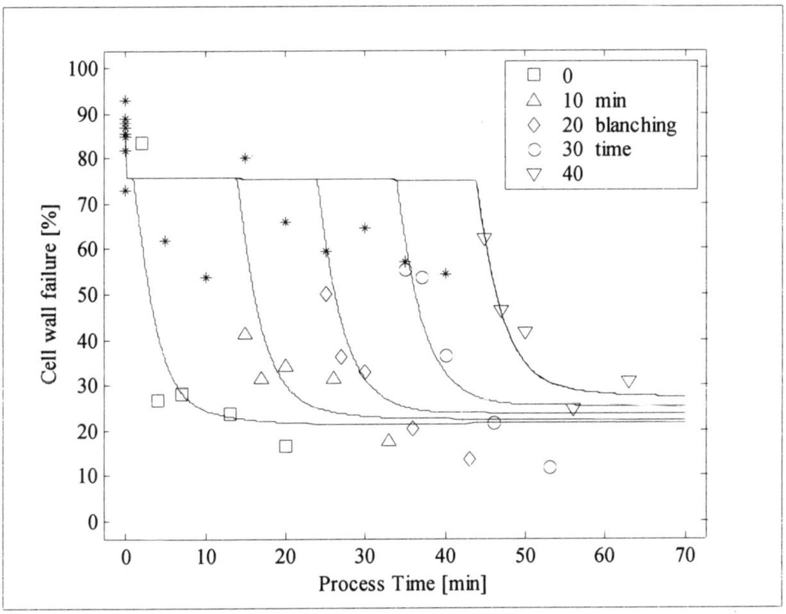

Figure 6. Experimental data and model of the tissue failure mode expressed as cell wall failure for blanching (55°C) and subsequent cooking (95°C) of carrots with different blanching time. The * indicates the blanching only process.

One important event is the loss of turgor at the start of the heating process resulting in a rapid phase of firmness degradation. The velocity of turgor degradation is dependent on the heating temperature and can be expressed using an Arrhenius type model (see Eq. (4)) (De Belie *et al.*,

2000b). Loss of turgor can be attributed to the increased permeability of membranes at higher temperatures resulting first in a passive efflux of solutes before the membranes are completely destroyed.

The heat treatment also has effect on the way the tissue ruptures. During the tissue failure the rupture occurs as a failure of the cell wall or rupture of the middle lamella. The cells on the ruptured surface were categorized by microscopic evaluation as ruptured through the cell wall or separated at the middle lamella. Experiments showed that when the raw vegetable tissue ruptures then the failure most likely occurs at the cells. While after the heat treatments the rupture occurs at the middle lamella (Fig. 6.)

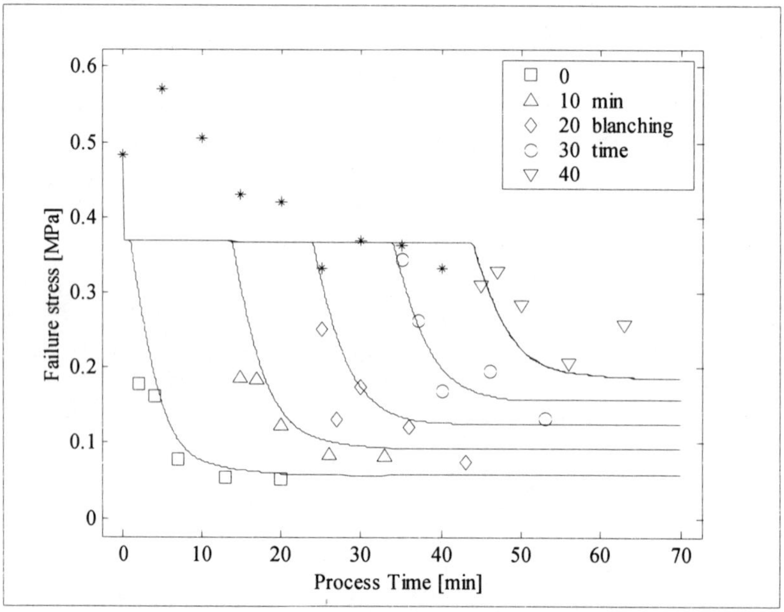

Figure 7. Experimental data and model of failure stress during blanching (55°C) and subsequent cooking (95°C) of carrots with different blanching time. The * indicates the blanching only process.

During blanching and subsequent cooling of potatoes and carrots changes in mechanical properties were different for the force- and deformation-based properties. (Verlinden and De Baerdemaeker, 1997, Verlinden *et al.*, 2000) The tensile rupture force measured with the apparatus shown on Fig. 1 showed a decrease during blanching as well as during subsequent cooking. However, the rupture deformation showed an initial increase during blanching with subsequent cooling, before decreasing during cooking. This initial decrease was attributed to the loss of turgor in the first phase of the blanching stage. It was also observed that the decrease in firmness during cooking was diminished when the samples were blanched first (Fig. 7).

These observations were modelled using three hypothetical chemical reactions of which the reaction rates were put into mathematical form as first order differential equations. The model is suitable to be used in computer aided food process design as a description of firmness.

4. CONCLUSIONS

The texture attributes of food materials are often described in terms of their mechanical properties. A number of measurement techniques for texture were discussed. Destructive techniques probably come closer to the processes that occur in the mouth during sensory perception. They are also most frequently used for studying texture kinetics during thermal processing of fruits or vegetables. Although not fully understood, the basic mechanisms of complex cell wall processes which affect firmness change during thermal processing can be described by a simplified model.

Firmness decrease due to turgor loss occurs very rapidly in these processes.

Texture measurements on the basis of rupture and chewing sound in the mouth offer a possibility for quantitative description of the mouth feel, but the sound analysis does not yet yield very conclusive results when compared to sensory tests.

Non-destructive texture measurements offer possibilities to sort fruits according to firmness and to bring products on the market that have a firmness that is desired for consumers. Such equipment is already being introduced on grading lines. The techniques also allow to measure repeatedly on the same fruit in the course of its storage or shelf life. This offers the possibility to construct models for firmness change as a part of chain design or management.

REFERENCES

Abbott, A.J., Lu, R., Upchurch, L.B., and Stroshine, L.R., 1997, Technologies for nondestructive quality evaluation of fruits and vegetables. *Hortic. Rev.* **20**: 1-120.

Barreiro, P., Ortiz, C., Ruiz-Altisent, M., De Smedt, V., Schotte, S., Andani, Z., Wakeling, I., and Beyts, P.K., 1998, Comparison between sensory and instrumental measurements for mealiness assessment in apples: a collaborative experiment. *Journal of Texture Studies* **29**: 509-525.

Cooke, J. R., and Rand, R.H., 1973, A mathematical study of resonance in intact fruits and vegetables using 3 media elastic sphere model. *J. Agric. Eng. Res.* **18**: 141-157.

Dacremont, C., Colas, B., and Sauvageot, F. 1991, Contribution of air- and bone- conduction to the creation of sounds perceived during sensory evaluation of foods. *Journal of Texture Studies* **22**: 443-456.

De Belie, N., De Smedt, V., and De Baerdemaeker, J., 2000a, Principal component analysis of chewing sounds to detect differences in apple crispness. *Postharvest Biology and Technology* **18**: 109-119.

De Belie, N., Herppich, W., and De Baerdemaeker J., 2000b, A kinetic model for turgor loss in red cabbage cells during mild heat treatment. *J. Plant Physiol.* **157**: 263-272.

De Smedt, V. 2000, Measurement and modelling of mealiness in apples. PhD thesis Katholieke Universiteit Leuven, Belgium.

De Smedt, V., Pauwel, E., De Baerdemaeker, J., and Nicolaï, B., 1998, Microscopic observation of mealiness in apples: a quantitative approach. *Postharvest Biology and Technology* **14**: 151-158.

Dewulf, W., Jancsók, P., Nicolaï, B., De Roeck, G., and Briassoulis, D., 1999, Development of an experimental method for determining the firmness of a pear using finite element modal analysis. *J. Agric. Eng. Res.* **74**: 217-224.

Drake, B.K., 1963, Food crushing sounds. An introductory study. *J. Food Sci.* **28**: 233-241.

Galili, N., and De Baerdemaeker, J., 1996, Performance of acoustic test methods for quality evolution of agricultural products. ISMA22 conference, Leuven, Belgium, September.

Harker, F.R., and Hallett, I.C., 1992, Physiological changes associated with development of mealiness of apple fruit during cool storage. *Hort. Science* **27**(12): 1291-1294.

Harker, F.R., Redgwell, R.J., Hallett, I.C., Murray, S.H., and Carter, G., 1997, Texture of fresh fruit. *Horticultural Reviews* **20**: 121-224.

Hatfield, S.G.S., and Knee, M. 1988, Effects of water loss on apples in storage. *International Journal of Food Science and Technology* **23**:575-583.

Hung, Y.C., Prussia, S.E., and Ezeike, G.O.I., 1999, Nondestructive firmness sensing using a laser air-puff detector. *Postharvest Biology and Technology* **16**: 15-25.

Ingham, L.M., Parker, M.L., and Waldron, K.W., 1998, Peroxidase: Changes in soluble and bound forms during maturation and ripening of apples. *Pysiologia Plantarum* **102**(1): 93-100.

Jancsók, P. Clijmans, L., Nicolai, B., and De Baerdemaeker, J., 2001, Investigation of the Effect of Shape on the Acoustic Response of Conference Pears by Finite Element Modelling. *Postharvest Biology and Technology* **23**: 1-12

Kapur, K., 1971, Frequency spectrographic analysis of bone conducted chewing sounds in persons with natural and artificial dentitions. *Journal of Texture Studies* **2**: 50.

Klein, J.D., Hanzon, J., Irwin, P.L., Benshalom, and N., Lurie, S., 1995, Pectin esterase-activity and pectin methyl esterfication in heated golden delicious apples. *Phytochemistry* **39** (3): 491-494.

Massiot, P., Baron, A., and Drilleau, J.F., 1994, Characterisation and enzymatic-hydrolysis of cell-wall polysacchrides from different tissue zones of apple. *Carbohydrate Polymers* **25** (3): 145-154.

Seymour, S.K., and Hamann, D.D., 1988, Crispness and crunchiness of selected low moisture foods. *Journal of Texture Studies* **19**: 79-95.

Siddiqui, S., Brackmann, A., Streif, J., Bangerth, F., 1996, Controlled atmosphere storage of apples: cell wall composition and fruit softening. *Journal of horticultural science* **71** (4): 613-620.

Studman, C.J., and Yuwana, 1992, Twist test for measuring fruit firmness. *Journal of Texture Studies* **23**: 215-227.

Thai, C.N., Shewfelt, R.L., and Garner, J.C., 1990, Tomato color changes under constant and variable storage temperatures: empirical models. *Transactions of the ASAE* **33** (2): 607-614.

Tu, K., Jancsók, P., Nicolai B., and De Baerdemaeker, J., 2000a, Use of laser-scattering imaging to study tomato-fruit quality in relation to acoustic and compression measurements. *Int. J. of Food Science and Technology* **35**: 503-510.

Tu, K., Jancsók, P., Nicolaï B., and De Baerdemaeker, J., 2000b, A study of apple fruit quality based on the analysis of laser scattering image. *International Agrophysics* **14** (1): 135-139

Verlinden, B., and De Baerdemaeker, J. 1997, Modelling low temperature blanched carrot firmness based on heat induced processes and enzyme activity. *J. Food Sci.* **62**(2): 213-218; 229

Verlinden, B., 1996, Modelling of texture kinetics during thermal processing of vegetative tissue. PhD thesis Katholieke Universiteit Leuven, Belgium.

Verlinden, B.E., de Barsy, T., De Baerdemaeker, J., and Deltour, R. 1996, Modelling the mechanical and histological properties of carrot tissue during cooking in relation to texture and cell wall changes. *J. Texture Stud.* **27**(1):15-28.

Verlinden, B.E., Yuksel, D., Baheri, M., De Baerdemaeker, J. and van Dijk, C. 2000, Low temperature blanching effect on the changes in mechanical properties during subsequent cooking of three potato cultivars. *Int. J. of Food Science and Technology* **35**: 331-340

Vickers, Z. 1987, Sensory, acoustical and force-deformation measurements of potato chips crispness. *J. Food Sci.* **52**: 138-140.

Vickers, Z. 1991, Sound perception and food quality. *J. Food Quality* **14**: 87-96.

Vickers, Z.M., and Bourne M.C. 1976, A Psycho-acoustical theory of crispness. *J. Food Sci.* **41**: 1158-1164.

Determination of a Drop Damage Index to Estimate Bruising Susceptibility of Apple Royal Gala by Logistic Model

PAOLO MENESATTI[1], GRAZIELLA PAGLIA[1], SILVIA SOLAINI[1] and ANGELO ZANELLA[2]
[1]*Agricultural Mechanization Research Institute, Roma, Italy*
[2]*Research Centre for Agriculture and Forestry, Laimburg - Ora (Bolzano), Italy*

1. INTRODUCTION

The analysis of impact damage of fruits is a subject of high economic value and it is widely discussed in literature (Brusewitz and Bartsch, 1989; Hung, 1993). A number of research techniques have been used to evaluate the resistance of fruit and cultivars to different kinds of mechanical stress (Brusewitz et al.,1991; Heap, 1994; Holt and Schoorl, 1977; Mohsenin, 1986; Schoorl and Holt, 1980; Shulte et al., 1991a and 1991b; Studman, 1997) in order to set limits that must not be exceeded during handling and transportation (Chen and Sun, 1981; Garcia et al., 1988; Holt et al., 1981; Kunze et al., 1975; Roudot et al., 1991; Zhang and Brusewitz, 1991). Despite the evidence for impact damage, the causes of physical damage are complex to analyse (Garcia et al., 1995; Menesatti et al.,1999; Menesatti et al., 1998).

These factors determine the property 'bruising susceptibility' of fruits (Banks, 1991; Hyde, 1997; Klein, 1987; Mowatt and Banks, 1994).

The definition of a standard method (Pang et al., 1996; Kuang, 1998; Studman and Banks, 1989a and 1989b) for determining a damageability index of fruit is of considerable interest for fruit production and quality improvement, variety selection, crop growing techniques, agronomic evaluation and postharvest treatment (Dela Rue, 1996; Siyami et al., 1988).

Authors (Menesatti and Paglia, 2001; Menesatti et al., 2001a and 2001b), have developed an instrumental and processing method to determine the drop damage index (DDI). In a logistic model (Bollen et al., 2001) this index allows to calculate the maximum drop height for which the probability of damage of the cultivar is equal to a fixed threshold (usually 10%).

The aim of this work was to analyse the statistical method based on logistic regression and to verify the calculation performance of the DDI for the Royal Gala apple.

2. MATERIALS AND METHOD

A large fruit sample (more than 400 fruits) of the apple cultivar 'Royal Gala' was used (Table 1). The apples were harvested in 2000 in Val Venosta (BZ), Italy. Trials were performed on apples treated with 1-MCP (1 $\mu l \cdot L^{-1}$, AgroFresh Inc. – Rohm&Haas Company, PA, USA) after 8 months of conservation in a controlled atmosphere (CA) (1% O_2 – 2% CO_2, 1°C).

Three different tests were performed to determine and to verify the Drop Damage Index estimation. Because temperature has a very important effect on bruising, the fruit temperature at the impact was constant and equal for all the fruits (22 °C). To obtain this situation, fruit were acclimatized before impact for 24 hours at room controlled temperature.

Table 1. Basic information on the determination and verification test of Drop Damage Index for the apple Royal Gala by a logistic model

Place of harvest	Val Venosta (BZ) - Italy
Date of harvest	September 2000
Type of conservation	
Treatment with 1-MCP (AgroFresh)	$1 \mu l \cdot L^{-1}$
September 2000 - April 2001	Controlled atmosphere
May 2001	normal air
Date of tests	
model	08/05/2001
1st verification	18/05/2001
2nd verification	31/05/2001
Sample amount	
total fruit number	423
impacted fruits for model determination	110
impacted fruits for the 1st verification	113
impacted fruits for the 2nd verification	200

A first sub-sample (110 randomly selected fruits) was subjected to a bruising susceptibility estimation method as described below (Model Test).

The best logistic models were determined and two specific values of the Drop Damage Index were derived.

To evaluate the estimation performance of the bruising susceptibility method, two verification tests were done using two specific drop height values corresponding to the two DDI values from the model test.

Statistical comparison (chi-squared) between expected probability and observed frequency of impact bruising was performed to verify the capacity of the Drop Damage Index to estimate bruising susceptibility.

2.1 Method to Estimate Bruising Susceptibility

Several distinct operating phases were involved in determining the fruit impact bruising probability:

1. drop impact procedure and instrumental measurement of the damage and associated variables (impact, morphological);
2. statistical analysis of acquired data by logistic regression;
3. drop damage index determination on logistic regression curve.

2.1.1 Drop Impact Procedure and Instrumental Measurement of Damage and Associated Variables

Drop tests were done in fruits with impact equipment for free drop tests onto a rigid surface. The distance between the steel plate and the lower edge of each single suspended fruit identified the drop height with an error of less than 0.2 mm. This value was randomised in a range of 20 and 50 mm, in 0.27 mm steps. Impacts were performed on the equatorial part of the fruit.

For each fruit, the following morphological, impact, maturity and damage variables (Fig. 1) were measured:

1. mass in g;
2. pulp firmness (Magness-Taylor index) in N/cm²: the maximum force measured by a digital dynamometer (Zwick 1KN) while piercing the fruit (ferrule Ø 11 mm) for 8 mm after epidermis removal;
3. skin firmness in N/cm²: the maximum force measured by a digital dynamometer while piercing the fruit (ferrule Ø 11 mm) for 3 mm without epidermis removal;
4. fruit diameter at impact point in mm: the diameter of the fruit at the point of impact and oriented parallel to the equatorial plane of the fruit;
5. maximum and minimum damage diameters in mm: the extensions of the damaged area approximately parallel to the fruit surface after epidermis removal;

6. damage depth in mm: the maximum extension of the damaged area along an axis (roughly) perpendicular to the fruit surface and to the core cavity.

In order to define single damage variables for the logistic regression model, where a binary response (damaged or undamaged) is required, expressions of different damage thresholds were performed. In all calculations they were coded with letters, YD or YT, followed by a number. Y represents the dependent variable of the logistic model, i.e. the probability of the event "fruit damaged". D indicates the average value of the minimum and the maximum damage diameters, while T is the average of the minimum and the maximum diameters and the depth of the damage. The numbers after D or T indicate the damage threshold value in mm beyond which the fruit is considered damaged. For example, YD7 indicates that only apples with an average value of damage (the average of the min. and max. damage diameters) greater than 7 mm were considered damaged. Eleven different damage thresholds were considered: YT0 = YD0, YT4, YT5, YT6, YT7, YD5, YD6, YD7, YD8, YD9, YD10.

2.1.2 Logistic Model

In order to directly estimate the fruit bruising probability a logistic regression model was applied. This model was used in the analysis of fruit and vegetable quality and provided interesting results (Bollen et al., 2001, Lammertyn et al., 2000), showing a relationship between a binary response variable and a set of continuous independent variables. Logistic regression analysis allows estimation of multiple regression models when the modelled response is dichotomous and can be scored 0 or 1. In fact, the absence or presence of damage was coded with 0 (undamaged) and 1 (damaged), after selecting a damage threshold. The equation of the fitted model is (Statgraphics plus 5.0 – Manugistics Inc.):

$$Y = \frac{\exp^{eta}}{1 + \exp^{eta}} \qquad (1)$$

where Y represents the probability of event 1, i.e. the damage occurrence and

$$eta = hH + aX_1 + bX_2 + \ldots nX_n + I \qquad (2)$$

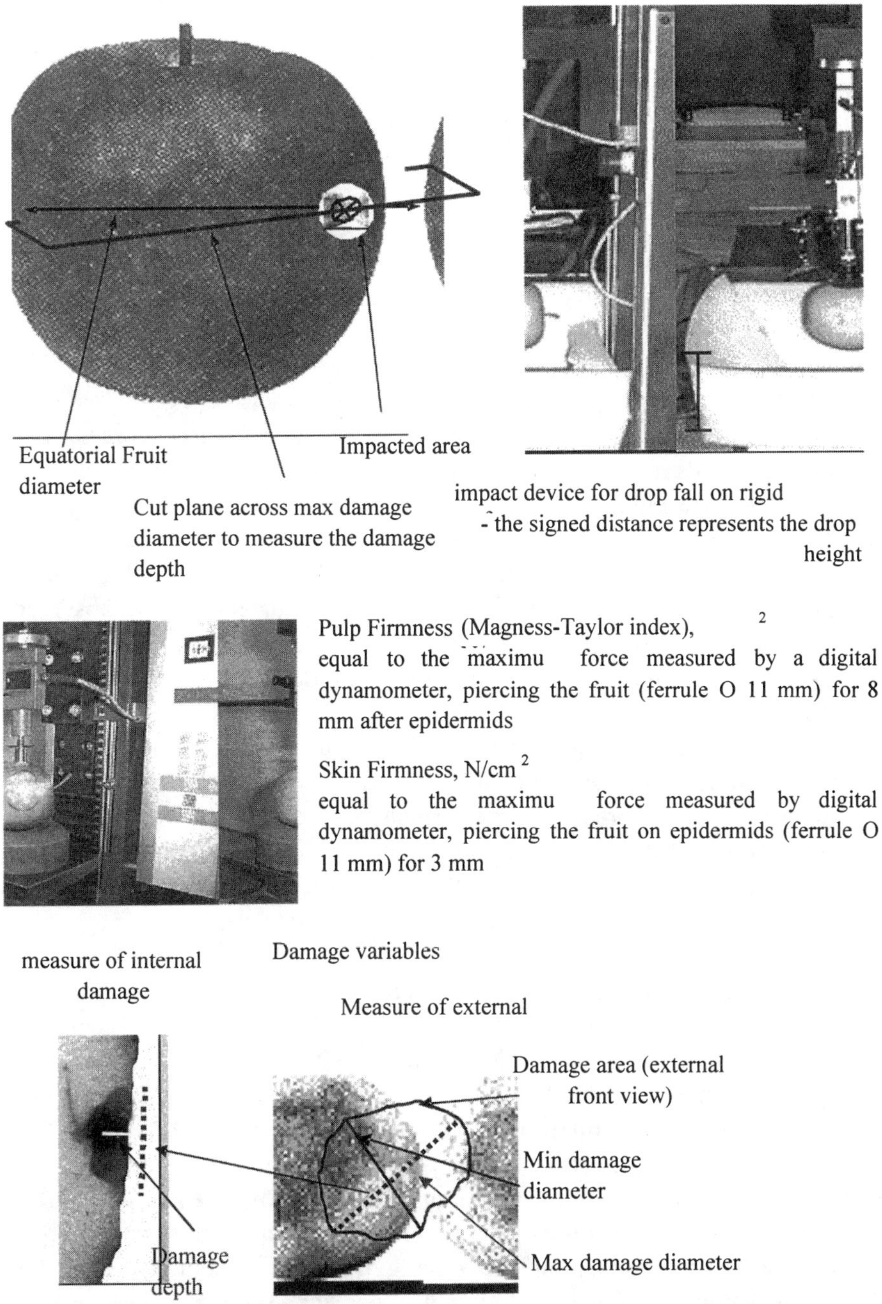

Figure 1. Impact device and variables measurement on the fruit sample used to determine the Drop Damage Index by logistic regression

where H is the drop height, h its coefficient, X the other independent variables (morphological, maturity) and I represents the constant.

To be accepted the models had to present significant coefficients of the variables ($P<0.05$) and not significance of the residual.

In logistic regression, the parameters of the model are estimated using the maximum-likelihood method. Since the logistic regression model is nonlinear, an iterative algorithm is necessary for parameter estimation. While in linear models, the goodness of fit is represented by the R^2 statistic, in logistic ones the Cox&Snell R^2 (Hosmer and Lemeshow, 1989) expressed only how is the variation explained by the final model, in respect to the first step calculation. In the logistic regression, a good way to asses how well the model fits is to compare the predictions to the observed outcomes. The percentages of damaged and undamaged fruits, correctly classified by models were calculated. The percentage of the observed data correctly classified, considering in the whole the two condition of damaged and the undamaged, were reported. Also the Chi-squared test, of the model goodness of fit, determines whether the logistic function adequately fits the observed data. When P-value of the chi-squared test is greater than or equal to 0.05, the model can be considered adequate to fit the data at the 95% or higher confidence level.

For each single fruit, the logistic model calculated a Y value that could range continuously between 0 and 1. The parameter indicated as cutoff was used to calculate the prediction performance of the model. Cutoff is the threshold for which the Y value indicates that the fruit has to be considered undamaged ($Y<$cutoff) or damaged ($Y\geq$cutoff). Cutoff can range between 0 and 1. For all models, a 0.5 (50%) cut-off was applied.

2.1.3 Drop Damage Index Determination

The logistic regression model directly estimates the probability of damage occurrence. Nevertheless, it could be very useful to express bruising susceptibility by a single numerical value instead of a complex function or a whole curve. By logistic model it is possible to calculate the value of a specified independent variable in the equation corresponding to a probability level. Probability was expressed as a function of the drop height (the impact factor) which was present in all the logistic models. The contribution of the other independent variables in models was considered equal to the average value of the observed data.

The Drop Damage Index was established as the maximum drop height, in mm, for which the probability of damage was equal to 0.1 (10%).

Eleven logistic models corresponding to different damage thresholds were analysed. Higher significant regressions were chosen and DDI values extracted. Two Drop Damage Index values were considered, corresponding to different damage thresholds.

2.2 Verification Tests of Drop Damage Index

In order to verify the damage probability associated with a specific drop height (DDI), two verification tests were performed on separate fruit samples.

For each verification test, the apples were impacted at a fixed value of the drop height corresponding to one DDI value. Only damage variables were measured. Damage frequency was expressed by applying damage thresholds equal to those used in the Drop Damage Index determination.

2.3 Statistical Validation of the Drop Damage Index

The statistical validation of the Drop Damage Index determined by the bruising susceptibility method and the logistic model was performed by comparing the chi-square tests of observed and expected damage frequency. Significance probability was considered.

3. RESULTS AND DISCUSSION

Table 2 shows the statistical values of the morphological, impact, maturity and damage variables. These were measured on the fruit sample used as a model to determine the Drop Damage Index by logistic regression analysis. Fruit mass values showed high variability in pulp firmness which represents the classical texture parameter to measure fruit maturity.

The other texture parameter, skin firmness, was found to be less variable. All variables were inserted in the logistic model input layout as independent while damage condition represented the dependent variable. Figure 2 shows the distribution of the percentage of the damaged fruits by different drop height classes for different damage thresholds.

The damaging percentage increases for higher drop height classes while it decreases for higher values of damage threshold. Nevertheless, it can be notice that the percentage of the damaged fruits is higher in the drop height class 20-25 then in 26-30. This fact depends on the contribution of the other independent variables not considered by a simple binary distribution of damage in function of the drop height. For this specific reason, it was necessary to refer to a complex statistical model that analyses the contribution of different variables to the damage.

Table 2. Statistical values of the variables measured on the fruit sample used determining the DDI by logistic regression

Variable	Mean	Standard deviation	Coefficient of variation	min	max
impact					
drop height, mm	35.2	8.7	24.8	20.3	50.0
morphological					
fruit mass, g	144.8	22.1	15.3	110.0	224.8
fruit diameter, mm	67.8	4.0	5.9	60.6	80.6
maturity					
pulp firmness (Magness-Taylor index), N/cm^2	68.3	15.2	22.3	38.8	104.8
skin firmness, N/cm^2	56.5	7.5	13.2	38.8	74.3
damage					
max diameter, mm	5.7	5.4	93.2	* 4.4	17.4
min diameter, mm	4.2	4.0	94.2	* 3.3	13.8
depth, mm	0.8	0.8	99.3	* 0.4	2.9
** T average, mm	3.6	3.3	92.8	* 3.1	11.2
*** D average, mm	5.0	4.6	93.1	* 4.2	15.6

* min values measured except zero

** mean damage obtained as average of the damage variables: max diam., min diameter and depth

*** mean damage obtained as average of the damage variables: max diam. and min diameter

Table 3. Coefficient values of the significant variables obtained by logistic regression applied to the different damage threshold models

Model	Constant	Drop Height	Fruit Diameter	Fruit Mass	Skin Firmness	Pulp Firmness
YT0 = YD0	-3.169	0.083			0.072	-0.049
YT4	-7.026	0.084	0.118			-0.058
YT5	-8.978	0.101	0.132			-0.059
YT6	-14.939	0.131	0.170			-0.036
YT7	-11.134	0.130		0.033		
YD5	0.690	0.083				-0.051
YD6	-7.223	0.082	0.117			-0.055
YD7	-10.361	0.099	0.152			-0.057
YD8	-12.142	0.148		0.040		
YD9	-12.088	0.142		0.036		

Differences were also found related to the calculation of the average damage used as threshold: for the same numerical value, YD presented higher damaging percentage than YT because the latter represents the average between two diameters (min and max) and the depth of damage normally ranged between 0.5-3 mm (Table 2).

Table 3 reports the coefficients of the independent variables that were found significant at the stepwise logistic regression applied to the considered damage threshold models. The impact variable, drop height, is always present. Among other variables pulp firmness and the two morphological variables, fruit diameter and mass, were often significant in damage determination.

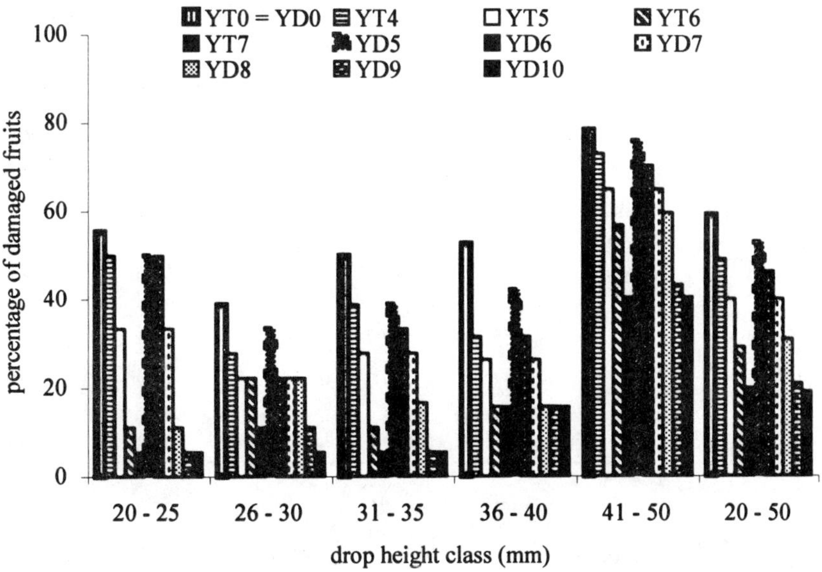

Figure 2. Distribution of the percntage of the damaged fruits for different damage thresholds by different drop height classes

All models were found significant at the logistic regression, but with different levels of inferential and prediction capability (Table 4).

It can be observed that three models, YT4, YD5 and YD6, present the P-value less than 0.05. Only in this case, we can reject the hypothesis that the fitted model is adequate at the 95% confidence level.

Through the logistic Eq. (1) using the variables and their coefficients indicated in Table 3, it was possible to estimate the probability to be damaged for every fruit of the sample. Applying a cut-off of 0.5 (50%), the percentage of undamaged fruit was estimated. Quite all the models present a high percentage of the undamaged fruits correctly classified. A lower percentage is shown for the prediction of the damaged ones. Nevertheless, to asses how well the model fits it is important to evaluate the total percentage of the observed data correctly classified, considering in the whole the two condition of damaged and the undamaged (total). The models present a good value of prediction percentage that, ranging from 64% to 83%, indicates that the logistic function adequately fits the observed data.

Table 4. Statistical parameters for the logistic models, comparison between observed and estimated values and determination of the DDI (Drop Damage Index) - ***Italicised*** and ***underlined*** rows correspond to the DDI values chosen for verification

model	Drop Damage Index (mm)	Chi-squared goodness of fit	P-value of the Chi-squared	% of the observed data correctly predicted at cutoff 0.5		
				damaged	undamaged	total
YT0 = YD0	2.85	3.839	0.279	72.3	53.3	64.6
YT4	8.85	10.176	0.017	66.7	75.0	70.9
YT5	*18.35*	*6.188*	*0.103*	*59.1*	*87.9*	*76.4*
YT6	27.70	2.773	0.428	53.1	91.0	80.0
YT7	*32.40*	*4.965*	*0.174*	*31.8*	*93.2*	*80.9*
YD5	6.96	8.291	0.040	67.2	59.6	63.6
YD6	10.00	8.451	0.038	54.9	79.7	68.2
YD7	*18.01*	*6.462*	*0.091*	*59.1*	*86.4*	*75.5*
YD8	27.80	1.757	0.624	52.9	90.8	79.1
YD9	*32.60*	*4.610*	*0.203*	*39.1*	*94.3*	*82.7*
YD10	33.60	6.316	0.097	33.3	93.3	81.8

The first column of Table 4 reports the determined value of the Drop Damage Index for each model. The ranges of DDI values are high, as a function of the damage threshold applied. They vary from about 3 mm, for YT0 or YD0, up to 33 mm measured for the highest threshold (YT7 and YD9).

Higher drop heights cause larger damages even considering the fixed probability of 10% of the damaged fruits. The whole logistic functions are shown in Fig. 3.

The damage probability is expressed only as a function of the drop height. The contribution of the other independent variables of Table 3 is considered constant and equal to the average values of the variable (Table 2) for each point. The vertical lines delimit the drop height range of the test while the horizontal line indicates the probability value to which the Drop Damage Index corresponds in abscissa. Some of these DDI values are included in the test range while the others are represented as a mathematical extrapolation on the logistic equation. The curves have different trends that present positive probability for zero drop height for models at lower damage threshold.

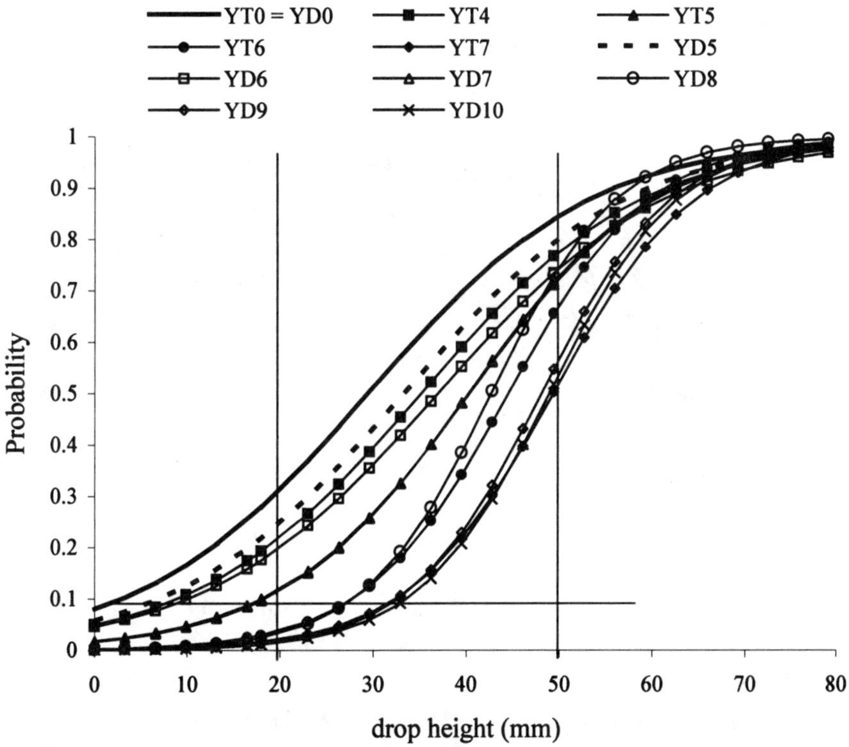

Figure 3. Logistic regressions for the different damage thresholds obtained by the model test.

The graph of Fig. 4 reports the percentage of fruit correct classified, both damaged than undamaged, in function of increasing value of cutoff. Using the cutoff value which maximizes the total percentage correct may provide a good value to use for predicting additional individuals. Moreover, it can be notice that all models provide a good level of prediction generally higher than 60%.

From the listed models in Table 4, only four models were used for the verification tests corresponding to two drop heights for each YT and YD threshold.

From the listed models in Table 4, only four models were used for the verification tests corresponding to two drop heights for each YT and YD threshold.

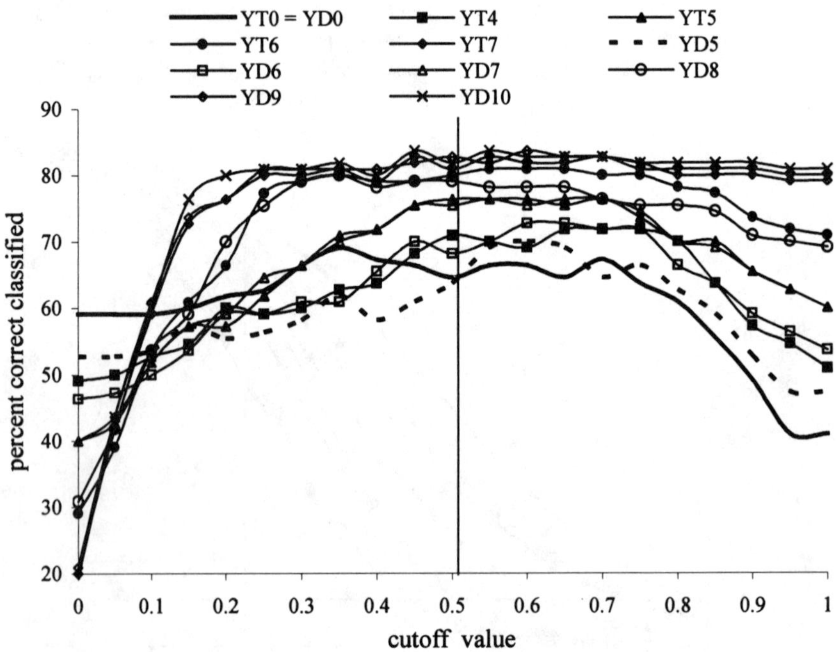

Figure 4. Logistic regressions for the different damage thresholds obtained by the model test.

The drop height of eighteen millimetres, corresponding to YT5 and YD7 models, were chosen for the first verification test. Thirty-three millimetres, corresponding to the DDI determined for YT7 and YD9 models, were chosen as the drop height for the second verification test.

Table 5 shows the results of the Drop Damage Index verification. For the first test, referring to the DDI value of 18 mm, differences between expected and observed numbers of damaged fruits were not significant in the chi-square statistical test. The observed frequencies of damaged fruits are 10% for YT5 and 7% for YD7.

The second DDI verification with a fixed drop height of 33 mm was found not significant only for the YT7 model in the chi-square test. This means that only in one out of four cases the Drop Damage Index determination method was not positively verified by an external sample of the model.

Table 5. Results of Drop Damage Index (DDI) verification by chi-squared test.

Logistic model	YT5	YD7	YT7	YD9
Determined DDI (mm)	18.4	18.0	32.4	32.6
Verification test	I	I	II	II
N° of impacted fruits	134	134	200	200
Fixed value of drop height (mm)	18	18	33	33
N° of observed Damaged Fruits	14	10	13	11
N° of Expected Damaged Fruits	13	13	20	20
Observed damage frequency (%)	10	7	7	6
2	0.09	0.77	2.72	4.50
P	0.770	0.381	0.099	0.034

4. CONCLUSION

Logistic regression was found to be a very powerful statistical approach to estimate bruising susceptibility of 'Royal Gala' apples in impact tests on free fall onto a rigid surface. Using a limited sample of impacted fruits (nearly 100), it is possible to estimate the damage probability in function of the impact factor, drop height, measuring few simple variables (fruit weight and diameter, firmness, damage diameter). The logistic regression model offers the possibility to extract the value of the drop height corresponding to a specified probability of damage to have a simpler value of evaluation instead of a whole curve or equation. The drop height corresponding to the 10% damage probability of the apples was chosen (Drop Damage Index or DDI).

Among the different damage thresholds to calculate the DDI, the most fitting models were obtained for those corresponding to a high percentage (60-80%) of undamaged impacted samples. This factor is important to correctly set up the drop height range of the impact tests. No particular difference was found between the two average values of the damage, one considering only the measure of the diameters (YD), the other also the depth (YT).

To simplify the trials, it could be possible to consider only YD thresholds of damage that reduces the error due to the manual measure of a very small distance as damage depth. A damage of 7 mm in average diameter measured on the fruit surface was found the most interesting damage threshold to discriminate undamaged fruit, presenting values lower than 7 mm, from the damaged ones. This value corresponds with a bruise spot of nearly 0.4 cm^2 which could be considered also a valid threshold in consumer acceptance of apple bruising.

At this threshold, the DDI value is 18 mm, this value was confirmed by the verification test made in another sample of more than one hundred fruits.

More uncertainty was found for the verification of a second value of DDI (33 mm) corresponding to higher damage thresholds (YT7 and YD9) where only one model (YT7) was validated.

The methodology proposed which includes both the measurement of physical-morphological and damage variables and logistic regression analysis, was positively verified and seems very powerful to estimate a Drop Damage Index for the quantification of the bruising susceptibility of apples.

REFERENCES

Banks, N. H., 1991, Identification of sources of bruises on Hawkes Bay grading equipment. *Project report to New Zealand Apple and Pear Marketing Board*, Massey University, New Zealand.

Bollen, A. F., Cox, N. R., Dela Rue, B. T., Painter, D. J., 2001, A descriptor for damage susceptibility of a population of produce. *Journal of Agricultural Engineering Research*, **78** (4): 391-395.

Brusewitz, G. H., Bartsch, J. A., 1989, Impact parameters related to post harvest bruising of apples. *Transactions of the ASAE*, **32**(3): 953-957.

Brusewitz, G. H., McCollum, T. G., Zhang, X., 1991, Impact bruise resistance of peaches. *Transactions of the ASAE*, **34**: 962-965.

Chen, P. and Sun, Z., 1981, Impact parameters related to bruise injury in apples. *Transactions of the ASAE*, Paper **81**: 3041.

Dela Rue, B. T., 1996, A hierarchical modelling procedure to determine factors in apple bruising. *Postharvest 96*, Conference proceedings, Taupo, New Zealand.

Garcia, C., Ruiz, M., Chen, P. 1988, Impact parameters related to bruising in selected fruits. ASAE Tech. Paper No. **88**-6027, *American Society of Agricultural Engineers*, St. Joseph, Michigan.

García, J. L., Ruiz-Altisent, M., Barreiro, P., 1995, Factors influencing mechanical properties and bruise susceptibility of apples and pears. *Journal of Agricultural Engineering Research*, **61**: 11-18.

Heap, R. A., 1994, Performance of recycled paper pulp trays in relation to impact damage in apple cartons. *M. Horticultural Science thesis*, Department of Agricultural Engineering , Massey University.

Holt, J. E., Schoorl, D., 1977, Bruising and energy dissipation in apples. *Journal of Texture Studies*, **7**: 421-432.

Holt, J. E., Schoorl, D., Lucas, C., 1981 Prediction of bruising in impacted multilayered apple packs. *Transactions of the ASAE*, **24**: 242-247.

Hosmer, D. W., Lemeshow, S., 1989, *Applied logistic regression*. New York, John Wiley and Sons.

Hung, Y. C., 1993, Latent damage - a systems perspective. In *Postharvest handling - a systems approach* (Shewfelt. R L and Prussia S E eds), chapter 10, Academic Press, San Diego.

Hyde, G. M., 1997, Bruising - impacts, why apples bruise, and what you can do to minimize bruising. *Tree fruit postharvest journal*, Washington State University, **8** (4): 9-12.

Klein, J. D., 1987, Relationship of harvest date, storage conditions, and fruit characteristics to bruise susceptibility of apple. *Journal of the American Society of Horticultural Science*, **112** (1): 113-118.

Kuang, M., 1998, Evaluation of alternative methods of bruise measurements in apple fruit. *Masterate thesis, Massey University*, New Zealand.

Kunze, O. R., Aldred, W. H., Reeder, B. D., 1975, Bruising characteristics of peaches related to mechanical harvesting. *Transactions of the ASAE*, **18**: 939-945.

Lammertyn, J., Aerts, M., Verlinden, B. E., Schotsmans, W., Nicolaï, B. M., 2000, Logistic regression analysis of factors influencing core breakdown in "Conference" pears. *Postharvest Biology and Technology*, **20**: 25-37.

Menesatti, P. and Paglia, G., 2001, Determination of a Drop Damage Index (DDI) of Fruit Resistance to Damage. *Journal of Agricultural Engineering Research*, in press.

Menesatti, P., Beni, C., Paglia, G., Marcelli, S., D'Andrea, S., 1999, Predictive statistical model for the analysis of drop impact damage on peach. *Journal of Agricultural Engineering Research*, **73**: 275-282.

Menesatti, P., Beni, C., Paglia, G., Marcelli, S., Gentile, A., 1998, Evaluation of pear and apricot drop impact bruises by image analysis. *13th International Congress on Agricultural Engineering (CIGR)* Proceedings Rabat (Morocco), **6**: 215-223.

Menesatti, P., Paglia, G., Solaini, S., 2001a, Comparison of Peach Cultivar Damageability by Drop Damage Index MODEL-IT *2nd Int. Symp. of Application of Modeling as an Innovative Technology in the Agri-Food Chain* 9-13 December 2001 – Palmerston North, NZ, accepted.

Menesatti, P., Paglia, G., Solaini, S., Zanella, A., Stainer, R., Costa, C., Cecchetti, M., 2001b, Non-linear Multiple Regression Models to estimate the Drop Damage Index of Fruit. *Journal of Agricultural Engineering Research*, submitted.

Mohsenin, N. N., 1986, *Physical properties of plant and animal materials.* Second Edition, Gordon and Breach, New York.

Mowatt, C. M., Banks, N. H., 1994, Factors influencing the bruise susceptibility of apples. *Final project report to the New Zealand Apple and Pear Marketing Board*, Hastings, New Zealand.

Pang, D. W., Studman, C. J., Banks, N. H., Baas, P. H., 1996, Rapid assessment of the susceptibility of apples to bruising. *Journal of Agricultural Engineering Research* **64**: 37-48.

Roudot, A. C., Duprat, F., Wenian, C., 1991, Modelling the response of apples to loads. *Journal of Agricultural Engineering Research*, **48**: 249-259.

Schoorl, D. and Holt, J. E., 1980, Bruise resistance measurements in apples. *Journal of Texture Studies*, **11**: 389-394.

Schulte, N. L., Timm, E. J., Armstrong, P. A., Brown, O. K., 1991a, Apple bruising - a problem during hand harvesting. *Proceedings of the American Society of Agricultural Engineers*, paper 911021.

Schulte, N. L., Timm, E. J., Ladd, J. D., Brown, G. K., 1991b, Peach and pear impact damage thresholds: a progress report. *ASAE paper*, 6620.

Siyami, S., Brown, G. K., Burgess, G. J., Gerrish, J. B., Tennes, B. R., Burton, C. L., Zapp, R. H., 1988, Apple impact bruise prediction models. *Transactions of the American Society of Agricultural Engineers*, **31** (41): 1038-1046.

Studman, C. J., 1997, Factors affecting the bruise susceptibility of fruit. *Proceedings of Conference on Plant Biomechanics* 1997, (Jeronimidis O; Vincent J F V eds), University of Reading, Reading: 273-281.

Studman, C. J., and Banks, N. H., 1989a, Indenter shape effects in the impact bruising of apples and nashi. Conference Proceedings, Paper A 14, *4th International Conference on the Physical Properties of Agricultural Materials*, 809, Rostock, East Germany.

Studman, C. J. and Banks, N. H., 1989b, The measurement of bruise susceptibility in apples and nashi. Conference proceedings, *International Symposium on Agricultural Engineering,* **3**: 31-36.

Zhang, X., Brusewitz, G. H., 1991, Impact force model related to peach firmness. *Transactions of the ASAE*, **34**: 2094-2096.

Color as a Quality Factor of Fruits and Vegetables

BOHDAN DOBRZAŃSKI, JR. and RAFAŁ RYBCZYŃSKI
B. Dobrzański Institute of Agrophysics, PAS, Lublin, Poland

1. INTRODUCTION

Apples grown for consumption should be crisp, have a lot of juice, taste good, have nice smell and have an appealing color (Kader, 1999). Disappointingly, most of the harvested apples do not meet quality requirements for the consumer market and are send for industrial processing. One of the basic improvements of quality of apples selected for consumption is proper sorting and handling (Bellon et al., 1992; Guyer et al., 1993; Harrell et al., 1988; Miller and Delviche, 1988; Studman, 1998; Thomason, 1986). For example, "sorting-for-quality" on harvest machines would go a long way towards maximizing food potential. Separating the ripe fruits from the over ripe and unripe fruits would allow the "good" food (with adequate shelf life) to be shipped to fresh market while the less desirable, the green and the over ripe fraction, could be send to a processing plant where quality could be enhanced by appropriate bio-processing techniques (McClure, 1995). Color and size of apple are most important criteria among other quality parameters estimated by consumer (Chen, 1996; Chen et al., 1986; 1981; 1993; Dobrzański and Rybczyński, 2000; 2001; Kader, 1983; Kappel et al., 1995; Francis, 1995; Paulus and Schrevens, 1997; Płocharski and Konopacka, 1999).

With the increasing diversity of pome fruits cultivars, fruit quality recognition becomes more and more important. Along with quality estimation, color is one of the major factors in creating a fruit image (Studman, 1994; Alchanatis and Searcy, 1995; Felföldi et al., 1996; Good,

2002; Kader, 1983; Kameoka et al., 1994; Kavano, 1994; Lancaster, 1992; Molto et al., 1996; Motonaga et al., 1997; Nielsen, 1996). Control of fruit color has a great effect on sales, however, in many cases is performed by visual evaluation, relying on the subjectivity of an individual in determining fruit color. Unfortunately, color perception of every individual is different. It is also difficult to accurately describe color in words, since each person interprets the described color differently.

Although, research in this chapter concentrates mainly on properties of apples. Other fruits and vegetables are also described, such as pears, lemons, oranges, paprika, mushroom, and carrots. Therefore, the objective of this research was to devise a numerical way to accurate identification of color for fruits and vegetables, without relaying on an individual's color perception and to apply it in the consumer quality estimation.

2. THEORY OF COLOR

The illuminated object reflects the light, which is perceived and interpreted by the persons. In physics, visible light is said to be composed of electromagnetic rays. The electromagnetic rays of visible light are different only in their frequency from the other rays such as: gamma rays, X-rays, ultraviolet, infrared, microwaves and rays carrying radio and television. The frequency range of rays reflected from an object is influenced by the physical and chemical properties of the object and by the frequency ranges of the rays that are absorbed. However, the color of an object is unknown, because electromagnetic rays are colorless (Hunt, 1987).

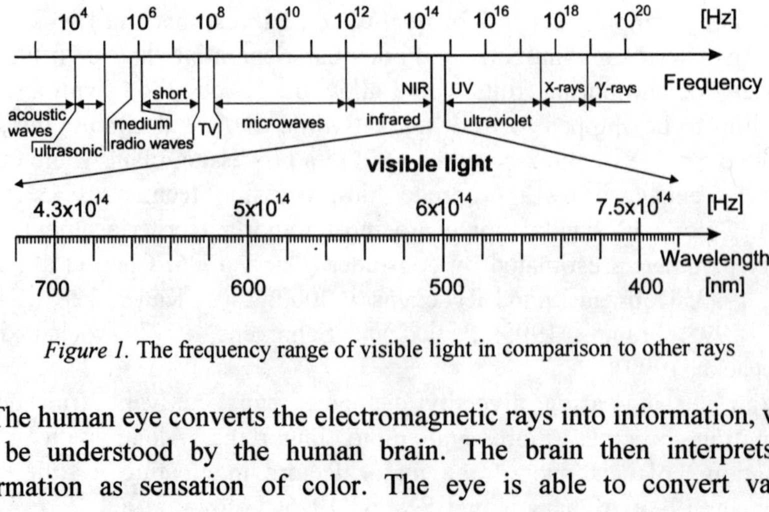

Figure 1. The frequency range of visible light in comparison to other rays

The human eye converts the electromagnetic rays into information, which can be understood by the human brain. The brain then interprets this information as sensation of color. The eye is able to convert varying frequencies of electromagnetic rays into differ information for the brain as different color. The eye is also able to convert the intensity of rays into information, which the brain interprets as a sensation of brightness. It is important to remember that all objects are colorless and the sensation of color originates only in the human brain (Hunt, 1987).

Embedded in the eye's retina are the staff cells and three different types of cone cells, which are responsible for daylight and color vision. The retina contains approximately 120 million staff cells and 6.5 million cone cells (Epson, 1995). Three different types of cone cells convert various wavelengths of electromagnetic rays. The perception of red color is allocated to the cone cell type with maximum sensitivity of 620 nm. Green is allocated to the cone cell type with maximum sensitivity of 520 nm and blue to the cell with maximum sensitivity of 450 nm (Hunt, 1987).

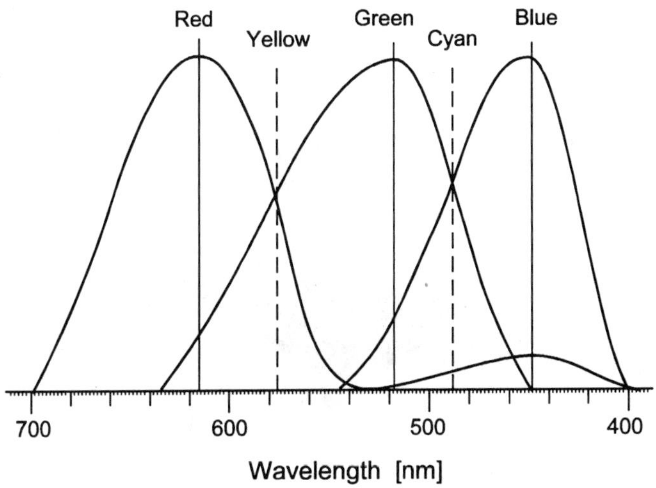

Figure 2. Color sensitivity of the cone cells of the human eye's retina

Our perception of color is dependent on which wavelengths are reflected by an illuminated object. The way in which these electromagnetic rays are reflected is partly responsible for the way of the human color perception. It highly depends on the available wavelengths (color composition) of the illuminating light. However, the composition of daylight changes throughout the day (Epson, 1995). The proportion of red is at its highest in the morning whereas the proportion of blue increases toward evening.

Based on the fact that the retina of the human eye contains three different types of cone cells (Achenbach, 2001) which are sensitive to the primary colors of red, green, and blue respectively, it is possible to formulate laws. These lows state that all colors can be derived from a mixture of three primary colors. The additive color mixing process states that by mixing red, green, and blue the color white is produced. This color mixing process is used wherever light passed directly into the eye without being reflected from an object, e.g. monitors and televisions. In the case of the subtractive color mixing from the primary colors cyan, magenta, and yellow using the process of subtracting or filtering. Subtractive color mixing processes are used when the reflection of light from object passes into eye, e.g. colored paper or fruit skin (Hunt, 1987).

Because human is able to distinguish between several hundred thousand shades of color (Epson, 1995 -- approx. 350000), it is necessary to introduce mathematical color models which enable each shade to be described exactly in terms of a numbered value. Because the number of colors, is high it is impossible to give each particular shade an individual name.

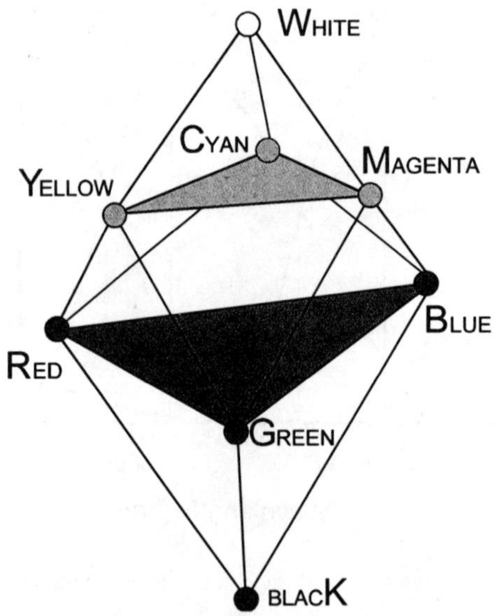

Figure 3. RGB and CMY color mixing model

To enable colors to be described as geometrical interpretation, there are various 3-D models for color description. Some of these, e.g. the RGB (red, green, and blue) color model are derived directly from the additive or CMY (cyan, magenta, and yellow) from the subtractive color mixing system (Hunt, 1987). The RGB color system is often used by software as an internal color model as it can be easily used for calculating and requires no conversion in order to display colors on computer screen (Epson, 1995). In practice when color is measured by color camera and RGB system is used the processing is often done by using the normalized rgb system. The rgb color system is a normalized version of the RGB system by dividing all the color coordinates with the intensity (R+G+B). The CMY color model enables any color to be created from the primary colors cyan, magenta, and yellow, which are converted the subtractive color mixing into a digital system. Each of the primary colors in both models is allocated to one of the eight corners of the cube. Therefore, each color in this cube is identified by its coordinates.

Compared with the RGB and CMY color models, the HSV (hue, saturation, and value) color model, as the hexagon pyramid, has the advantage that the colors correspond most closely to our description of color. It made the bases to elaborate many devices, which used the systems of tristimulus coordinates of color and brightness factor; corresponding to the sensitivity of human eye.

Most of new colorimeters allow measurements of absolute color to be displayed in any of five color systems: Yxy, L*a*b*, L*C*Ho, Hunter Lab, or tristimulus values XYZ. Measurements of color difference can be displayed in any of four systems: Δ(L*a*b*)/ΔE*ab, Δ(L*C*Ho)/ΔE*ab, Δ(Yxy), and Hunter Δ(Lab)/ΔE (Good, 2002). Two of these systems are frequently applied in any quality estimation of fruit color.

1. The Yxy color system was defined by the CIE (Commission Internationale de I'Eclairage) in 1931, and forms the base for other CIE color systems. In this system, Y is the lightness factor expressed as a percentage based on a perfect reflectance of 100 %; x and y are the chromaticity coordinates in the CIE x, y diagram, and are defined by the following equations:

$$x = \frac{X}{X+Y+Z} \quad (1)$$

$$y = \frac{Y}{X+Y+Z} \quad (2)$$

where:

X, Y, Z: Tristimulus values based on the color-matching functions of the CIE 2^0 Standard Observer

2. The L*a*b* color system is one of the uniform color spaces recommended by CIE in 1976 as a way of more closely representing perceived color and color difference. In this system, L* is the lightness factor; a* and b* are the chromaticity coordinates (Good, 2002).
    - L* (lightness) axis -- 0 is black; 100 is white.
    - a* (red-green) axis -- positive values are red; negative values are green; 0 is neutral.

- b* (yellow-blue) axis -- positive values are yellow; negative values are blue; 0 is neutral.

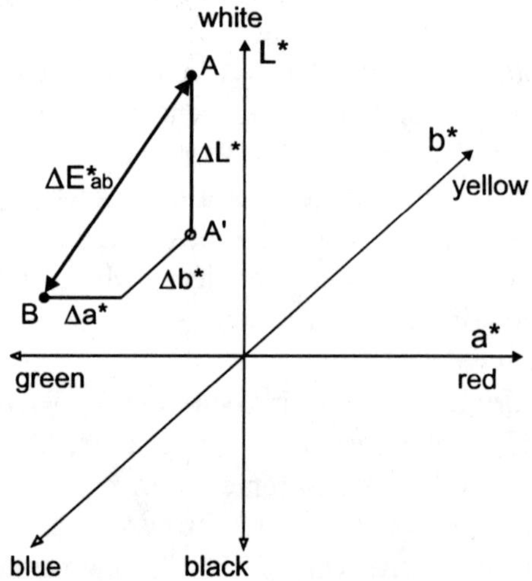

Figure 4. L*a*b* color mixing model. A – target color, B – specimen's color, A' – target color at same lightness as specimen's color

The lightness factor L* and chromaticity coordinates a* and b* are defining as follows:

$$L^* = 116\left(\frac{Y}{Y_0}\right)^{\frac{1}{3}} - 16 \tag{3}$$

$$a^* = 500\left[\left(\frac{X}{X_0}\right)^{\frac{1}{3}} - \left(\frac{Y}{Y_0}\right)^{\frac{1}{3}}\right] \tag{4}$$

$$b^* = 200\left[\left(\frac{Y}{Y_0}\right)^{\frac{1}{3}} - \left(\frac{Z}{Z_0}\right)^{\frac{1}{3}}\right] \tag{5}$$

where: X_0, Y_0, Z_0 - Tristimulus values of luminance
X_0=98.072, Y_0=100, Z_0=118.225 for standard illuminant C (2^0 observer)

X_0=95.045, Y_0=100, Z_0=108.892 for standard illuminant D65 (2^0 observer)

Above formulas apply only when X/X_0, Y/Y_0, and Z/Z_0 are greater than 0.008856.

ΔE*ab is the strait-line distance between two colors in the L*a*b* system and is defined as follows: $\Delta E^*_{ab}=[(\Delta L^*)^2+(\Delta a^*)^2+(\Delta b^*)^2]^{1/2}$

3. PRACTICAL APPLICATION OF L*a*b* COLOR SYSTEM IN QUALITY ESTIMATION OF FRUIT AND VEGETABLES

Studman (1994) observed, that consumers of the 1990's are more conscious of quality than any previous generation. There is no doubt that at present time and during last decade in most developing countries, including the East-European countries, the market has changed. Therefore, the appearance of fruits and vegetables has a major influence on perceived quality. However, color as one of most important quality parameter is influenced by cultural and consumer preferences. The preferences of color depend on: uniformity of external color, repeatability of fruit color in crop, differences between high color of blush and ground color, intensity of high color (saturation of red), intensity of ground color, surface area of high color, brightness-darkness, whiteness, physical defects, dents, browning, bruising, stage of maturity (ripeness).

Seven apple cultivars (Delikates, Holiday, Granny Smith, Spartan, James Grieve, Empire, Šampion) were taken into consideration for color uniformity and to estimate the differences between high color of blush and ground color of the fruit skin. The color of each apple was determined at six points around the stem-axis of fruit, to allow to measure surface area of high color. Each measuring area covered 60 degree of the circumference of fruit.

The measurements were performed with Braive Instruments 6016 supercolor™ colorimeter (Braive, 1994). The measuring system employed by Braive colorimeter is designed to provide accurate readings and uniform response. The light received by the meter is divided three ways and passed through special filters whose light absorbing characteristics combine with the spectral response of the photocells. Upon striking the silicon photocells, light energy is converted into electrical signals and sent to the microprocessor, where it is adjusted for the illuminating condition desired and then converted into coordinates fore chosen color space. Readings are displayed in the LCD panel and can be transferred to a separate computer or processor through the data output terminal. For color readings, these values are translated into Yxy coordinates or in color L*a*b* standard. This device allows changing and selecting the following illuminant:

A - incandescent (tungsten) lamplight
C - daylight (filtered tungsten)
D65 - daylight, color temperature (65000 K)
F2/CWF - cool white fluorescent lamplight (42000 K)
F11 - narrow band fluorescent lamplight (42000 K)

and Observer:

CIE 2^0 (1931) Standard Observer
CIE 10^0 (1931) Supplementary Observer

3.1 High Color and Ground Color, Uniformity, Darkness-Brightness

The lightness parameter L* of the base color and high color dependent on exposure to sun rays at growing. On the other hand, the low value of L* parameter indicates dark skin of observed fruit.

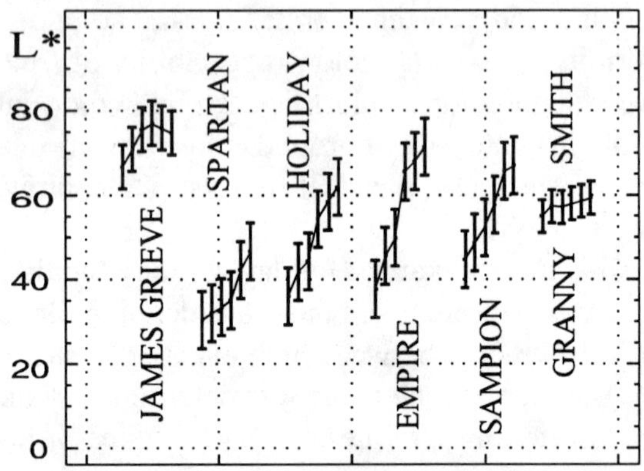

Figure 5. Distribution of brightness factor L* around the fruit of six apple cultivars

Note that the L* data ranged from 30 to 70 for studied cultivars (Fig. 5). Interestingly, the James Grieve and Granny Smith cultivars were the only fruit that produced a nearly constant L*. However, the James Grieve fruits were brighter and the darkest was Spartan. The Spartan apples had dark-purple skin and, therefore, gave low L* values.

The area of high color of Empire apple is smaller than for other cultivars and more than 180^0 around the fruit is differ to ground. In this case, the ground color sharply changed in to the high color. Darkness of these apples ranged from 36 to 70. The differences between high color and ground of

Holiday and Šampion apples were significant, however, gradually change of darkness from ground color to high color of blush was observed.

Most uniform are Granny Smith and James Grieve apples, where average color of the skin are at range from 66 to 76 and from 56 to 59 respectively. In both cases the high color was not observed.

3.2 Intensity of Color (Saturation of Red and Yellow), Repeatability in Crop

The chromaticity parameter a*, with value over 20 (up to 60) indicates the high red color affecting the cosmetic's appearance of fruits.

The saturation of red color as the index of chromaticity or a*, ranges from -15 to +40 for the cultivars tested (Fig. 6). James Grieve apples, having no red color, gave a* values very near to zero. Empire apples had a slight hue of green, while Granny Smith apples are completely pure green. Intensity of green color steady in this case indicates that the ground color is the same as high color of blush.

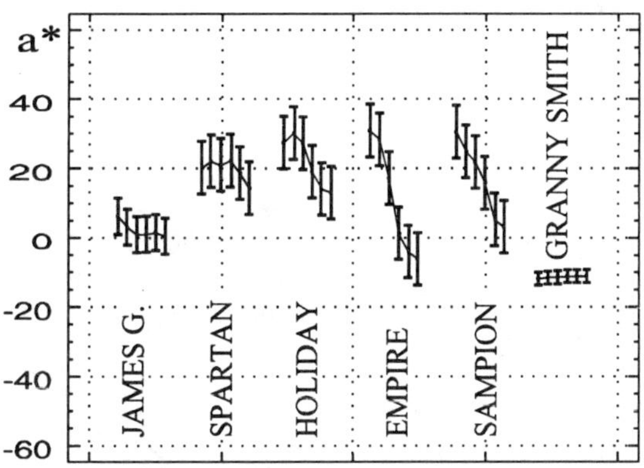

Figure 6. Distribution of chromaticity coordinate a* around the fruit of six apple cultivar

Repeatability in crop for Granny Smith apples is high and differences in population between fruits are low, indicating, good quality for marketing and handling. On the other hand, cultural preferences in most countries based on the opinion, that green color indicated not ripe, bitter and sour fruits, generally distasteful. However, good storability and high resistance to transport, shipping, and handling improves validity of this apple.

Spartan apples were saturated with red color that passed through the purple. The average of high color of Holiday, Empire, and Šampion apples range from 23 to 36 and intensity of red is similar. Any significant negative a* values indicate presence of green color in apples. Mature apples, except Granny Smith, exhibited no green color.

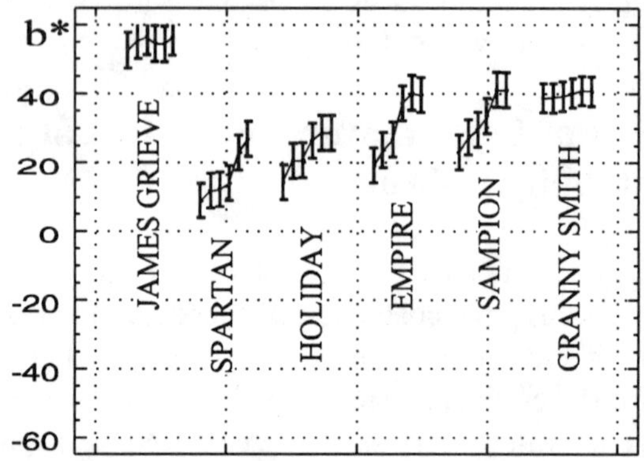

Figure 7. Distribution of chromaticity coordinate b* around the fruit of six apple cultivar

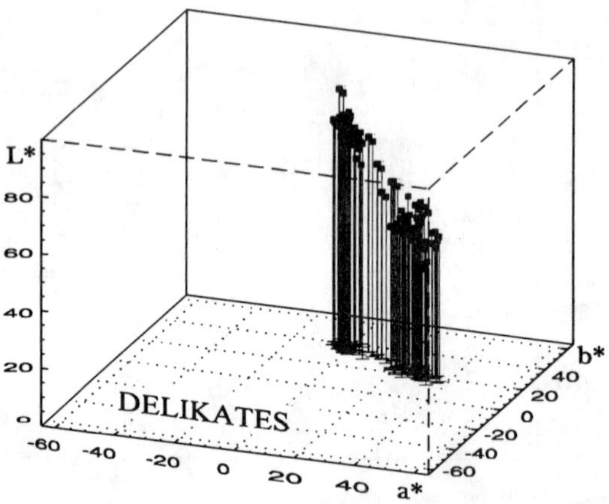

Figure 8. The cube of color coordinates of brightness factor L* and chromaticity a* and b* of Delikates apple

For the chromaticity parameter b*, all fruits are saturated with yellow color, with values of coordinate b* in the range from 18 to 58 (Fig. 7). James Grieve apples were significantly different from other apples in terms of index b*. The values of index b* for James Grieve apples were close to 60 indicating that the skin of the James Grieve apples was practically pure yellow in color. Granny Smith apples were uniformly close to 40 indicating same chromaticity factor b* of skin color.

Constant value of parameter b* as well as a* of this cultivar and any differences observed from high color to ground gave the color very uniform. Interestingly, a high value of b* coordinate observed for ground color is represented with low value of high color of coordinate a*.

Difference between cultivars is easy to observed on the cube of color coordinates, where factors of brightness L* and chromaticity coordinates of a* and b* are represented in 3-D figures.

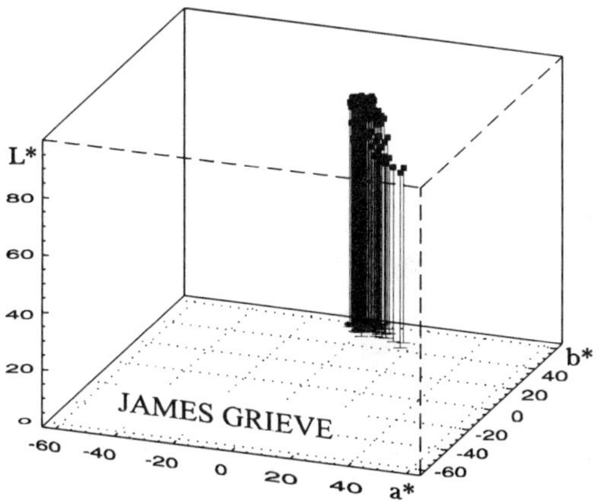

Figure 9. The cube of color coordinates of brightness factor L* and chromaticity a* and b* of James Grieve apple

Wide range of color coordinates of Delikates apples are shown in the figures 8. In comparison, figure 9 presents concentrated values of chromaticity coordinates represented yellow color and bright, what is exposed with high vertical position of measuring points. All parameters of color obtained in 3-D figures shows that L*a*b* indexes for several apples were not significantly different. Nonetheless, representation in 3-D figures based on the L*a*b* system should be useful for marketing since it would more clearly define consumer color preferences.

4. MATURITY, RIPENESS STAGE

Trough the years, authors have concentrated on the estimation of fruit quality based on maturity stage (Birth et al., 1983; Delwiche and Baumgartner, 1983; 1985; Fornes et al., 1994; Hoehn et al., 1996; Ihl et al., 1999; Jonhson and Luton, 1996; Saks et al., 1990. It was found that maturity stage was related to the fruit color. The L*a*b* system allowed to estimate a fruit color of Conference and Lukasówka pears at various ripening stages (kd2, kd4, and kd5). The brightness index L* showing intensity of Conference pear's color was for most fruits in the range close to 45. The pears of Lukasówka cultivar were bright at all stages of maturity, reaching value close to 60. The saturation of red color, presented as the index of chromaticity a*, ranges from 0 to 10 for pears harvested at earlier term. Only for later harvested Lukasówka pears a slightly red color was observed (black spot in figure 10). For mature pears the green color was not observed. The values of index b* for Lukasówka pears were close to 50, which indicates almost pure yellow color of fruit skin at maturity stage. However, the chromaticity coordinate a* is the most important factor of maturity appearance describing color of fruits.

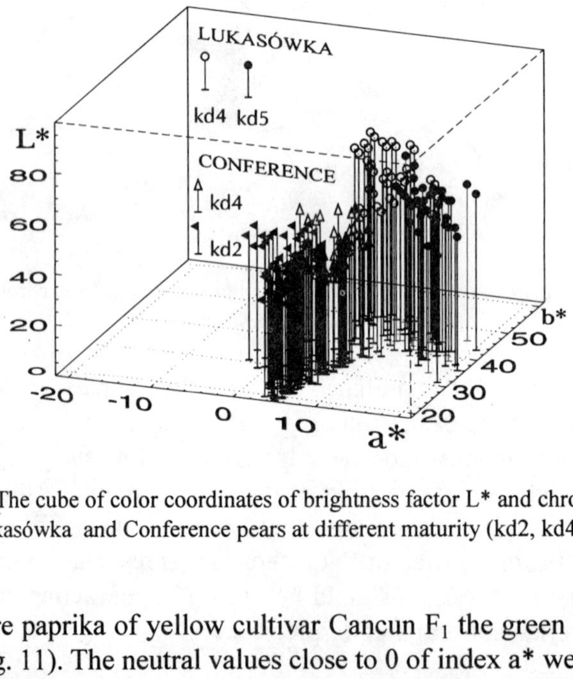

Figure 10. The cube of color coordinates of brightness factor L* and chromaticity a* and b* of Lukasówka and Conference pears at different maturity (kd2, kd4, and kd5)

For mature paprika of yellow cultivar Cancun F_1 the green color was not observed (Fig. 11). The neutral values close to 0 of index a* were noticed for unripe fruits what was connected with slightly green observed in the skin. On the other hand, this parameter for overripe yellow paprika has reached, on this scale, the level up to 20. It describes maturity stage with the orange color, which is composed of red and yellow.

Immature red paprika shows low value of parameter a*, what indicate unripe stage of Renegade F_1. Ripe and overripe paprika of this cultivar was characterized by high value of chromaticity parameter a* in the range from 33 to 38 (Fig. 11).

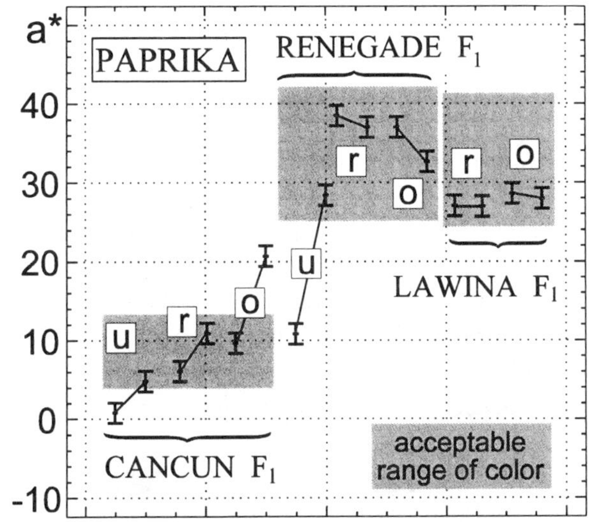

Figure 11. The chromaticity factor a* of Cancun, Lawina, and Renegade paprika F_1 at different maturity stage (unripe – u, ripe – r, and overripe – o)

The intensity of red color normally indicates full maturity and high ripeness. The parameter a* for Lawina F_1 cultivar ranges from 25 to 28. It seems that red color component of Lawina F_1 fruits is not high, however, the visual perception is contradictory. In this case, the human assessment of a deep red color is connected with darkness of fruits represented by low value of brightness parameter L*.

5. COLOR CHANGE AS A RESULT OF STORAGE, SHELF-LIFE, AND BRUISING

When apples are grown, it is possible to eliminate the negative influence of some hygiene and quarantine factors affecting fruit quality. After harvest, cosmetic appearance seems to be the most important quality factor. Frequently, the storage has great influence on final quality of products, affected by its appearance and color changes (Dobrzański et al., 2001; Henze, 1995; Johnson and Luton, 1996; Kameoka et al., 1994; Saks et al., 1999; Turk et al., 1997). Thus, sorting to select the fruit with the same level of high color of blush or for the same base color, could be most important factor in improving product quality and in influencing its price.

Some of products are more affected by storage conditions than others. Especially, shelf-life is the period of storage, where the conditions are very unsuitable and rudest i.e. a high temperature and not adequate humidity for keeping products in good quality. At this time, the quality of the product decreases and one of the most influential parameters of quality observed by consumer, such as external color of product changes, browning and darkening of skin occurs.

Frequently, the predicted color of mushroom is white, indicating freshness of product and being a factor of acceptability for consumption. Figure 12 presents coordinates of mushrooms' skin color determined with L*a*b* system. It is easy to conclude, that brightness parameter L* indicated darkening of mushroom after every day. All mushrooms' whiteness was not acceptable after 2 days of shelf-life, while after one day only half of the crop was slightly dark. After 5 days the pure white color has changed into brown, which brightness (darkness) is in the range close to 60.

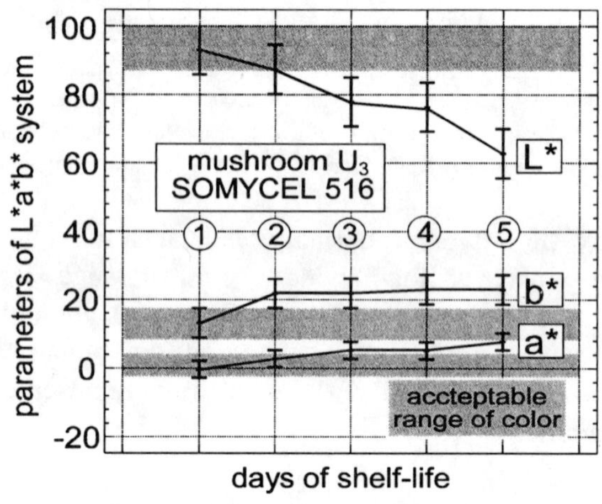

Figure 12. The coordinates of L*a*b* system and color changes of Somycel 516 mushroom U_3 at shelf-life

The chromaticity parameter b* changed value up to 20, after two days of shelf-life, however, remaining almost unchanged for all remaining days of shelf-life. In comparison, this value after harvest and one day of storage was low at 12.

The chromaticity parameter a* influenced the color of mushroom skin with red color changing the white color into pink-brown, thus, after 3 days of shelf-life the difference was significant.

Throughout the years, authors concentrated on finding an automatic method for detecting bruise damages on apples (Bennedsen and Qu, 1996, Studman and Li, 1997; Blahovec, 1999; Blahovec et al., 1997; Brusewitz and Bartsh, 1989; Dedolph and Austin, 1961; Garcia et al., 1994; Sinobas et al., 1991). Some of the methods based on visual light reflection (Ayuso et al., 1996; Chen and Mohri, 1997), light transmission (Kuczyński et al.,

1992; 1994; Worthington et al., 1976), acoustic impedance (Chen et al., 1992), vibration response (De Baerdemaeker et al., 1982), CO_2 development etc. The most promising results have been reported using visual light and digital image processing (Bennedsen and Qu, 1996). The bruises are observed, while the oxidation process produces brown color. However, visible volumetric increase after impact is related to the laps of time and properties of the skin (transparency and saturation of color).

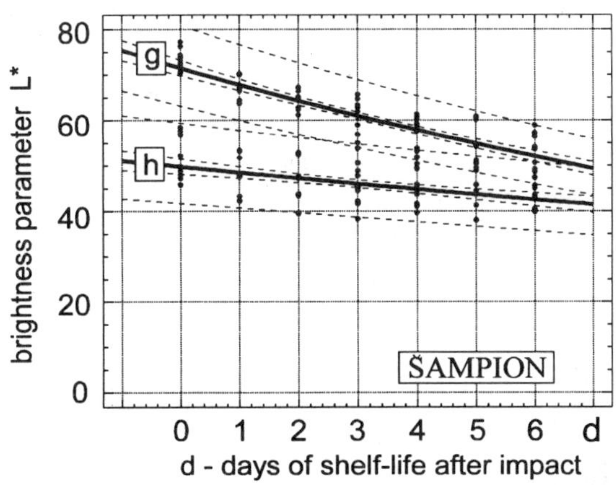

Figure 13. The changes of brightness factor L* of Šampion apple as a result of bruising represented by exponential function $L^* = 51.14e^{-0.026d}$ (R = -0.52) for high color of blush – h, and $L^* = 75.42e^{-0.0528d}$ (R=-0.87) for ground color - g

The impact caused bruising, what involve darkening of fruit skin of Šampion apple (Fig. 13). The average parameter of brightness L* of the ground color (g) decreases at shelf-life from 73.36 to 53.33, while from the opposite side the dark side of the blush changes the high color from 52.1 to 43.78. It is easy to conclude, that only bright side of fruit changes its color significantly (R = -0.87).

At that time, the red color represented by chromaticity parameter a* increases from 7.75 to 15.83 for ground and decreases from 35.49 to 27.8 for the high color (Fig. 14). The ground color of the skin after 3 days of shelf-life was significantly darker (R = 0.81).

The yellow color of both sides of fruit, represented by chromaticity parameter b*, decreases from 52.6 to 38.91 for ground and from 32.97 to 28.36 for the high color (Fig. 15). The ground yellow color of skin decreases more rapidly at shelf-life (R = -0.79) in comparison to high color (R = -045).

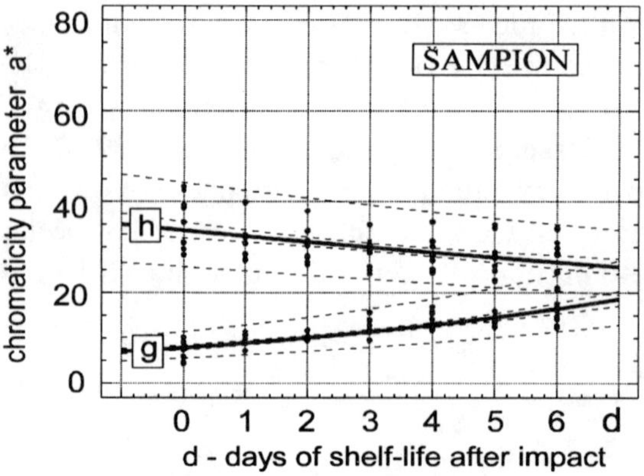

Figure 14. The changes of chromaticity factor a* of Šampion apple as a result of bruising represented by exponential function $a^* = 35.09\ e^{-0.039d}$ (R= - 0.51) for high color of blush – h, and $a^* = 7.11e^{0.121d}$ (R = 0.81) for ground color - g

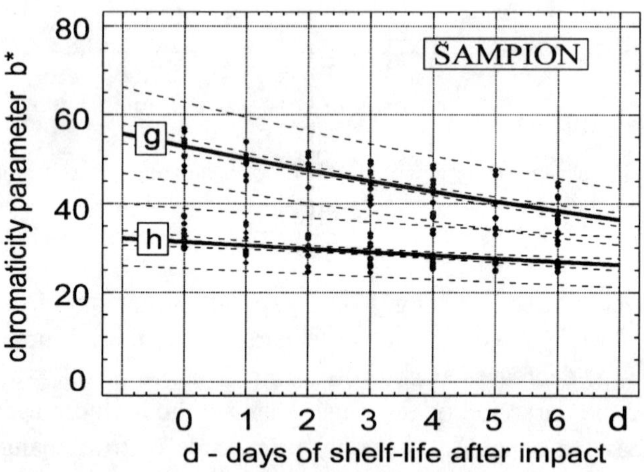

Figure 15. The changes of chromaticity factor b* of Šampion apple as a result of bruising represented by exponential function $b^* = 32.25e^{-0.026d}$ (R = -0.45) for high color of blush – h, and $b^* = 55.70e^{-0.053d}$ (R = -0.79) for ground color - g

The brightness parameter L*, as well as chromaticity coordinates a* and b* for the ground and high color after that time were at similar level. The color of bruised apples at shelf-life is more uniform, however, after 6 days of storage the appearance quality was unsatisfactory.

6. IMPROVING CUSTOMER SATISFACTION AND PERCEPTION OF COLOR

Labeling, packing and preparing agricultural products for market are important techniques in product handling operations. Fruits and vegetables are crops, which are harvested frequently at different time and different maturity stage.

As with other fruits, lemons and oranges are growing differently exposed to sun and are differ colored. Citrus fruits are frequently harvested as unripe, and reach full maturity just at handling or transport (Harrell et al., 1988). However, some fruits are still not ripe, have non-uniform color or skin appears green. Other fruits, well exposed to sun at growing, and over ripe, in short time are over saturated with red color showing as dark skin.

Because fruits of different maturity are characterized by non-uniform color, the sellers prepare special packing (plastic or vinyl color bags and nets) to improve appearance. Varietal color skin spectrum of lemon range from green to intensively saturated yellow. Oranges cover the range from slightly yellow to dark orange color. To improve customer satisfaction and to create more uniform perception of color, lemons are packed with yellow nets and oranges with red nets.

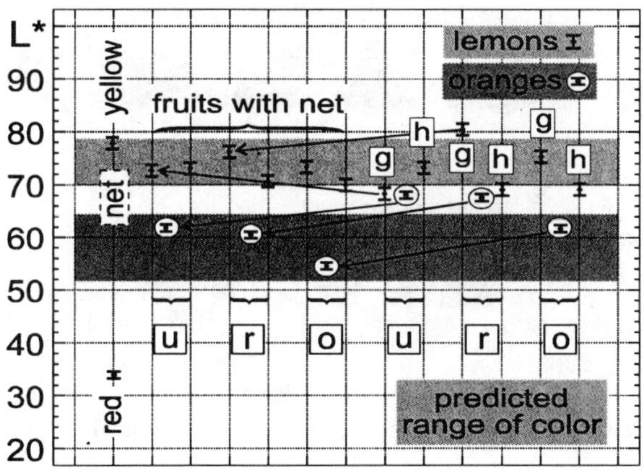

Figure 16. The influence of net color on predicted range of brightness factor L* of lemons and oranges (u – unripe, r – ripe, o – overripe, g – ground color, h – high color)

The red net used for packing of oranges causes decrease of brightness parameter L* for all maturity fruits (Fig. 16). However, the some applies too both bright fruits and unripe oranges, through red net they look more maturated, with parameter L* values in predicted range. On the other hand, unripe lemons with green color present become more bright, reaching predicted range of parameter L*.

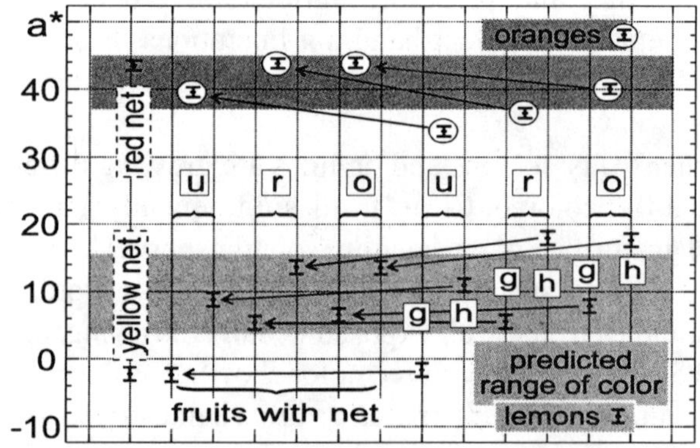

Figure 17. The influence of net color on predicted range of chromaticity parameter a* of lemons and oranges (u – unripe, r – ripe, o – overripe, g – ground color, h – high color)

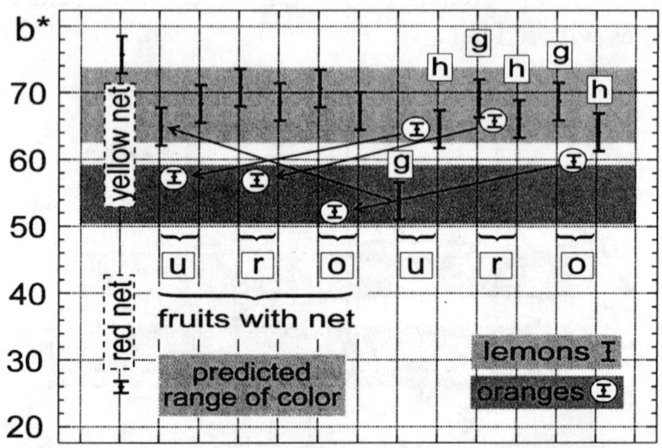

Figure 18. The influence of net color on predicted range of chromaticity parameter b* of lemons and oranges (u – unripe, r – ripe, o – overripe, g – ground color, h – high color)

The chromaticity parameter a* is the most influencing factor affecting the human perception of fruit color. Red net (Fig. 17) improves this parameter significantly, such unripe yellow oranges become more saturated changing perception of color in to predicted range. Ripe and overripe fruits become more red. The fruit color is insignificant to net color, however, is acceptable for human classification and quality assessment. Yellow net used for ripe and overripe dark lemons decreases this parameter into predicted color.

Using yellow net the brightness parameter b* describing coordinate of lemon's color did not change the perception for all ripeness stage except for the green side of unripe fruits (Fig. 18). On the other hand, using red net unripe oranges look less yellow and are classified as more mature.

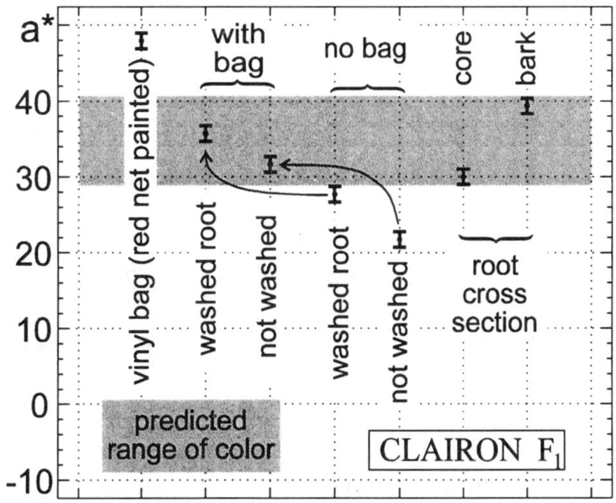

Figure 19. The changes of chromaticity parameter a* improving perception of carrot color

Carrot is a vegetable mostly used in salads and in cooked dishes, therefore, predicted color is based on visual assessment of cross section of carrot root (Fig. 19). The spectrum of red color in this case ranges from 30.0 (core) to 39.4 (bark). However, carrot is served frequently as tarnish roots, which chromaticity parameter a* reach merely 21.7. The surface of carrot root changes after washing, which improves perception of red color up to 27.7, being near the color of core. Using vinyl bag painted with red net, results in increase of chromaticity coordinate a* of unwashed roots to 31.6 and of washed roots to 35.7, which satisfies consumer perception of carrot's red color.

7. CONCLUSIONS

Estimation of fruit quality based on the system describing color should be useful for marketing allowing to check consumer preferences and assess quality of products after storage and at shelf-life. The L*a*b* system as additional measurement allows determination of the following parameters which characterize external properties of fruits and vegetables:

Base color and high color are dependent on exposure to sun rays at growing. Lightness parameter L* describes skin darkness and represents freshness of the product. Low value of L* parameter indicates dark skin of fruit. The change of this parameter as a result of storage or shelf-life depends on storage conditions or bruising development.

White color of mushroom represented by brightness parameter L* indicates freshness of product being a factor of acceptability for consumption.

Parameter a* shows the saturation of red affecting size of blush and intensity of high color, which affects the cosmetic' appearance and maturity stage.

Parameter b* indicates yellow color of skin, frequently observed in most mature pears, apples, lemons, oranges, and fruits without high color present. The negative value of this parameter indicates darkness of the skin, which results in the purple color on apples or green color on lemons and yellow paprika.

Frequently, during the range of storage the skin color of fruits and vegetables remains unchanged, however, as an important factor it must be included in any consumer quality estimation.

Determination of fruit quality based on L*a*b* system color should be useful for making decisions pertaining to the marketability of fruits and vegetables. This system, if probably integrated into a marketing plan, could improve appearance of fruits and vegetable, making consumers more aware of true quality factors.

REFERENCES

Achenbach, J. , 2001, The power of light. National Geographic, 10, 2-29.

Ayuso, M.C., Bravo, F.J., Ruiz-Altisent, M., and Barreiro, P., 1996, Sensing colour stability and mixtures of powder paprika using of optical reflectance and image analysis. *AgEng'96*, Madrid, **2**: 903-904. (paper 96F-049)

Alchanatis, V., Searcy, S.W., 1995, A selectable wavelength imaging sensor for multispectral inspection of agricultural products. *ASAE Paper* No. 95-3210. St Joseph, MI.

Birth, G.S., Dull, G.G., Magee, J.B., Chan, H.T., and Covaletto, C.B., 1983, An optical method for estimating papaya maturity. *J. Amer. Soc. Hort. Sci.* **109**(1): 62-66.

Bellon, V., Rabatel, G., and Guizard, C., 1992, Automatic sorting of fruit: sensors for the future. *Food Control*, 49-54.

Bennedsen, B.S. and Wei Qu., 1996, Detection of bruise damages on apples by thermal properties. *AgEng'96*, Madrid, **2**: 860-861. - (paper 96F-026)

Blahovec, J., 1999, Bruise resistance coefficient and bruise sensitivity of apples and cherries. *Int. Agrophysics*, **13**(3): 315-322.

Blahovec, J., Patocka, K., and Bares, J., 1997, Low-level bruising of stored apples due to quasi-static loading up to constant compression strain. *J. Texture Stud.*, **28**: 87-89.

Braive, 1994, Supercolor user's guide. Braive Instruments, Liège, Belgium.

Brusewitz, G.H. and Bartsh, J.A., 1989, Impact parameters related to post harvest bruising of apples, *Trans. of the ASAE*, **32**: 953-957.

Chen, H., De Baerdemaeker, J., and Vervaeke, F., 1992, Acoustic impulse response of apples for monitoring texture change after harvest. *Agric. Eng.& Rural Development*. Beijing, China, Vol. 1, **IV**: 30-38.

Chen, P., 1996, Quality evaluation technology of agricultural products. *Proc. of ICAME'96*, Seul, Korea, Vol. **1**: 171-190.

Chen, P., Ruiz, M., Lu, F., and Kader, A.A., 1986, Study of impact compression damage on Asian pears., *Trans. of the ASAE*, 30(4): 1193-1197. ASAE Paper no. 86-3025

Chen, P.M. and Mellenthin, W.M., 1981, Effects of harvest on ripening capacity and postharvest life of d'Anjou pears. *J. Am. Soc. Hort. Sci.*, **106**: 38-42.

Chen, P.M., Varga, D.M., and Facteau, T.J., 1993, Ripening behavior of „Gebhard" strain of Red d'Anlou pears after cold storage as influenced by harvest maturity and ethylene. *6th Int. Symp. on Pear Growing*, Medford, Oregon, 12-14.

Chen, Y.B. and Mohri, K., 1997, Image analysis of bruised Oorin apples. *5th Int. Symp. on Fruit, Nut, and Vegetable Production Engineering*, Davis, California, USA, **S12**(2): 1-6.

De Baerdemaeker, J., Lemaitre, L.L., and Meire R., 1982, Quality detection by frequency spectrum analysis of the fruit impact force, *Trans. of the ASAE*, 175-179.

Dedolph, R.R. and Austin, M.E., 1961, The evaluation of impact bruises on apple fruit. *J. Amer. Soc. Hort. Sci.*, **80**: 125-129.

Delwiche, M.J. and Baumgardner, R.A., 1983, Ground color measurements of peach. *J. Amer. Soc. Hort. Sci.* **108**: 1012-1016.

Delwiche, M.J. and Baumgardner R.A., 1985, Ground color as a peach maturity index. *J. Amer. Soc. Hort. Sci.*, **110**: 53-57.

Dobrzański, jr.B. and Rybczyński, R., 2000, Physical description of the fruit colour in apple quality grading. *Acta Agrophysica*, **37**: 17-27. - (in polish)

Dobrzański, jr.B. and Rybczyński R., 2001, Mechanical and optical properties of pears as parameters of stage maturity. *Acta Agrophysica*, **45**: 61-68. - (in polish)

Dobrzański, jr.B., Rybczyński, R., Dobrzańska, A., and Wójcik, W., 2001, Some physical and nutritional quality parameters of storage apple. *Int. Agrophysics*, **15**(1): 13-18.

Epson, 1995, *The theory and practice of color*. 5th ed., Düsseldorf: Epson Deutschland GmbH

Felföldi, J., Fekete, A., and Győri, E., 1996, Fruit colour assessment by image processing. *AgEng'96*, Madrid, **2**: 869-870 - (Report 96F-031).

Fornes, I., Ruiz, L.A., Juste, F., Molto, E., and Mehouachi, J., 1994, Study of variation of physicochemical properties of citrus fruits related to maturity. *XII CIGR, AgEng'94*, Milano, **2**: 907-908, (Report N. 94-G-072).

Francis, F.J., 1995, Quality as influenced by color. *Food Quality and Preference*. **6**: 149-155.

Garcia-Fernandez, J.L., Ruiz-Altisent, M., and Barreiro, P., 1994, Factors influencing apple and pear physical properties and bruise susceptibility. *AgEng'94*, Milano, **2** - (Report N. 94-G-062).

Good, H., 2002, Measurement of color in cereal products. *Cereal Foods World*. **47**(1): 5-6.

Guyer, D., Brook, R., and Timm, E., 1993, Lighting systems for fruit and vegetable sorting., USDA, Michigan State University: East Lansing, *Bulletin AEIS*, **618**.

Harrell, R.C and Levi, P., 1988, Vision controlled robots for automatic harvesting of citrus. *AgEng'88*, - Paper No. 88.426.

Harrell, R.C., Adsit, P.D., Pool, T.A., and Hoffman, R., 1988, The Florida robotic grove-lab. *ASAE Paper* No. 88-1578, St. Joseph, MI.

Henze, J., 1995, Storage and after ripening of pears. *Acta Horticulturae*, **398**: 159-166.

Hoehn, E., Dätwyler, D., and Gasser, F., 1996, Maturity indices to predict optimum harvest date for the storage of Conference pears in Switzerland. *Proc. of Work. Group on Optimum Harvest Date*, Lofthus, Norway, 149-156.

Hunt, R.W.G., 1987, *Measuring colour*. London: Ellis Horwood Limited, John Wiley&Sons.

Ihl, M., San Martin, A., and Bifani, V., 1999, Preliminary report on colour quality measured as chlorophylase activity in strawberries at different stages of maturity. *Acta Horticulturae*, **485**: 181-185.

Johnson, D.S. and Luton, M.T., 1996, Maturity indices to predict optimum harvest date for the storage of Conference pears in UK. *Proc. of Work. Group on Optimum Harvest Date*, Lofthus, Norway, 133-147.

Kader, A.A., 1983, Postharvest quality maintenance of fruits and vegetables in developing countries. *Postharvest Physiology and Crop Production*, Plenum: New York, pp.455-470.

Kader, A.A., 1999, Fruit maturity, ripening, and quality relationships. *Acta Horticulturae*, **485**: 203-208.

Kameoka, T., Hashimoto, A., and Motonaga, Y., 1994, Surface color measurement of agricultural products during post-ripening, *Color Forum Japan'94*, 11-14.

Kappel, F., Fisher-Fleming, R., and Hogue, E.J., 1995, Ideal pear sensory attributes and fruits characteristics. *HortScience*. **30**: 988-993.

Kavano, S., 1994, Present condition of nondestructive quality evaluation of fruits and vegetables in Japan. *JARQ*, **28**: 212-216.

Kuczyński, A., De Baerdemaeker, J., and Oszmiański, J., 1994, An optical reflectance method for studying the enzymatic browning reaction in apple. *Int. Agrophysics*, **8**: 421-425.

Kuczyński, A., Varoquaux, P., and Varoquaux, F., 1992, Reflectometric method to measure the initial colour and the browning rate of white peach pulps. *Sci. des Aliments*, **12**: 213-221.

Kappel, F., Fisher-Fleming, R., and Hogue, E.J., 1995, Ideal pear sensory attributes and fruits characteristics. *HortScience*. **30**: 988-993.

Lancaster, J.E., 1992, Regulation of skin color in apples. *Crit.Rev. Plant Sci.* **10**: 487-502.

Miller, B. K. and Delwiche, M. J., 1988, A color vision system for peach grading. *ASAE Paper* No. 88-6025.

McClure, W.F., 1995, Biological measurements for the 21st century. Keynote paper at *Agricultural and Biological Engineering Conference, "New Horizons, New Challenges"*, Newcastle, **3**: 1-9.

Molto, E., Aleixos, N., Ruiz, L.A., Vazquez, J., and Juste, F., 1996, An artificial vision system for fruit quality assessment. *AgEng'96*, Madrid, **2**: 956-957. - (paper 96F-078)

Motonaga, Y., Kameoka, T., and Hashimoto, A., 1997, Constructing color image processing system for managing the surface color of agricultural products. *J. Japanese Soc. Agric. Machin.*, **59**(3): 13-21.

Nielsen, H.M. and Paul, W., 1996, Colour image analysis for quality assessment of tomatoes. *AgEng'96*, Madrid, **2**: 863-864. - (paper 96F-028)

Paulus, I. and Schrevens, E., 1997, A methodology to study the interaction of external features of apples on human quality classification. *5th Int. Symp. on Fruit, Nut, and Vegetable Production Engineering*, Davis, California, USA, **S2**(1): 1-6.

Płocharski, W. and Konopacka, D., 1999, The relation between mechanical and sensory parameters of apples and pears. *Acta Horticulturae*. **485**: 309-317.

Saks, Y., Copel, A., and Barkai-Golan, R., 1996, Improvement of harvested strawberry quality by illumination: colour and Botritis infection. *Postharvest Biol. Technol.*, **8**: 19-27.

Saks, Y., Hofman, P.J., and Meiburg, G.F., 1999, Potential for improvement of mango skin colour during storage. *Acta Horticulturae*, **485**: 325-329.

Saks, Y., Sonego, L., and Ben-Arie, R., 1990, Artificial light enhances red pigmentation but not ripening of harvested „Anna" apples. *HortScience*, **25**: 547-549.

Sinobas, L.R., Ruiz-Altisent, M., and de la Plaza Perez, J.L., 1991, Bruise development and fruit response of pear (c.v. 'Blanquilla') under impact conditions. *J. Food Engineering*, **14**: 289-301.

Studman, C., 1994, Quality in fresh fruit meaning, measurement and maintenance. *XII CIGR, AgEng'94*, Milano, **2**: 897-898. - (Report N. 94-G-067)

Studman, C. and Ouyang, Li., 1997, Bruise measurement by image analysis. *5th Int. Symp. on Fruit, Nut, and Vegetable Production Engineering*, Davis, California, USA, **S12**(5): 1-7.

Studman, C., 1998, Ergonomics in apple sorting: a pilot study. *J. agric. Engng Res.* **70**, 323-334.

Thomason, R.L., 1986, High speed machine vision inspection for surface flaws, textures and contours. *Proc. Vision '86 Conf.* Detroit, Michigan. **5**: 51-61.

Türk, R., Eris, A., and Akbudak, B., 1997, Physiological and biochemical changes occur in pears (cv. „Deveci") stored at different temperatures. *Book of Abstracts of Int. Symp. on Effect of Preharvest and Postharvest Factors on Storage of Fruit*, Warsaw, 69.

Worthington, J.T., Massie, D.R., and Norris, K.H., 1976, Light transmission technique for predicting ripening time for intact green tomatoes. pp. 46-49. In: *Quality detection in foods*. ASAE Publication, 1-76. Amer. Soc. Agric. Eng., St. Joseph, Michigan.

Uncertainty of Measurements of the Fruit Impacts
by Artificial Fruit (PMS-60) in Tomato Handling Chains

VEERLE VAN LINDEN, MICHELE DESMET, PAUL DARIUS, JOSSE DE BAERDEMAEKER and HERMAN RAMON
Katholieke Universiteit Leuven, Leuven, Belgium

1. INTRODUCTION

During harvest and handling, fruits and vegetables suffer from different impacts, which may cause bruises and other quality degradations. Besides impacts, bruises can result from vibrations, static loads and friction. Today's high quality demands urge a screening of the entire handling chain from grower to consumer in order to locate critical parts responsible for fruit damage (Hilton, 1998).

A widely used method to measure the mechanical load exerted on fruits and vegetables, is the use of an artificial fruit. This fruit is a device, resembling biological commodity in shape, dimensions, weight and mechanical properties. The device is placed in the processing chain as a regular fruit. In general, two systems are used for research purposes: an instrumented sphere (IS) measuring accelerations in three directions and a pressure measuring sphere (PMS) registering static and dynamic forces acting on the fruit surface.

Zapp *et al.* (1990) used an IS to estimate fruit bruising resulting from various impacts during handling and transportation. Brown *et al.* (1990) characterized cushioning material in packing lines by means of an IS. Jourdain *et al.* (1992) compared several apple packing lines with an IS, as did Ragni & Berardinelli (2001). Herold *et al.* (1993) compared two types of electronic fruit with respect to their mechanical stress-strain behaviour and the characteristics of the obtained data. García *et al.* (1994) evaluated fruit damage along the handling process utilizing two types of electronic fruit.

This research focuses on the PMS-60 (Institute of Agricultural Engineering, Bornim, Germany).

Forces during impact are mainly dependent on the energy of impact, (electronic) fruit properties and characteristics of the impacting body. It is assumed that the PMS-60 is a homogeneous fruit with isotropic fruit characteristics, which implies the same forces to be measured for equal impact characteristics. This would imply that the place on the artificial fruit where an impact is given should not be of importance with regard to the force output.

Since one cannot control the impact location on the fruit during on-line measurement, the main goal of this report is to get insight into spatial variation of force readings measured by the PMS-60. Further, the accuracy of the electronic fruit for measuring fruit impacts will be investigated as well.

This report comprises two separate topics: (1) the investigation of the uncertainty of measurement of fruit impacts by a PMS-60 artificial fruit, and (2) analysis of a tomato sorting line by means of a PMS-60 artificial fruit.

Both the experiment on the uncertainty of measuring impacts and the on-line use of the PMS-60 in a tomato handling chain will be described, analyzed and discussed.

2. MATERIALS AND METHODS

2.1 Materials

The PMS-60 is a pressure measuring device capable of measuring both static and dynamic loads above a pre-set threshold. It is a 62 ± 1 mm diameter ball and weighs 181 ± 5 g. Its outer layer is a 4 mm thick rubber skin. An inner 42 mm diameter electronic unit is centered by means of 16 conical steel springs. The space between the inner and outer ball is filled with silicon oil. The inner ball contains all the electronic parts, including: a pressure sensor, a central processing unit (CPU), 4 NIMH (Nickel Metal Hydride) accumulators and a communication port. The sphere is connected to a power supply device that is connected to the PC to act as an interface between PC and PMS-60 (Herold, 1994).

Measuring conditions can be specified such as sampling frequency, primary sampling frequency, duration of measurement, threshold value, delay before measurement... Sampling frequency is raised to primary sampling frequency once the threshold is surpassed. Sampling frequency can be up to 3.33 kHz while primary sampling frequency can be up to 10 kHz. Impact data are registered as force in Newton. The measuring range for static load is 0-100 Newton, for dynamic load it is 0-400 Newton (corresponding to a drop height of 1 m onto a concrete surface).

Forces on the sphere surface cause an internal hydraulic pressure that is measured by the sensor and stored in the enclosed data logger if a pre-set threshold is exceeded. After data collection the measured data are downloaded to a Personal Computer and analyzed. The analyzed data can be used as indicators of damage risk.

The pendulum as described by Baheri (1997) is a device designed to conduct controlled impact experiments. The main body of the pendulum is a heavy metal construction with a platform and a wooden arm that is able to oscillate about a pivot shaft. The platform is equipped with a metal plate that can be moved up and down and on which the test specimen is mounted and kept stationary by means of an elastic band. The upper end of the arm is tooled up with an angle sensor; the lower end - the so-called arm head - is equipped with an impactor, a force sensor and an accelerometer.

For the experiment, an aluminum spherical impactor was used. The electronic fruit was fixed in such a way that it just touched the center point of the pendulum arm head when the arm head was at its lowest point of oscillation. Variation of the angle of impact corresponds to a variation of impact energy.

2.2 Methods

In order to investigate the uncertainty of measurement of fruit impacts with the PMS-60, measurements at 5 levels of impact were repeated 3 times at 12 different spots on the fruit by means of a pendulum.

An equidistant grid of 12 points was laid on the fruit, according to the vertices of an icosaeder, to mark the test spots. Levels of impacts are chosen in the range of impacts that occur in fruit handling chains. Impact angles of 5°, 10°, 15°, 20° and 30° corresponding to impact forces of approximately 10, 20, 30, 40 and 60 Newton were used. For each impact level, the 36 measurements were completely randomized.

The experiment was designed to permit only one source of variation at a time, i.e.: for a given impact level, impact characteristics and energy of impact were kept constant. Variation in measured force could only result from spatial variation in fruit characteristics.

In the remainder of the article, forces measured by the PMS-60 will be referred to as *PMS-60 output*; forces measured by the pendulum force sensor will be referred to as *pendulum output*. Following tasks were carried out and will be discussed:

1. Forces measured by the artificial fruit are analyzed with respect to the place of impact.
2. The 3 repeated measurements per spot are averaged and minimum and maximum deviations (%) from the average PMS-60 output based on 36 observations are calculated for all 5 levels of impact.
3. 95% confidence intervals (CI's) are determined per level of impact (Eq.(1)). New PMS-60 outcomes for impacts at that particular level, which are randomly distributed over the fruit surface, will be within the interval limits with 95% certainty.

$$95\%CI = \bar{x} \pm 1.96\sqrt{\frac{MSE}{n}} \tag{1}$$

<u>Explanation of symbols:</u> *$\bar{x}$ = average PMS-60 output for a particular impact level, based on 36 values (12 spots, 3 repetitions); $\sqrt{MSE}$ = standard deviation of PMS-60 outputs for a particular impact level, based on 36 values; n = number of impact measurements.*

1. PMS-60 outputs and pendulum outputs are compared in order to explain the variance in PMS-60 outputs throughout the surface. Since the pendulum sensor measures the 'real force' determined by the energy of impact and fruit and impact characteristics, a variation in pendulum outputs can only be due to a local variation in electronic fruit characteristics (cfr. experiment design). No variation would imply that the fruit characteristics of the artificial fruit are homogeneous.
2. An on-line measurement of fruit impacts in a tomato sorting line was carried out in order to locate parts in the chain where fruit damage might occur. Critical events were recorded 10 times to obtain a reliable average impact value. Pre-set thresholds varying between 10 and 12 Newton were selected. Maximum and average impact forces are plotted against impact events. Impact measurements below the threshold were represented by the threshold value.

   In the sorting line, following events can be distinguished:

- **A:** emptying a box with freshly picked tomatoes on a conveyor belt, made up of cushioned rollers

- **B**: changeover of the fruits to a subsequent conveyor belt, associated with a 10 cm drop
- **C**: a 10 cm drop from the conveyor belt onto a separating belt, i.e.: a belt specifically designed to separately lead the fruits onwards
- **D**: drop from single fruits into plastic cups that pass through the sorting machine where fruits are weighed and assigned to appropriate weight classes
- **E**: while being weighed the cups are lifted slightly
- **F**: cups are turned over and fruits fall from a height of 25 cm onto a small rubber conveyor belt
- **G**: fruits finally tumble from small conveyor belt onto the table where they are put into boxes.

3. RESULTS

Fig.1 shows the average PMS-60 output of the 12 spots for each level of impact. The average PMS output for a particular impact energy is based on 36 observations. Observations at spots 1, 3, 7 and 12 are systematically higher than the average output while observations at spots 2, 5, 10 and to a lesser extent 9 are systematically lower. Observations at spots 4, 6, 8 and 11 are situated around the average output.

Minimum and maximum average deviation from the average PMS-60 outputs are shown in Fig.2. The 0 % deviation line represents the average PMS-60 output around which average minimum and maximum deviations are scattered. Most extreme output values differ approximately 19, 23, 40, 30 and 24% for the respective impact levels.

Table 1 lists 95% confidence intervals for all five levels of impact for on-line impact measurements by means of the PMS-60 (n = 10). A new observation at that particular impact level will fall within the interval with 95% certainty. Obviously, CI's for lower impacts (10-20N) do not contain that imposed impact value while higher impacts have CI's that are centered around these values. The higher the impact, the broader the 95% confidence interval. This implies a bigger range of possible outputs, and hence a larger variation among measurements.

Fig.3 shows the PMS-60 output in function of the pendulum output for each measurement. Per impact level, pendulum outputs differ and are distributed around their average. All pairs are scattered around the diagonal, but for lower impacts, they are situated somewhat above it. In general, for low/high PMS-60 outputs, pendulum outputs tend to be low/high as well.

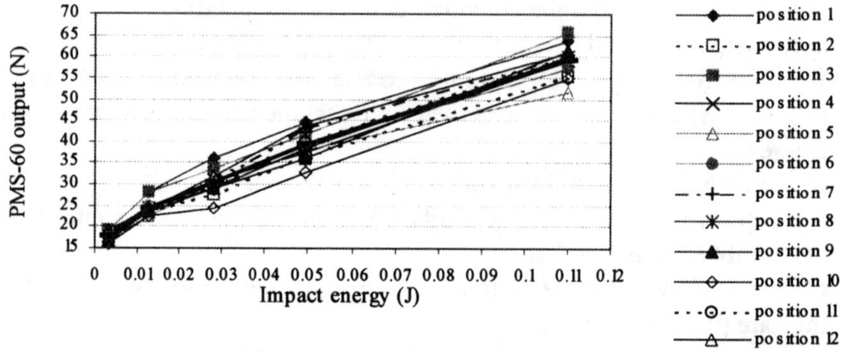

Figure 1. Average measured impact force for 5 levels of impact at 12 separate locations on the fruit.

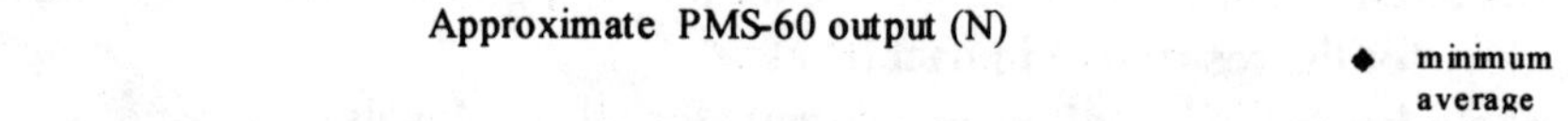

Figure 2. Minimum and maximum deviation from average PMS-60 output per impact level. Difference in impact location accounts for different PMS-60 outputs.

Table 1. 95% confidence intervals for on-line measurements with the PMS-60.

Approximate impact level (N)	95% confidence interval (N)	CI width (N)
10	[16.89; 18.72]	1.83
20	[23.11; 26.05]	2.94
30	[28.42; 32.14]	3.72
40	[37.00; 41.78]	4.78
60	[56.27; 62.34]	6.07

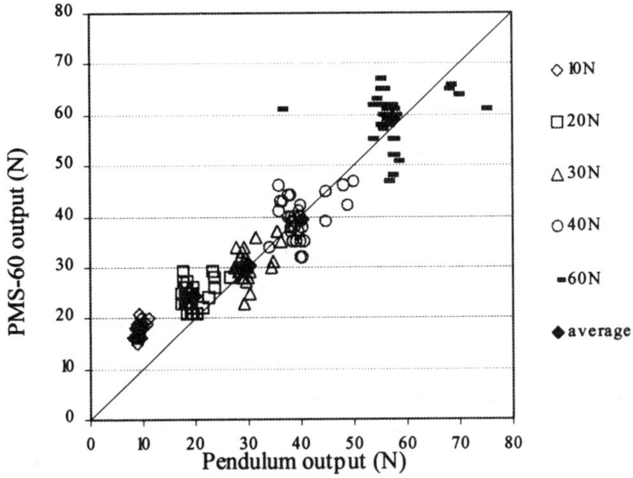

Figure 3. Outputs of PMS-60 compared to outputs of pendulum force sensor.

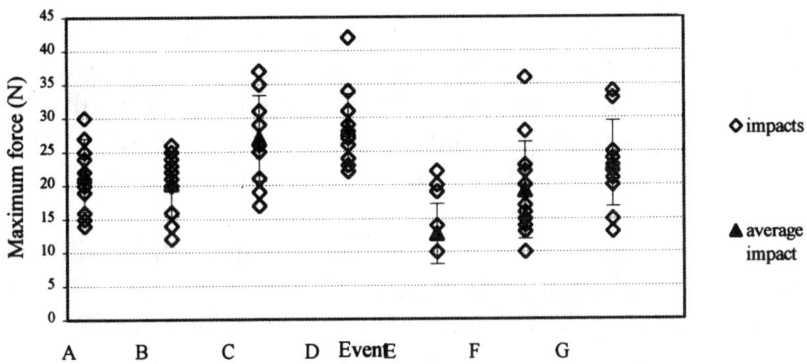

Figure 4. Maximum impact forces for critical events in a tomato sorting line.

On-line measured maximum impacts for critical events in a tomato sorting line are shown in Fig.4. Average impact and standard deviation are plotted for each event. 95% confidence intervals for the impact forces relevant to the actual sorting line are shown.

4. DISCUSSION

According to Fig.1, locations on the fruit surface can be divided into three classes: spots with respectively higher, lower and average force readings. All spots per class are more or less equally distributed over the surface. The experiment clearly reveals that the PMS-60 is a non-homogeneous fruit with anisotropic mechanical properties. According to the inner construction of the fruit, there are areas on the fruit surface that are connected to the center by means of a spring. It is possible that these areas are responsible for higher force readings because the displacement of the spring is there at the highest and so is the measured force. Fig.3 seems to support this hypothesis. Pendulum outputs and PMS-60 outputs follow the same trend: lower PMS-60 values correspond mainly with lower pendulum outputs. The same holds true for higher PMS-60 values. What the fruit experiences as higher impacts is also registered as higher impact forces by the pendulum force sensor. Both different force-measuring systems give outputs that vary in a similar way over the fruit surface. In conclusion, the source of variation of PMS-60 outputs is not a coincidence but a consequence of the anisotropy of the inner construction of the fruit.

Fig.2 illustrates the sometimes large deviation of the extreme PMS-60 outputs from the average output. It needs no comment that multiple measurements are necessary in order to reliably characterize a critical event in terms of maximum impact force. Similar conclusions were drawn by Herold *et al.* (1996).

Repetition of measurements in a tomato sorting line will lead to a variation in force outputs. Two main factors are responsible for this variation: the anisotropic fruit properties and the natural variation in impact force of a certain event. A single event after all is rarely related to a particular impact force. For example: when a tomato drops, it either falls freely or on top of another fruit. In both cases, the event-related impact value ('event-impact') will be different. The natural variation in event-impact brings about the largest source of variation in on-line measurements.

While natural variation in event-impact force cannot be predicted, variation due to anisotropic fruit properties can be taken into account. Given a fixed event-impact, its 95% CI represents a range of possible PMS-60 outputs that all refer to that impact force. From Eq. (1) it is clear that more measurements ($n>10$) will narrow the CI and will give a more accurate approximation of the event-impact.

Lower forces have CI's that do not contain the force itself. It must be mentioned that these forces do not belong to the measuring range of the fruit. Hence, the accuracy of measurement is much less. Nevertheless, impacts of 10 to 20 N frequently occur in fruit and vegetable processing chains and can be harmful as well. Sufficient attention must be paid when it comes to the interpretation of these lower PMS-60 outputs.

Higher forces have broader confidence intervals. This could be due to the fact that higher forces were sometimes inevitably associated with a small displacement of the fruit upon impact. It is a plausible explanation for the relative larger standard deviations and resulting broader CI's.

Fig.4 reveals that the on-line measurement is characterized by a large variation in fruit output for each event. Different measurements for a single event seem to belong to different CI's. Again, the large natural variation in event-impact is responsible for this. The average PMS-60 output is a reliable approximation for the event-impact.

Higher impacts occur at the changeover from the conveyor belt to the separating belt (C) and when the fruit is dropped into a plastic cup (D). Event E, where the cups are lifted slightly for weighing the tomato, has an almost non-detectable event-impact force; several measurements had forces less than the threshold. It must be noted that most of the event-impacts are no part of the measuring range of the fruit. As a consequence, translating the on-line measured impact forces in real event-impact forces is very difficult.

5. SUMMARY

Equal forces exerted on the fruit surface differ with location on the fruit. It seems that areas on the surface that are connected to the center by means of a spring give systematically higher readings for an equal impact force while areas with systematically lower readings are present as well. The idea is that the local visco-elastic properties of the fruit are somewhat different throughout the surface, causing deviation from homogeneity and isotropic behaviour of the fruit.

Impact measurements with the PMS-60 in a tomato sorting line approximate the real event-related impact forces if each event is measured several times. The average PMS-60 output for an event is a reliable value for the event-impact force that has taken into account the variation in output due to inhomogeneity of the fruit.

6. CONCLUSIONS

The PMS-60 is an appropriate evaluation tool for measuring impacts on-line. Attention must be paid when it comes to the interpretation of lower force readings that refer to impact forces that are no part of the fruit's measuring range. Inhomogeneity of fruit characteristics can give rise to large variation between the measurements. This can easily be compensated for by conducting multiple, independent measurements and averaging them.

ACKNOWLEGDEMENTS

This research was financed with a specialisation bursary of the Flemish Institute for the Promotion of the Scientific-Technological Research in Industry (IWT). The authors also wish to express their gratitude to the Ministry of Small Trade and Agriculture of Belgium.

REFERENCES

Baheri, M., 1997, *Development of a method for prediction of potato mechanical damage in the chain of mechanized potato production.* Doctoral thesis Nr. 342 at the Faculty of Agricultural and Applied Biological Sciences, K.U.Leuven, Belgium, pp. 1-301.

Brown, G.K., Schulte-Pason, N.L., Timm, E.J., 1990, Impact classification using the instrumented sphere. *Meeting Presentation of the American Society of Agricultural Engineers*, Paper No. 90-6001, pp. 1-8.

García, J.L., Barreiro, P., Ruiz-Altisent, M., Vicente, M., 1994, Use of electronic fruits to evaluate fruit damage along the handling chain. *Report of Agricultural Engineering* No. 94-G-045: 1-7.

Herold, B., Siering, G., Truppel, I., 1993, Comparison of measuring spheres as means to determine mechanical stress on fruits during handling. *IV International Symposium on Fruit, Nut and Vegetable Production Engineering,* March 22-26, 1993, Valencia-Zaragoza (Spain), pp. 35-42.

Herold, B., Truppel, I., Siering, G., Geyer, M., 1994, *User Manual, PMS-60.* Institute of Agricultural Engineering, Bornim, Federal Republic of Germany, pp.1-26.

Herold, B., Truppel, I., Siering, G., Geyer, M., 1996, A Pressure Measuring Sphere for monitoring handling of fruit and vegetables. *Computers and Electronics in Agriculture*, **15**: 73-88.

Hilton, D.J., 1998, *Impact and vibration damage to fruit during handling and transportation.* National Centre for engineering in agriculture. Faculty of engineering and surveying, University of Southern Queensland, Toowoomba, Queensland 4350, Australia., pp 116-126.

Jourdain, J.M., Vaysse, P., Mora, P., Cammal, P., Villes, C., 1992, *Apple bruise damage evaluation on packing lines using an instrumented sphere.* Centre Technique Interprofessionnel des fruits et légumes, 22 rue Bergère, 75009 Paris, France, pp. 99-106.

Ragni, L., Berardinelli, A., 2001, Mechanical behaviour of apples, and damage during sorting and packaging. *Journal of American Engineering Research*, **78** (3): 273-279.

Zapp, H.R., Ehlert, S.H., Brown, G.K., Armstrong, P.R., Sober, S.S., 1990, Advanced instrumented sphere (IS) for impact measurements. *Transactions of the American Society of Agricultural Engineers*, **33** (3): 955-960.

Thermal Imaging

A Promising High-Tec Method in Agriculture and Horticulture

HANS J. HELLEBRAND[1], HORST BEUCHE[1] and MANFRED LINKE[2]
[1]*Department of Technology-Assessment and Substance Flow*
[2]*Department of Horticultural Engineering*
Institute of Agricultural Engineering Bornim (ATB), Potsdam, Germany

1. INTRODUCTION

Thermography has become an important tool in engineering, in medical research, and in other disciplines in the course of the last decades. There are many engineering problems and research tasks where temperature and temperature distributions are the source of information and control. New infrared sensors and powerful computers changed thermal imaging from extraordinary military method into a routine tool in science and daily engineering practice. And this development continues. Night vision will increase traffic safety of cars within the next years and young engineers promise $1000 IR-cameras on the market in the near future (Grietens 2001).

What kind of applications are visible and practicable in agriculture and agricultural engineering?

All agricultural and horticultural objects or processes, where heat is generated or lost can be analysed by thermography regarding temperature gradients in space and time. Such applications include:

- Health and fertility diagnostics in animal husbandry (Mastitis, Oestrus, etc.)
- Design and control of heat insulation, air conditioning and air flow in animal houses, storage houses, detection of cold storage cooling loss

- Thermal inspection of electrical installations to find overheating in connections and components, or faulty components, especially at places with inflammable materials
- Thermal inspections of motors, pumps, fans, bearings, compressors, heating elements, humidifiers and similar components can be analysed for faults and decrease in efficiency.
- Moisture contamination evaluations in buildings and stored materials (e.g., grain, hay)
- Site specific classification of soil moisture content
- Site specific evaluation of field plant water stress and disease
- Evaluation of forced and natural drying of plant materials and plant products
- Evaluation of transpiration loss and changes in turgor pressure of sensitive horticultural products.

The water status and water content of fruits and vegetables can influence the properties of biological products. The loss of water due to transpiration depends on the type and state of fruit or plants (size, shape, skin, stage of development, etc), the temperature and humidity of the surrounding air, and on the airflow around and against the object. A temperature decrease of the surface will be observed when the heat transmitted from the product for the transformation of water into vapour exceeds metabolic heat generation. Thermal imaging can measure local surface temperature changes.

2. BASIC PRINCIPLES

An imaging camera is utilised normally as a "two-dimensional" thermometer. In reality, it measures the spatial distribution of the thermal radiant flux (radiation flux or radiation intensity) at a limited wavelength band (3 - 5 μm or 7 -14 μm, dependent on type of sensor). This two-dimensional distribution of the radiant flux of a three-dimensional object is then converted into a temperature distribution. At least two main error sources limit the accuracy of the temperature determination:

- The total radiant flux E of an object with the surface area A (or the radiant flux density $\Phi = E\ A^{-1}$) depends on the emissivity ε and on the temperature T (Stefan-Boltzmann law)

$$E = \sigma\ \varepsilon\ T^4\ A \qquad or \qquad \Phi = \sigma\ \varepsilon\ T^4 \tag{1}$$

with σ a proportionality constant (dimension factor). Therefore, the emissivity of the object in the waveband of interest is needed to convert radiation data into temperature. Usually a "global" emissivity of $\varepsilon = 1$ or $\varepsilon = 0.95$ is used. The value of $\varepsilon = 1$ (blackbody radiation) is chosen in cases, where the measurement of temperature gradient is of importance and not the exact temperature. A more realistic approximation uses $\varepsilon = 0.95$, since many non-metallic surfaces show an emissivity above 0.8. The thermal results based on global emissivity settings without local corrections are called apparent temperatures. For accurate measurements, the emissivity of the objects must be known. Until now, only few data on emissivity of agricultural products was available from literature. Standard handbooks like Ebert 1962, King 1987, LaRocca 1996, Schuster or Kolobrodov 2000 list the emissivity of water, wood, paper and some other organic materials. The emissivities of these materials are in the range between 0.80 and 0.95.

- The thermal radiation of a body is a superposition of direct (specular) radiation and of diffuse radiation. The radiation intensity depends on the angle between measuring direction and the normal vector to the surface under consideration. If the surface would emit thermal radiation with the intensity distribution according Lambert's (cosine) law, then even an arbitrary shaped body would give equal temperatures at all pixels at the projection on the measuring sensor plane. Because of the superposition of direct and diffuse radiation, the surface parts with strongly deviating normal vectors from the measuring direction appear to be cooler. The share of direct radiation of water, the main component of fruits and vegetables, depends on the thickness of the water layer (Haußecker 1996). Only those surfaces of fruit will be measured correctly that are nearly parallel to the sensor plane of the camera. This effect is visible in thermal images of spherical fruits, e. g. apples or tomatoes. The temperature at the outer circumference seems to be lower than at the centre of the fruit. Additionally, thin laminar air layers superpose this effect. Across these air layers we find the temperature gradient caused by heat exchange between objects under consideration. Since several effects (type of radiation, surface roughness and reflectance/absorption at 10 μm wavelength, and airflow rate) produce the final apparent temperatures, an exact temperature calculation is hardly possible for image pixels standing for radiation data at surface parts with normal vectors deviating strongly from the measuring direction. This difficulty can be overcome in thermography applications by using evaluation areas and spots excluding these critical parts of the thermal images.

2.1 Literature Review

Canopy and plant leaf temperatures have been studied for about 160 years. The application of thermography for the study of plants started about 40 years ago (Monteith and Szeicz 1962, Tanner 1963). There is an increasing number of publications in which plants or plant materials are studied by thermal imaging. Infrared vision technique was applied for the determination of physiological depression in crop plants (Inoue 1990), leaf transpiration rates (Inoue et al. 1990), stomatal conductance (Jones 1999) and single cell studies (Zohar et al. 1998). Differences in the flag leaf temperature of cereals, caused by diseases and wind gusts, were observed by thermography measurements (Nilsson 1995, Daley 1995). Thermal imaging was used for visualisation of plant-virus interactions. The presymptomatic temperature increase of plant leaves, as resistance to tobacco mosaic virus, was detected thermographically before any disease symptoms became visible on tobacco leafs (Chaerle et al. 1999).

The viability of seedlings of Scots pine (Pinus sylvestris L.) and Norway spruce (Picea abies (L.) Karst.) was tested by infrared thermography. Significant positive correlation was found between seedling temperature and degree of damage. The warmest seedlings had a lower survival rate as a group when compared with the remaining seedlings. There was significant negative correlation between seedling temperature and annual height increment in the first growing season (Egnell and Orlander 1993).

Infrared video thermography has recently been used to visualise ice nucleation and propagation in plants. The patterns of ice propagation are largely explained by the presence of particular anatomical features in the leaf and fruit. Cranberry leaves are hypostomatous; therefore, ice penetration into the plant is prohibited on the adaxial surface of the leaf. Ice penetration is possible on the abaxial surface, presumably via stomata. Mature fruits appear to be able to supercool by decreasing or eliminating the vascular connections in the pedicel, as well as by developing a compact and pigmented cell layer at the calyx end of the fruit (Wisniewski et al. 1997 & 1999; Workmaster et al. 1999).

First attempts have been performed to analyse the quality of horticultural products and to detect bruises in apples (Danno et al. 1978, Miyazato et al. 1981, Beverly et al. 1987, Tollner et al. 1993, Varith 1999), and to estimate the health state of plants (James and Aldous 1997, Moran 2000). At present, new applications in agriculture and horticulture are visible like designing and control of storage houses and vegetable displays, quality evaluation and control of sensitive biological products (Linke et al. 2000).

2.2 Materials and Methods

For calibration and emissivity studies, the IR imaging camera Varioscan™ 3021-ST (Jenoptik Technologie GmbH) serves for the measurement of the radiant flux. The software IRBIS Plus V2.0™ (InfraTec GmbH Dresden) enables inputs of emissivity as "global" for the whole image, and as "local" for a definite area or a free definable pixel range. A Stirling engine cools the sensor of this camera. The experimental examinations of different kinds of horticultural produce were carried out earlier by the infrared thermography system Varioscan 2011™ (Jenoptik Technologie GmbH) with nitrogen cooling and software IRBIS V2.0™ and ThermoView V1.0™ (InfraTec GmbH Dresden). The infrared sensor of both systems (MCT - *M*ercury *C*admium *T*elluride, a photodetector alloy) works in the 8-12 μm wavelength range with a germanium optic. Temperature resolution is about 0.03 K at 300 K. All three versions of image analysis software determine the temperature of single points as well as the average temperature along lines or adjustable surfaces. Temperature profiles in two perpendicular axes can be made visible, and different isothermal areas can be analysed within freely adjustable temperature limits.

3. RESULTS AND DISCUSSION

3.1 Emissivity of Fruits

The principle for emissivity measurements is to use a calibrated marker at the surface of the fruit to get the true surface temperature. Knowing the true temperature and the radiant flux, the emissivity can be derived. For calibration, markers of different materials are measured at the surface of a temperature-controlled aluminium ingot containing a blackbody radiator (black hole C01 in Fig. 1).

The temperature of the aluminium ingot is regulated by an internal water bath. The emissivity of the blackbody radiator is measured by dynamic temperature measurements (heating curves and cooling curves), as it is necessary to have temperature contrasts. The measurement error increases at low contrast (near room temperature) because of small signal-to-noise ratio (Fig. 2).

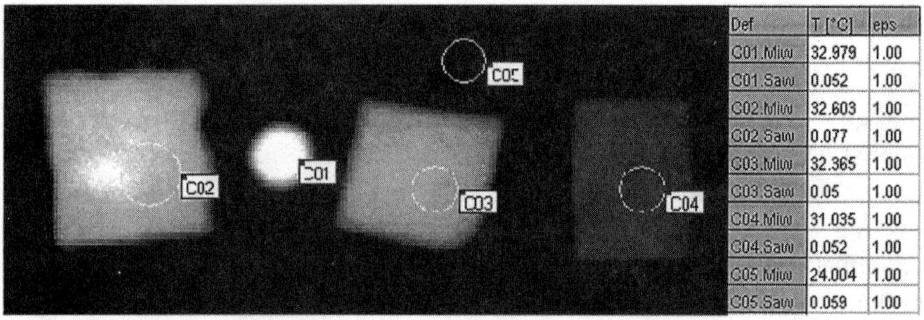

Figure 1. Thermal image and apparent temperatures (ε = 1) of calibration measurements (C01: blackbody radiator; C02: white paper; C03: adhesive paper; C04: adhesive aluminium film; C05: aluminium ingot).

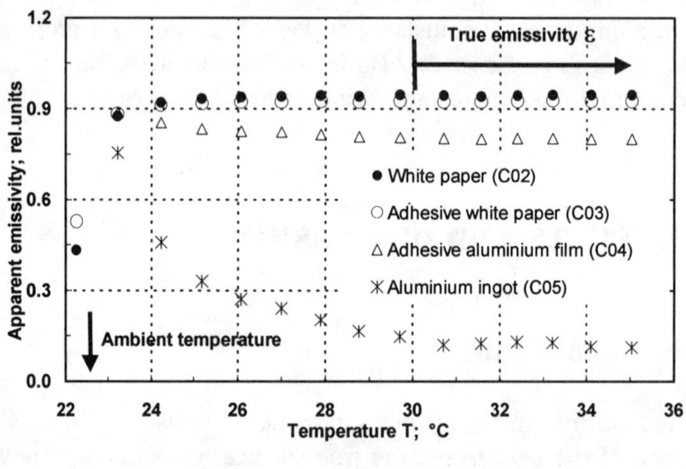

Figure 2. Emissivity of different materials (heating curve, ambient temperature 22.5 °C).

Thermal equilibrium at room temperature produces no image. The reason is shown here by the following simplified consideration:

Having an object and background, the image pixels of the object are generated by the superposition of thermal radiation flux density of the object Φ_o, and the background thermal radiation flux density reflected by the object Φ_r.

$$\Phi_o = \varepsilon_o \, \sigma \, T_o^4 \tag{2}$$

with ε_o representing the emissivity of the object, σ the Stefan's constant, and T_o the temperature of the object,

$$\Phi_r = \rho_o \, \sigma \, T_b^4 \tag{3}$$

with ρ_o the reflectivity of the object and T_b the background temperature. If absorption at the object is neglected then the reflectivity ρ_o equals $1 - \varepsilon_o$. The image pixels of the background are caused by its temperature and emissivity:

$$\Phi_b = \varepsilon_b \, \sigma \, T_b^4 \tag{4}$$

with ε_b the emissivity of the background (or global emissivity). An image will be obtained, if the difference between image and background pixels is not zero.

$$(\Phi_o + \Phi_r) - \Phi_b < 0 \;\; or \;\; (\Phi_o + \Phi_r) - \Phi_b > 0 \tag{5}$$

This difference of thermal radiation flux densities produces the image. The image will be visible, if the corresponding signal Δ is above the noise signal and thermal resolution limit. As Δ is proportional to the difference $(\Phi_o + \Phi_r) - \Phi_b$, different objects with different temperatures T_{oi} or emissivities ε_{oi} give different signals Δ_i.

$$\Delta_i = c_i \, (\Phi_{oi} + \Phi_{ri} - \Phi_b) \tag{6}$$

This equation can be evaluated in terms of ε_{oi} :

$$\varepsilon_{oi} = (a_i \, \Delta_i - b) \, / \, \{(T_{oi} / T_b)^4 - 1\} \tag{7}$$

with $a_i = (c_i \; \sigma \; T_b^4)^{-1}$ and $b = (1 - \varepsilon_b)$ being parameters, depending on measuring conditions c_i, ε_b, T_b, and on the object *i*. Now it can clearly be recognised that emissivity measurements in the course of heating curves (heating of the object) will show irregularities, if T_o approaches and passes T_b, the background or ambient temperature (Fig. 2 and Fig. 3).

The ingot temperature is measured with thermocouples and the radiant flux with the imaging camera. The emissivity of the blackbody radiator is found in such a way to be 0.99. With this value, the emissivity of the markers is determined (Fig. 2).

Additional systematic deviations may result from small local temperature differences caused by the different heat conductivity and heat transport at the contact ingot to marker during dynamic measurements. Therefore, the constant values (above 30 °C) were chosen as emissivity data for the markers.

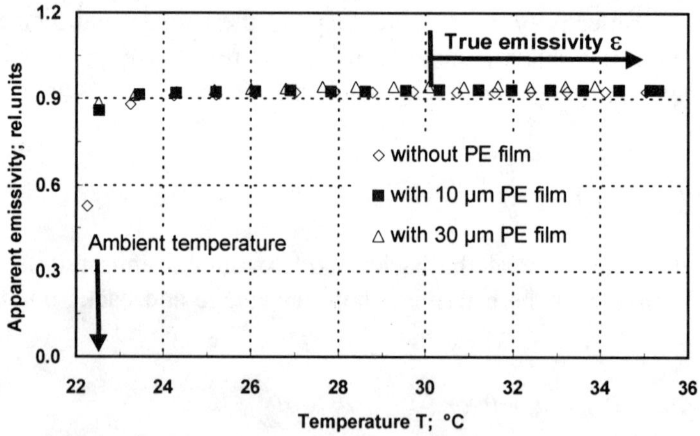

Figure 3. Emissivity of adhesive white paper without film and covered by polyethylene films of different thickness (heating curve, ambient temperature 22.5 °C)

Small pieces of adhesive paper are used as markers for the fruit experiments, as they show little variation in emissivity ($\varepsilon = 0.93$), and can easily be handled. Since a marker adheres locally to the surface and stops transpiration with local changes in the temperature, the solution is a protecting film (10 μm polyethylene), which covers parts of the fruit near the marker too and thus guaranties similar ambient conditions during the time of measurement. In order to know the influence of transmitting films on the measurement, the properties of films had to be checked. Measurements of the calibration device without film and with two different films (10 μm and 30 μm) did not change the calibration (Fig. 3). Hence it is possible to use transparent films as cover in emissivity measurements of fruits.

Assuming a similar temperature near the marker C01, the known temperature of the calibrated marker is taken as the temperature of the measuring circles C02 and C03 at the fruit (Fig. 4).

Then the emissivity at the circles C02 and C03 is found by changing the inputs for the software until the measuring circles C02 or C03 show the same temperature as the marker. The measurement of fruit emissivity has to be performed dynamically to have sufficient contrast. Therefore, the fruits are warmed up for few hours (sealed in plastic films) at 32 °C, then placed on a support and thermal images are taken every 30 s. After 120 s, the fruit-with-marker is pressed against the protecting film and imaging is continued up to 270 s (Fig. 5).

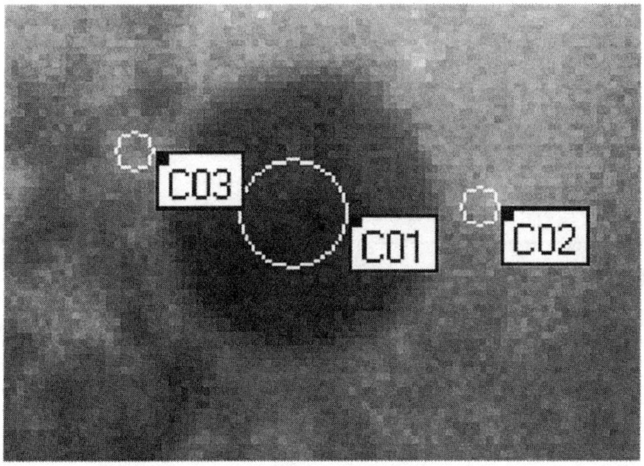

Figure 4. Part of a thermal image of an apple (Golden Delicious) with adhesive paper marker (C01: $\varepsilon = 0.93$, T = 30.726 °C) and measuring circles (C02, T = 31.128 °C; C03, T = 31.102 °C). These temperatures are mean values of pixels inside the circles. The apparent temperatures correspond to a global emissivity of the images of $\varepsilon = 1$ (ambient temperature 22.5 °C).

The decline of the apparent emissivity of C02 and C03 till 120 s (Fig. 5) results mainly from transpiration-caused temperature differences between marker and measuring circles. After stopping transpiration by the protecting film, the temperature increases and the true emissivity is measurable. There are several error sources. Changing air conditions give deviations in the radiation flux measurements. Temperature gradients between marker and measuring circles result in incorrect emissivity calculations. Thus the placing of the measuring circles is important. From Fig. 4 it can be seen that there is a local variability of temperature in the surrounding of the marker.

This variability might be due to insufficient contact between film and fruit surface, resulting in thin insulating air layers or water vapour layers. Further careful experiments must be performed to analyse this situation. The first measured emissivities are in the range 0.90 to 0.97 (apple: 0.94 - 0.97; pear: 0.95 - 0.96; nectarine: 0.93 - 0.95; tomato: 0.90 - 0.95). Since these numbers are from only a few fruits, more measurements must be performed for statistical verification of these preliminary results.

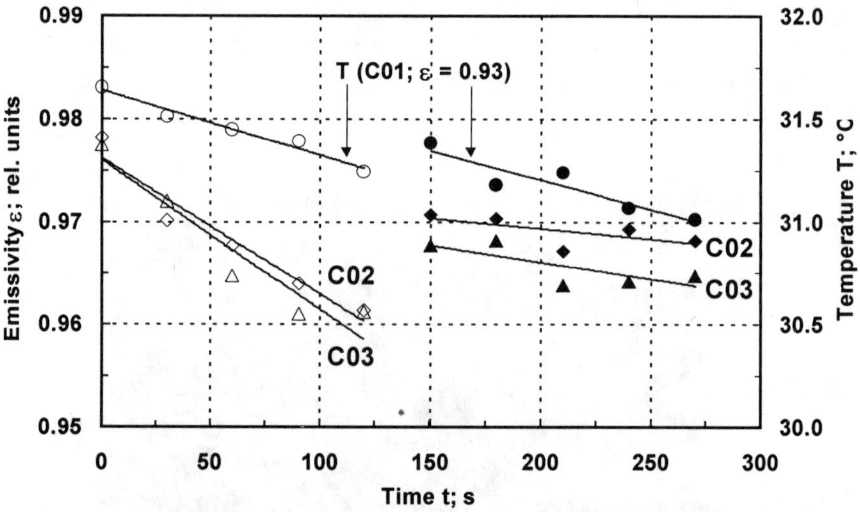

Figure 5. Time-dependent measurement (temperature decay) of apple emissivity (Golden Delicious; ambient temperature 22.5 °C). The Temperature T(C01) is calculated using the marker emissivity of ε = 0.93.

3.2 Temperature Distributions and Temperature Profiles

An apple's temperature distribution will be determined using the measured emissivity and temperature profiles (Fig. 6).

A circle C01 is set with ε = 0.93, the emissivity of the marker. The effect of direct radiation is visible; the "true temperature" is valid for the centre of the image only. "Lambert's cosine law radiation" causes this apparent temperature decrease from the centre to the circumference. Additionally, at the circumference we have a superposition with the laminar boundary layer, which results from transpiration and heat exchange with ambient air.

3.3 Transpiration Measurements

Under controlled flow conditions (e g natural convection around single produce items), the temperature difference, which can be evaluated using the thermal imaging system, depends upon the water status of the produce and the water vapour partial pressure difference between the produce surface and the ambient air. The transpiration resistance values of horticultural produce extend over a very wide range. While values of 0.25 s/cm were measured for young radish tubers, the values for stored apples amount to more than 500 s/cm. Under comparable ambient conditions, this results in temperature differences between six degrees and less than a tenth of a degree (Linke 1999).

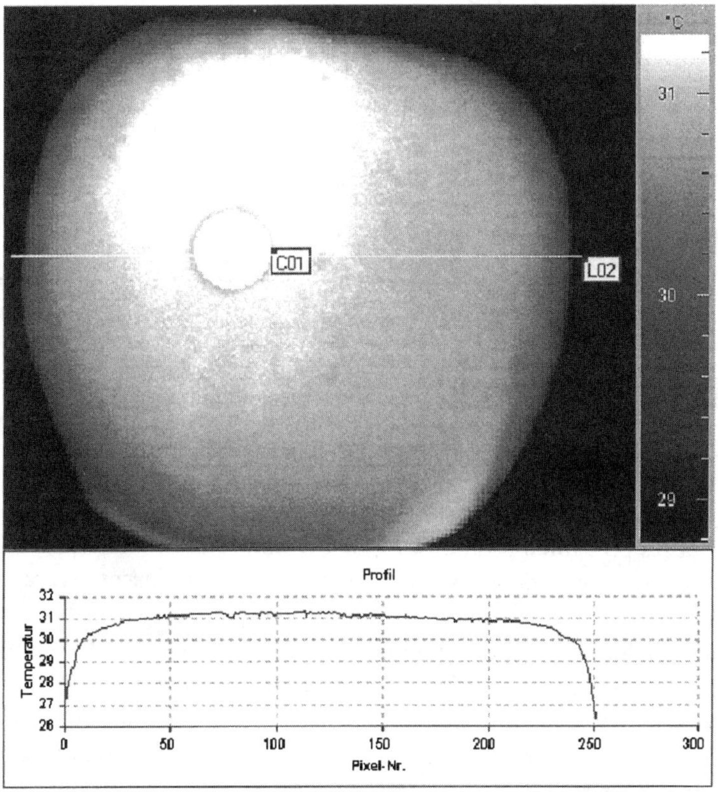

Figure 6. Thermal image and temperature profile across an apple (Golden Delicious; $\varepsilon = 0.965$) and marker part C01 ($\varepsilon = 0.93$) without protecting film.

The transpiration behaviour of many kinds of produce changes as the natural development of the produce progresses. It exhibits a characteristic course, which applies to the pre-harvest stage (cultivation period) and (with certain restrictions), also to the post-harvest behaviour. As the physiological age increases, transpiration resistance grows as well before diminishing more slowly after a maximum value has been reached (Pieniazek 1943). In many kinds of produce, this maximum value coincides with the stage of ripeness. As ageing begins, parts of the epidermal tissue obviously atrophy, which also leads to a reduction of transpiration resistance.

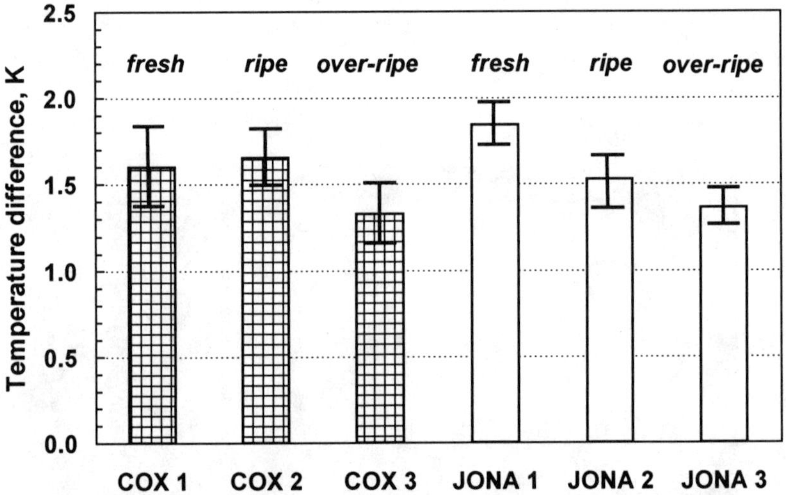

Figure 7. Temperature difference between indoor air (22 °C) and apples (varieties Cox Orange and Jonagold at three storage classes; difference of mean values of 8 apples and 8 simultaneously measured air temperature values with humidity in the range of 38 ± 2%; the standard deviation 2σ is indicated).

In studies promoted by the European Union (ASTEQ 2000, Hellebrand et al. 2000), two apple varieties (each at three storage classes) were examined after the appropriate storage period using thermal image analysis and other methods. Under comparable indoor climate conditions (relative air humidity of 38 % and temperature of 22 °C), a significant decrease in transpiration was observed for both varieties Cox Orange and Jonagold at the stage "over-ripe" (Fig. 7). The storage classes do not stand for "developmental stages - fresh, ripe, over-ripe" (mature, fresh picked apples have the lowest relative transpiration). The storage classes here give a "shelve-life information". There is nearly no difference between "fresh" and "ripe" Cox Orange, which shows similar physiological stage for this two Cox classes. The "over-ripe" apples have lost more water with the result of reduced transpiration. The monotone decrease of the air to surface temperature difference for the variety Jonagold can serve as an indicator for the shelve-life of each storage class. As a result it can be stated that IR imaging is suitable for relative classification of apples regarding transpiration and physiological stage.

3.4 Evaluation of Microbial Infestation

In two cases, locally restricted microbial infestation can be detected with the aid of thermal images. Using peaches as an example, it was possible to show that destruction of the surface structure due to microbial activities leads to locally different transpiration intensities, which cause divergent surface temperatures (Fig. 8). Studies on the microbial infestation of green vegetables without visible destruction of the surface structure detected (through heat emission) certain cultures with relatively strong respiration activities, while other cultures were not. In such cases, it is not yet possible to clearly assess the suitability of thermal image analysis.

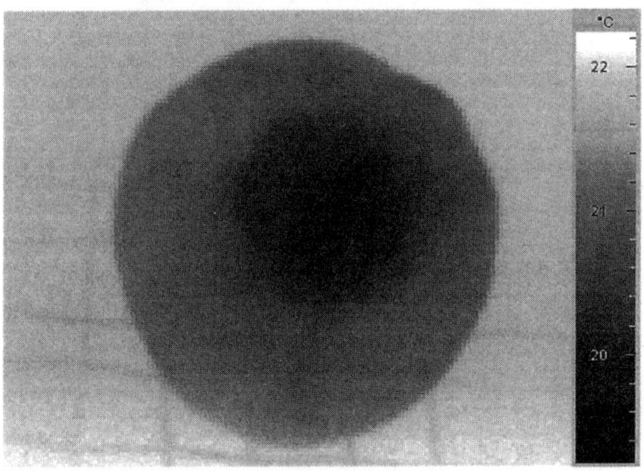

Figure 8. Thermal image of a rotting peach

3.5 Evaluation of the Display Conditions

Infrared thermography can also be used directly to analyse flow conditions near the produce, if appropriate measures are taken to make the temperature field above the produce visible (Fig. 9). For this purpose, a thin sheet of paper is spread in the plane under investigation. A camera situated at right angles to the paper can record the temperature distribution on the paper sheet. The figure shows the temperature distribution over lettuce, which is displayed in two cartons in a chilled food display cabinet. If the indoor climate parameters are determined at a sufficient distance to the produce, this also enables humidity distribution over the produce to be calculated by approximation. Such investigations were carried out at a commercially manufactured chilled food display cabinet with the goal of adapting the operating characteristics of the cabinet to the properties of the produce and, ultimately, maintaining produce quality as long as possible (Geyer and Linke 2000).

Figure 9. Thermal image of the temperature distribution over lettuce in a chilled food display cabinet

4. CHANCES AND SHORTCOMINGS

In principle, thermal imaging systems are suitable for examining horticultural produce at the post-harvest stage. They may lead to qualitatively new insights and thus make a contribution towards the maintenance of quality. For several reasons, however, the commercial application of thermal imaging systems very quickly reaches its limits. Not only the costs of these systems are still relatively high, but the possibilities of use are also restricted due to unsteady temperatures. For scientific studies and in other areas, which offer laboratory conditions, this technology provides numerous, very suitable possibilities of use. For different applications, it will be appropriate to carry out pre-investigations under laboratory conditions in order to prepare the use of inexpensive infrared thermometers (e g infrared thermocouples), which are available in a wide spectrum for different conditions (distance from the produce, measurement surface, etc). Detailed analyses of the transpiration behaviour of the produce and its interrelation with the environment allow qualified conclusions to be drawn regarding the optimal harvest date and the adapted design of post-harvest machinery and technological processes on the way from the producer to the consumer. The essential advantage of such a perspective is that greater value can be attached to the produce characteristics.

5. CONCLUSIONS

Thermal imaging allows the visualisation and quantitative determination of the surface temperature distribution of horticultural products. For exact temperature measurements, the emissivity of the objects must be known. For qualitative differentiation and evaluation of fruits, the influence of the emission factor can be neglected. The emissivity of most of the horticultural products differs less than 10 % to each other ($\varepsilon = 0.94 \pm 0.04$). Infrared vision was successfully tested for examinations of quality changes of horticultural produce, for discrimination of products, and designing storage and shelve conditions. Thermal imaging is able to measure all those produce properties connected with thermal processes (transpiration, respiration). Infrared imaging may lead to qualitatively new insights and thus make a contribution towards quality maintenance. When interpreting the results, the interrelation between the produce and its environment must always be taken into account.

REFERENCES

ASTEQ, 2000, **A**rtificial **S**ensing **T**echniques for **E**valuating **Q**uality, *Concerted Action No FAIRS-CT97-3516* (http://www.inapg.inra.fr/ens_rech/siab/asteq/index.htm)

Beverly, D., Hung, Y.-C., Prussia, S. E., Schwefelt, R. L., Tollner, E. W., Garner, J. C., and Robinso, D., 1987, Thermography as a non-destructive method to detect invisible quality damage in fruits and vegetables. *HortScience* **22**: 5, 1056

Chaerle, L., Van Caeneghem, W., Messens, E., Lambers, H., Van Montagu, M. and Van der Straeten, D., 1999, Pre-symptomatic visualisation of plant-virus interactions by thermography. *Nature Biotechnology* **17**: 8, 813-816

Daley, P. F., 1995, Chlorophyll fluorescence analysis and imaging in plant stress and disease. *Canadian Journal of Plant Pathology* **17**: 167-173

Danno, A., Miyazato, M., and Ishiguro, E., 1978, Quality evaluation of agricultural products by infrared imaging method. I. Grading of fruits for bruise and other surface defects. *Memoirs of the Faculty of Agriculture, Kagoshima University* **14**: 23, 123-138

Danno, A., Miyazato, M., and Ishiguro, E., 1980, Quality evaluation of agricultural product by infrared imaging method. III. Maturity evaluation of fruits and vegetables. *Memoirs of the Faculty of Agriculture, Kagoshima University*, **16**: 25, 157-164

Ebert, H., 1962, Physikalisches Taschenbuch (in German, Handbook of Physics). Vieweg&Sohn Braunschweig, pp 349-354

Egnell, G. and Orlander, G., 1993, Using infrared thermography to access viability of Pinus sylvestris and Picea abies seedlings before planting. *Canadian Journal of Forest Research* **23:** 9, 1737-1743

Geyer, M. and Linke, M., 2000, Display Cabinets for Perishables with Atmospheric Evaporation Cooling. In: Integrated View of Fruit & Vegetable Quality (W.J. Florkowski, S.E. Prussia, R.L. Shewfelt, Eds.), Technomic Publishing Co., Inc. Lancaster. 259-267, ISBN 1-58716-019-6

Grietens, B., 2001, Xenics - Best European Small Enterprise Initiative. http://www.xenics.com/Article3X.asp

Haußecker, H., 1996, Messung und Simulation von kleinskaligen Austauschvorgängen an der Ozeanoberfläche mittels Thermographie (in German, Measurement and simulation of small scale exchange at the surface of ocean by thermography,). PhD thesis, Ruprecht-Karls-University, Heidelberg,
http://klimt.iwr.uni-heidelberg.de/PublicFG/ProjectB/CFT/disshhaus/node20.html#transemiss

Hellebrand, H. J., Linke, M., Beuche, H., Herold, B., and Geyer, M., 2000, Horticultural Products Evaluated by Thermography. *AgEng2000, University of Warwick, UK, 2-7 July 2000, Abstracts Part 2,* Paper No. 00-PH-003, 26-27

Inoue, Y., 1990, Remote detection of physiological depression in crop plants with infrared thermal imagery. *Japan Jour Crop Sci* **59**: 4, 762-768

Inoue, Y., Kimball, B. A., Jackson, R. D., Pinter, P. J., and Reginato, R. .J 1990, Remote estimation of leaf transpiration rate and stomatal resistance based on infrared thermometry. *Agricultural and Forest Meteorology* **51:** 21-33

James, K. R. and Aldous, D. E., 1997, Thermal imaging: Tool for measurement of turfgrass stress. 4th ATRI Turf Research Conference, Proceedings of the 4th ATRI Research Conference, 80-83, Concord West, NSW: Australian Turfgrass Research Institute Ltd.

Jones, H. G., 1999, Thermographic studies of stomatal conductance over leaf surfaces. *Plant, Cell and Environment* **22:** 1043-1055

King, W. J., 1987, Emissivity and absorption, in R C Weast ed, Handbook of Chemistry and Physics. CRC Press, Boca Raton (Florida), pp E-393 - E-395

LaRocca, A. J., 1996, Artificial Sources, in "Sources of Radiation", editor Zissis, G. J., vol. 1 of "The Infrared & Electro-Optical Systems Handbook" (8 volumes), exec. eds. Accetta, J. S. and Shumaker,D. L., ERI Ann Arbor (Michigan) and SPIE Optical Engineering Press Bellingham (Washington), pp 49-135

Linke, M., Beuche, H., Geyer, M., and Hellebrand, H. J., 2000, Possibilities and limits of the use of thermography for the examination of horticultural products. *Agrartechnische Forschung* **6:** 110-114

Linke, M., 1999, Measurement and modelling of produce responses in post-harvest effected by pre-harvest parameters. In: Control applications in post-harvest and processing technology (Farkas, I. ed.). Pergamon Press, Oxford UK, pp 1-6

Monteith, J. L., Szeicz, G., 1962, Radiative temperature in the heat balance of natural surfaces. *Q J Roy Meteorol Soc* **88:** 496-507

Moran, M. S. 2000, Thermal infrared measurement as an indicator of plant ecosystem health. USDA-ARS Southwest Watershed Research Center, Tuscon Arizona, 42 pages, June 21, 2000, http://webpages.acs.ttu.edu/smaas/asa2000/moran.pdf

Miyazato, M., Ishiguro, E., and Danno, A., 1981, Quality evaluation of agricultural products by infrared imaging method. IV. Grading of fruits and root vegetables for bruise and other surface defects. *Bulletin of the Faculty of Agriculture, Kagoshima University*, **31:** 149-156

Nilsson, H.-E., 1995, Remote sensing and image analysis in plant pathology. *Canadian Journal of Plant Pathology* **17:** 154-166

Pieniazek, S. A., 1943, Maturity of apple fruits in relation to the rate of transpiration. *J Amer Soc Hort Sci* **44:** 231-237

Schuster, N. and Kolobrodov, V. G., 2000, Thermal imagigng. Viley-VCH Berlin Weinheim New York , 339 p, p 59

Tanner, C. B., 1963, Plant temperatures. *Agronomy Journal* **55:** 210-211

Tollner, E. W., Brecht, J. K., and Upchurch, B. L., 1993, Nondestructive evaluation: Detection of external and internal attributes frequently associated with quality or damage, in "Postharvest-handling: a system-approach". *Food Science and Technology Series* **7**: 225-255

Varith, J., 1999, Non contact bruise detection in apples by thermal imaging. http://www.wsu.edu/~jatuphon/research/thermal_im.htm

Wisniewski, M., Lindow, S., and Ashworth, E., 1997, Observations of ice nucleation and propagation in plants using infrared video thermography. *Plant Physiol.* **113**: 327-334

Wisniewski, M. and Fuller, M., 1999, Ice nucleation and deep supercooling in plants: new insights using infrared thermography. In: Cold-adapted Organisms, eds. Margesin, R. and Schinner, F. Springer Verlag Berlin Heidelberg

Workmaster, B., Palta, J. P., and Wisniewski, M., 1999, Ice Nucleation and Propagation in Cranberry Uprights and Fruit Using Infrared Video Thermography. *J. Amer. Soc. Hort. Sci.* **124** : 6, 619-625

Zoha, O., Ikeda, M., Shinagawa, H., Inoue, H., Nakamura, H., Elbaum, D., Alkon, D., and Yoshika, L., 1998, Thermal imaging of receptor-activated heat production in single cells. *Biophysical Journal* **74**: 1, 82-89

Sensing Taste Attributes of Plums Using Near Infrared (NIR) Reflectance Spectroscopy

NAWAF ABU-KHALAF and BENT S. BENNEDSEN
Department of Agricultural Sciences, The Royal Veterinary and Agricultural University (KVL), Taastrup, Denmark

1. INTRODUCTION

In recent years, research has been focused on the development of non-destructive techniques for measuring quality parameters of different agricultural commodities. The aim is development of techniques which include fast execution, can easily be used in process control and grading system and requires limited sample pre-processing (Lammertyn *et al*., 2000). NIR spectroscopy is one of these non-destructive techniques. NIR methods have already been used to detect bruised on apples (Upchurch *et al.,* 1994) and to study dry matter content in onions (Birth *et al.,* 1985) and potatoes (Dull *et al.,* 1989). Kawano *et al.* (1992) used an optical fibre with interactance mode to study the sugar content in peaches. Slaughter (1995) managed to measure non-destructively the internal quality of peaches and nectarines as characterised by their soluble solid content (SSC), sorbitol and chlorophyll contents using visible and NIR spectroscopy. Bellon *et al.* (1993) used the wavelength region between 800-1050 nm and developed a NIR instrument coupled with optical fibres to detect sugar at the speed of three apples per second with a standard error of prediction of 2.4 g/l of glucose. A relationship between NIR spectra and apple fruit quality parameters such as pH, acidity, sugar content and texture parameters was established by Lammertyn *et al.* (1998) and Moons *et al.* (2000).

According to Ozanich (1999), NIR spectroscopy is uniquely qualified for analysis in food and related industries. NIR spectroscopy is particularly sensitive to molecules containing C-H, O-H, and N-H groups. These bonds interact in a measurable way in the NIR portion of the spectrum. Hence, constituents such as starch and sugars (C-H), alcohols, moisture and acids (O-H), and protein (N-H) can be quantified in solids, liquids and slurries. In addition, the analysis of gases is possible. NIR is not a trace analysis technique and is generally used for measuring components that are present at concentrations greater than 0.1 %. Kim *et al.* (2000) stated that "spectral data of a fruit can be treated as a signature allowing fruits to be grouped on the basis of their similarities". The use of NIR technology provides for a faster, safer work environment and does not require chemicals (Ozanich, 1999).

Previous researches reported use of different wavebands. In this research, wavelengths between 700-1100 nm were used. NIR in this range is promising and more useful for intact foods due to the following (Carlini *et al.*, 2000; McGlone and Kawano, 1998;Walsh *et al.*, 2000):

1. Penetration of the radiation into the fruits was found to be deeper than for other ranges;
2. Water absorbance peaks are less strong and broad than they are at other ranges, and the risk to mask spectral information correlated to low concentration constituents is low;
3. The instrumentation that can be used at this range has low cost, suitable for process control and is portable enough for in situ field measurements;
4. The bands are ascribed to the third and forth overtones of O-H and C-H stretching modes and are expected to be separated due to anharmonicity; and
5. Lower absorbance at these wavelengths allows for transmission optics.

Optical measurements can be done in different modes: transmittance, absorbance and reflectance. Chen (1978) stated that reflectance is generally easier to use for quality evaluation of agricultural products due to:

1. Its relative high intensity: Reflectance in the visible and infrared regions ranges up to 80 % of the incident energy; and
2. Reflectance measurement is not adversely affected by low-intensity background light.

Taste is an important internal quality parameter of a commodity. Fruit's taste, as the major asset of fruit quality, is dominated by the sugar to acid ratio (Blanke, 1996). Apart from the sweet and sour taste, provided by sugar and acid, fruits have a distinctive taste, which originates from contents of other chemicals. This taste is characteristic for the variety. Auerswald *et al.* (1999) quoted a number of authors indicating that the overall flavour of fruits is a combination of sugars, acids, aroma volatiles, amino acids and amines. This research aims at measuring these taste parameters, using NIR technology. The idea is that, if it is possible to distinguish two different varieties of the same commodity, having the same ratio of soluble solids content (SSC) and titratable acidity, then both components (sugar and acid) are eliminated, leaving only the characteristics taste (specific chemical compounds) parameters of each variety to be detected, which indicates that NIR technology can sense taste characteristics parameters of each variety. The ability of NIR spectroscopy to detect these parameters will open a wide range for applying NIR sensors for online grading of fruits according to their taste quality.

2. MATERIALS AND METHODS

2.1 Fruits

Two varieties of plums, Reine Claude (*Prunus domistica*) and Blackamber (*Prunus silicina)* were used. Eighty plums of each variety were selected for uniform maturation and surface. They were obtained from a local retail market in August. After being in cold store, plums were kept at room temperature (20 ^{o}C) for 24 hours for equilibration before the experiment.

2.2 Reflectance Measurements

A scanning Zeiss MMS1 NIR enhanced spectrometer was used to collect reflectance readings over a wavelength range of 700-1100 nm in 2 nm increments, yielding 200 values per spectrum. For each variety (80 samples), the reflectance measurements were done in one day. For each plum, three reflection spectra were taken at three equidistant positions around the equator in order to eliminate the spatial variability. The light source consisted of 12V/100W tungsten halogen lamp. The lamp was calibrated before and after reflectance measurement of each variety, using a standard reflectance plate made of $BaSO_4$. Light from the lamp passed through a bundle of optical fibres to the fruit, and reflected light was transferred to a photo diode array (PDA) detector through another bundle of fibre optic. A holder was designed to support plums and to direct the light in a 45 degree angle to the plums (to avoid specular reflectance), and to maintain a distance of 1 cm, according to manufacture recommendation, between the probe and the plums. The integration time (time needed for a spectrum to be acquired) was 181 milli second (ms). The three spectra from each plum were averaged for further treatment using multivariate analysis.

2.3 Taste Attributes

After reflectance measurements, SSC and titrable acidity were measured on juice extracted from plums. A digital refractometer (RFM 90-Struers) was used to measure SSC. The SSC was expressed in Brix. Titrable acidity (citric acid) was measured using A 719 S Titrino, by titrating the juice with 0.1 *M* NaOH to a pH 8.1 endpoint.

2.4 Multivariate Analysis

The calculations were carried out using 'Unscrambler' v. 7.5 (CAMO, ASA, Oslo, Norway), a statistical software package for multivariate analysis. Matlab R.12 (The Math Works Inc., Natick, MA) was used as a bridge program between the spectrometer outputs and Unscrambler to transfer the reflectance data to be analyzed.

The modelling and classification of the spectra was based on full multiplicative scatter correction (MSC) pre-processing technique. This technique corrects the additive and multiplicative effects in the spectra and improves the predictive ability (Mobley *et al.,* 1996). In this work, full MSC pre-processing yielded the lowest root mean square error of prediction (RMSEP) compared to raw data or first and second derivative pre-processing techniques. Moreover, distinguished groups (two varieties) were explained very clearly along the first principal component (PC) in the spectra score plot.

Partial least square (PLS) technique was used for modelling plums' SSC. 100 samples were used for calibration, and 60 samples were used for validation. Test set validation was used for building PLS model.

Soft independent modelling of class analogies (SIMCA) classification technique was used for classification plums according to their reflectance spectra. In SIMCA classification technique, the classification succeeds when there are very clear distinguished groups of the targeted classes in the different SIMCA's pertinent graphs (Esbensen *et al.*, 2000). Samples are considered wrongly classified when they are not belonging to any of the models of the targeted classes. SIMCA's classification tables (not shown in the results) were used for expressing classification percentage rate of different verities. For the classification of all the plum samples, a training set containing 50 samples and test set containing 30 samples from each variety were used. 63 samples, having the same values of SSC/acidity, within ± 5 %, were selected, representing both varieties. 35 of them were used in the training set and 28 samples in the test set. Coomans plots were used as SIMCA's pertinent graph. These graphs shows the object to model distance for both the new objects (test set) and the calibration objects (train set). The numbers on x and y-axis are the residuals to two groups. One coordinate is given for each sample, which is the summed square over all variable residuals at that component.

3. RESULTS AND DISCUSSION

The stability of the light source in the spectrometer was quite acceptable, with a variation of less than 5 % during the reflectance measurements. Fruits had a uniform surface temperature after being kept at 20 °C for 24 hours. The averaged values of SSC and titrable acidity for each plums' variety is tabulated in Table 1.

Table 1. SSC and titratable acidity values of Reine Claude and Blackamber plums

Variety	*SSC (Brix)*				*Titratable acidity (g/100g)*			
	Average	*SDev**	*CV* (%)*	*Range*	*Average*	*SDev*	*CV (%)*	*Range*
Reine Claude	11.54	2.21	19	7.5-16.5	1.05	0.43	41	0.1-1.97
Blackamber	8.55	1.77	21	4.5-12.3	1.06	0.20	19	0.53-1.57

**SDev: Standard Deviation*
**CV: Coefficient of Variation = SDev / Average*

The SSC differed significantly between the two varieties, while difference in titrable acidity was insignificant at 5 % significant level. The correlation between acidity and SSC was very low.

The reflectance spectra scores plot of all plum samples (from both varieties) is shown in Fig. 1. The first principal component (latent variable) describes 68 % of the variation between all samples. Very clear distinguished groups (two varieties) are explained along the first principal component (PC).

Partial least square (PLS-1) method was used for modelling SSC of both varieties. The SSC had a validation correlation of 0.8 and a RMSEP of 1.56 with four principal components (PCs) (see Fig. 2). The most important ranges for modelling plums' SSC are 772-810, 890-946, 956-1000 and 1030-1070 nm, since these ranges have the highest regression coefficients (Fig. 3). The relative error (relative error = standard error of prediction (SEP)/ range) in SSC model is 12.5 %, which is acceptable. Acidity modelling was not satisfactory. The correlation was very low. This may be due to a high variation and limited range of the plums' titrable acidity.

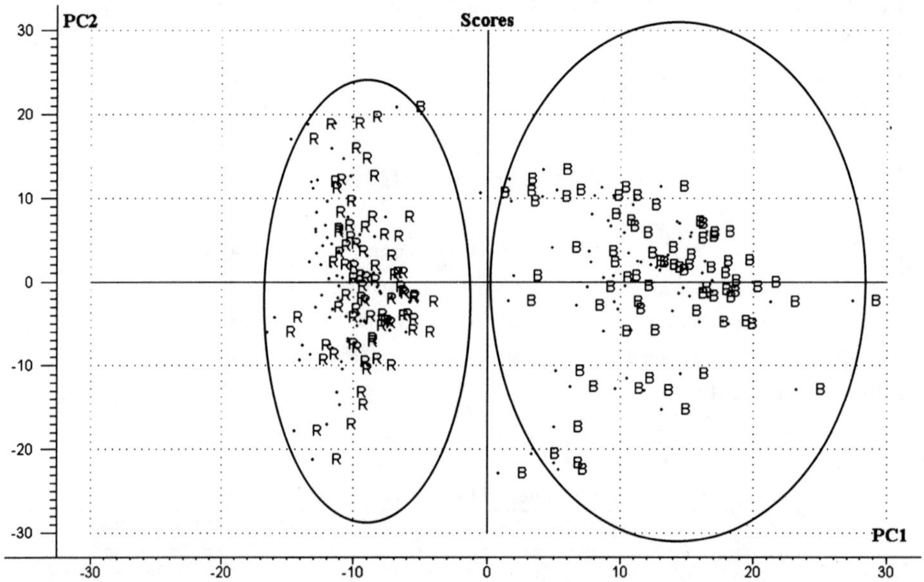

Figure 1. The reflectance spectra scores plot of all plums samples (R: Reine Claude, B: Blackamber)

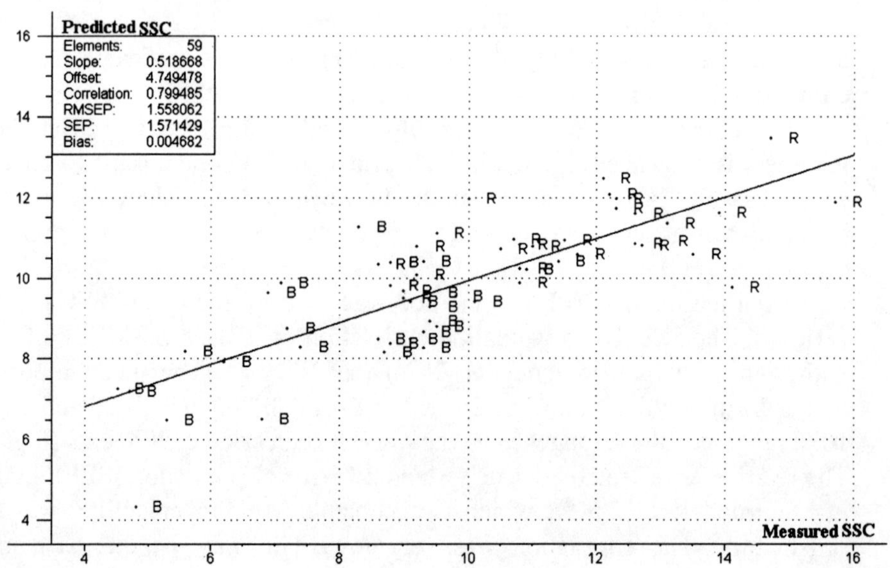

Figure 2. The PLS model for plums' SSC

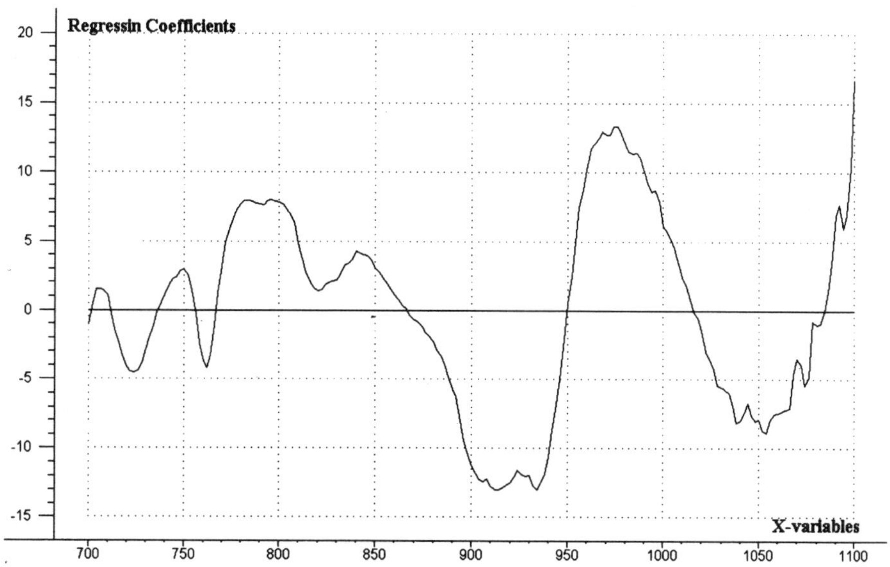

Figure 3. The regressions coefficient of the wavelengths variables used for plums' SSC modelling

The classification rate using SIMCA technique was quite high. Reine Claude samples were correctly classified in 90 % of the cases, and Blackamber in 96.6 % of the cases, at 5 % significant level (Table 2).

Table 2. Classification result of NIR reflectance spectra for all plums' samples

Variety	*Reine Claude (%)*	*Blackamber (%)*	*Error (%)*
Reine Claude	90	0	10
Blackamber	0	96.6	3.4

The Coomans plot is shown in Fig. 4. It shows clearly separated groups of both varieties. This indicates that the plums can be classified on the basis of the chemical components in the juice, measured by NIR spectroscopy.

The next step was to investigate the ability of NIR spectra to sense characteristic taste parameters, other than SSC and acidity. For that, 63 samples having the same ratio of SSC and titrable acidity, and representing both varieties were gathered. The reflectance spectra scores plot is shown in Fig. 5. The first PC describes 90 % of the variation between these samples. Very clear distinguished groups (two varieties) are explained along the first PC. SIMCA classification technique was used to classify these samples. The result is shown in Table 3. Both Reine Claude and Blackamber were correctly classified in 92.8 % of the cases at 5 % significant level. Coomans plot (Fig. 6) shows clearly separated groups of both varieties. The wavelengths that have the highest discrimination power are in the ranges of 715-740 and 820-850 nm.

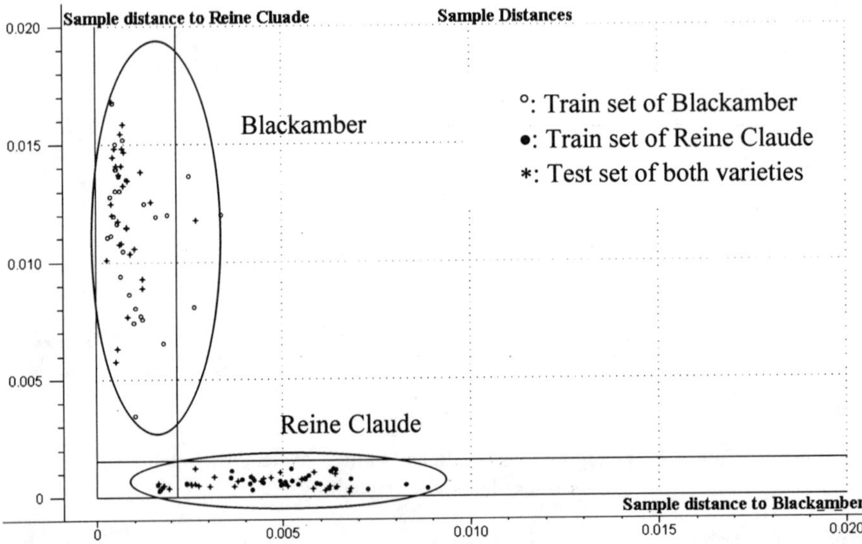

Figure 4. Coomans plot for all plums' samples at 5 % significant level

Information about the discriminating power of the variables is provided by the Unscrambler software. The classification of these samples were based on the characteristics taste parameters, since the effect of SSC and titrable acidity were eliminated by analysing samples having the same values of SSC/acidity. Further, it was found that plums with the same SSC but of different variety, still end up in the right variety group, when classifying according to the spectral information. This indicates that the classification was based on other factors than SSC.

From taste attributes measurements, it was noticed that there was a high variation in SSC of both plums' varieties (CV average is about 20 %), this variation is also seen in Fig. 2. Moreover, titrable acidity in both varieties had a high variation (CV average is about 30 %). This indicates that the samples were heterogeneous in terms of SSC and titrable acidity. Anyhow, the classification was based on plum samples having the same values of SSC/acidity, which means that the effect of SSC and acidity were already eliminated, and all of the other parameters (i.e. stability of light source, plums' surface temperature, etc.) were equal, taste is the only remaining parameter that plums classification was based on.

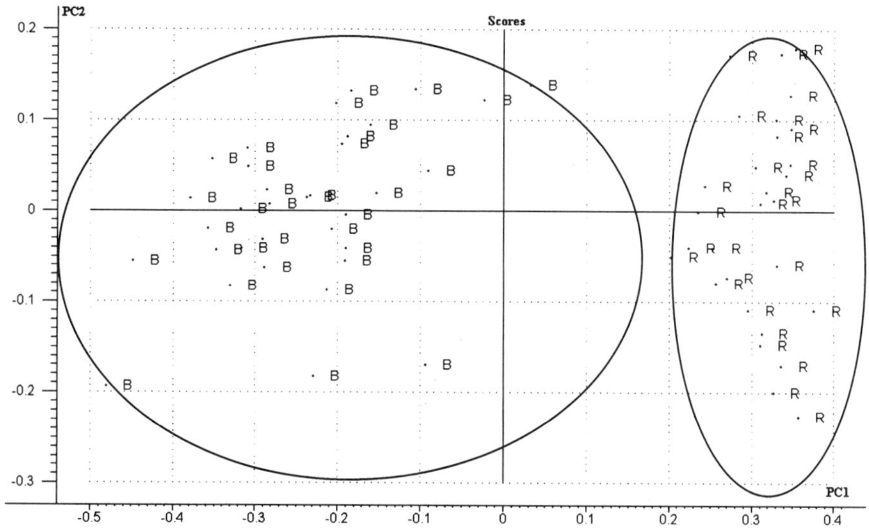

Figure 5. The reflectance spectra scores plot of plum samples having the same ratio of SSC and acidity (R: Reine Claude, B: Blackamber)

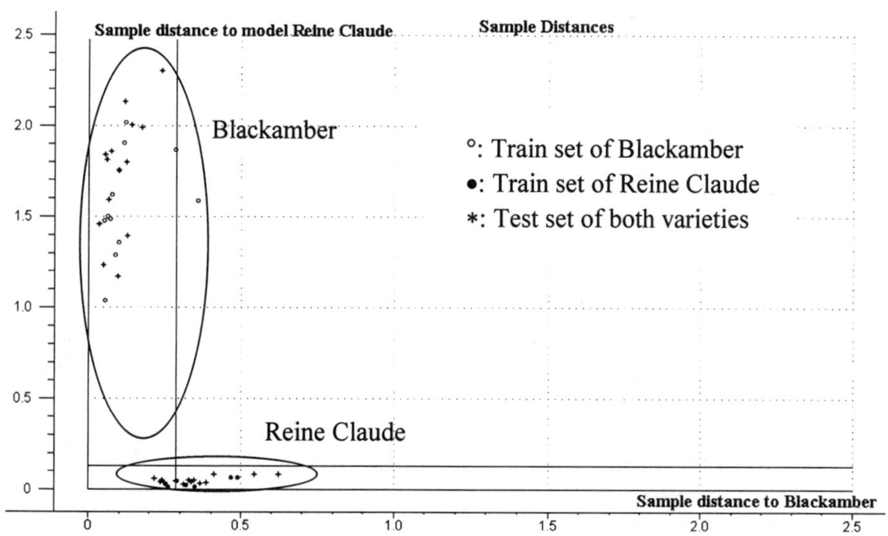

Figure 6. Coomans plot of plums' samples having the same ratio of SSC and acidity at 5 % significant level

Table 2. Classification result of NIR reflectance spectra for plums' samples having the same ratio of SSC and titratable acidity

Variety	*Reine Claude (%)*	*Blackamber (%)*	*Error (%)*
Reine Claude	92.8	0	7.2
Blackamber	0	92.8	7.2

Fig. 1 and Fig. 5 show that the MSC pre-processed reflectance spectra scores produce very clear distinguished groups, which are separated according to their variety along the first PC, this PC can be interpreted as a variety factor.

4. CONCLUSION

Based on the results presented in this paper, it can be concluded that taste attributes; SSC, acidity and characteristics taste parameters can be detected using NIR spectroscopy reflectance. Plums' SSC can be determined with a reasonable correlation. NIR technology has a high potential to sense plums' overall taste attributes and NIR is able to sense the characteristic taste parameters of plum samples, when the influence of SSC and acid has been eliminated.

REFERENCES

Auerswald, H., Peters, P., Bruckner, B., Krumbein, A., and Kuchenbuch R., 1999, Sensory analysis and instrumental measurements of short-term strored tomatoes (Lycopersicon esculentum Mill). *Postharvest Biology and Technology* **15**: 323-334.

Bellon, V., Vigneau, J. L., and Leclercq, M., 1993, Feasibility and performance of a new multiplexed, fast, low cost fiber optic NIR spectroscopy for the on-line measurement of sugar in fruits. *Applied spectroscopy* **47**: 1079-1083.

Birth, G. S., Dull, G. G., Refroe, W.T., and Kays, S. J., 1985, Non-destructive spectrophotometric determination of dry matter on onions. *Journal of the American Society of Horticultural Science* **110**: 297-303.

Blanke, M.M., 1996, Fruit photosynthesis and pome fruit quality. *Acta Hort.* **466**: 19-22.

Carlini, P., Massantini, R., and Mencarelli, F., 2000, VIS-NIR measurement of soluble solids in cherry and apricot by PLS regression and wavelength selection. *J. Agric. Food Chem.* **48**: 5236-5242.

Chen, P., 1978, Use of optical properties of food materials in quality evaluation and material sorting. *Journal of Food Process Engineering*, **2**: 307-322.

Dull, G. G., Birth, G. S., and Leffler, R. G., 1989, Use of near infrared analysis for non-destructive measurement of dry matter in potatoes. *Amer. Potato Journal* **66:** 215-225.

Esbensen, K., Guyot, D., and Westand, F., 2000, *Multivariate data analysis in practice*, 4th edition, Camo, Norway. 600 pp.

Kawano, S., Watatnabe, H., and Iwamoto, M., 1992, Determination of sugar content in intact peaches by near infrared spectroscopy with fiber optics in interactance mode. *J. Japanese Soc. Hort. Sci.*, **61:** 445-451.

Kim, J., Mowat, A., Poole, P., and Kasabov, N., 2000, Linear and non-linear pattern recognition models for classification of fruit from visible-near infrared spectra, *Chemometrics and intelligent laboratory systems* **51:** 201-216.

Lammertyn, J., Nicolai, B., Ooms, K., De Smedt, V., and De Baerdemaeker, J., 1998, Non-destructive measurements of acidity, soluble solids, and firmness of Jonagold apples using NIR spectroscopy. *Trans. Am. Soc. Ag. Eng.* **41**: 1089-1094.

Lammertyn, J., Peirs, A., De Baerdemaeker, J., and Nicolai, B., 2000, Light penetration properties of NIR radiation in fruit with respect to non-destructive quality assessment. *Postharvest Biology and Technology* **18**: 121-132.

McGlone, V. A., and Kawano, S., 1998, Firmness, dry-matter and soluble-solids assessment of postharvest kiwifruit by NIR spectroscopy. *Postharvest Biology and Technology* **13**: 131-141.

Mobley, P. R., Kowalşki, B. R., Workman, J. Jr., and Bro, R., 1996, Review of chemometrics applied to spectroscopy: 1985-95, part 2. *Applied Spectroscopy Reviews* **31**: 347-368.

Moons, E., Sinnaeve, G., and Dardenne, P., 2000, Non destructive visible and NIR spectroscopy measurement for determination of apple internal quality. *Acta Hort.* **517:** 441-448.

Ozanich, R. M., 1999, Near-Infrared spectroscopy: background and application to tree fruit quality measurements. *Tree Fruit Postharvest* **10**: 18-19.

Slaughter, D. C., 1995, Non destructive determination of internal quality in peaches and nectarine. *Transactions of the ASAE* **38**: 617-623.

Upchurch, B. L., Throop, J. A., and Aneshansley D. J., 1994,Influence of time, bruise-type and senverity on near-infrared reflectance from apple surfaces for automatic bruise detection. *Tran. Am. Soc. Ag. Eng.* **37**: 1571-1575.

Walsh, K. B., Guthrie, J. A., and Burney, J. W., 2000, Application of commercially available, low-cost, miniaturized NIR spectrometers to the assessment of the sugar content of intact fruit. *Aust. J. Plant Physiol.* **27**: 1175-1186.

Index